T0228016

IET PROFESSIONAL APPLICATIONS OF COMPUTING SERIES 18

Modeling and Simulation of Complex Communication Networks

IET Book Series on Big Data—Call for Authors

Editor-in-Chief: Professor Albert Y. Zomaya, University of Sydney, Australia

The topic of Big Data has emerged as a revolutionary theme that cuts across many technologies and application domains. This new book series brings together topics within the myriad research activities in many areas that analyze, compute, store, manage, and transport massive amounts of data, such as algorithm design, data mining and search, processor architectures, databases, infrastructure development, service and data discovery, networking and mobile computing, cloud computing, high-performance computing, privacy and security, storage, and visualization.

Topics considered include (but not restricted to) Internet of Things and Internet computing; cloud computing; peer-to-peer computing; autonomic computing; data centre computing; multicore and many core computing; parallel, distributed, and high-performance computing; scalable databases; mobile computing and sensor networking; green computing; service computing; networking infrastructures; cyberinfrastructures; e-science; smart cities; analytics and data mining; Big Data applications, and more.

Proposals for coherently integrated International coedited or coauthored handbooks and research monographs will be considered for this book series. Each proposal will be reviewed by the editor-in-chief and some board members, with additional external reviews from independent reviewers. Please email your book proposal for the IET book series on Big Data to Professor Albert Y. Zomaya at albert.zomaya@sydney.edu.au or to the IET at author_support@theiet.org.

Modeling and Simulation of Complex Communication Networks

Edited by
Muaz A. Niazi

The Institution of Engineering and Technology

Published by The Institution of Engineering and Technology, London, United Kingdom

The Institution of Engineering and Technology is registered as a Charity in England & Wales (no. 211014) and Scotland (no. SC038698).

© The Institution of Engineering and Technology 2019

First published 2019

This publication is copyright under the Berne Convention and the Universal Copyright Convention. All rights reserved. Apart from any fair dealing for the purposes of research or private study, or criticism or review, as permitted under the Copyright, Designs and Patents Act 1988, this publication may be reproduced, stored or transmitted, in any form or by any means, only with the prior permission in writing of the publishers, or in the case of reprographic reproduction in accordance with the terms of licences issued by the Copyright Licensing Agency. Enquiries concerning reproduction outside those terms should be sent to the publisher at the undermentioned address:

The Institution of Engineering and Technology
Michael Faraday House
Six Hills Way, Stevenage
Herts, SG1 2AY, United Kingdom

www.theiet.org

While the authors and publisher believe that the information and guidance given in this work are correct, all parties must rely upon their own skill and judgement when making use of them. Neither the authors nor publisher assumes any liability to anyone for any loss or damage caused by any error or omission in the work, whether such an error or omission is the result of negligence or any other cause. Any and all such liability is disclaimed.

The moral rights of the authors to be identified as authors of this work have been asserted by them in accordance with the Copyright, Designs and Patents Act 1988.

British Library Cataloguing in Publication Data
A catalogue record for this product is available from the British Library

ISBN 978-1-78561-355-5 (hardback)
ISBN 978-1-78561-356-2 (PDF)

Typeset in India by MPS Limited
Printed in the UK by CPI Group (UK) Ltd, Croydon

Contents

Preface

Thank you for choosing "Modeling and Simulation for Complex Networks." This book offers a unique set of chapters and case studies employing the use of various disparate techniques for the modeling and simulation of complex communication networks. Rather than focus on simplistic models using simple numerical simulations, the book focuses instead on tools and techniques which can be used for the realistic modeling of large-scale and complex communication networks—termed collectively as Complex Adaptive COmmunicatiOn Networks and environmentS (CACOONS).

The book has been logically sectioned in three parts. The first part focuses on the importance of modeling and simulation and also gives two varied examples of unconventional but powerful tools which can be useful for any type of modeling and simulation, in general, and modeling and simulation of CACOONS, in particular.

This is followed by the second part which presents three critical reviews and surveys in the domain of modeling and simulation. The third part of the book focuses on practical case studies of modeling and simulation using different techniques. Of interest here is a focus on the use of the Cognitive Agent-based Computing (CABC[1]) framework. CABC framework can be used to model any type of complex physical system or complex adaptive system (CAS). As such, it can be a very useful approach to model and simulate any type of CACOONS. The third part presents several practical case studies employing the use of CABC framework in various areas of CACOONS. Next, I give an overview of the various chapters in a bit more detail.

The first part starts with the essence and importance of "Modeling and Simulation," presented by Ören et al. The chapter not only gives an overview of modeling and simulation but also presents taxonomies. The chapter first gives an overview of why simulation could be needed instead of an actual system. Then it moves on to taxonomies and ontologies for use in modeling and simulation.

The next chapter in part I presents a detailed overview of using Simio, a modern, sophisticated object-oriented tool for developing simulations of complex real-world systems. The chapter first starts out with an overview of the Simio object framework. This is followed by a description of Simio object classes. Next, concepts related to modeling movement are presented. This can be transformed to model not only messages in the Internet of Things but also people, mobile devices, and more. A description of modeling physical components is also presented. This is followed by techniques for modeling processes. The chapter concludes by giving an overview of process tables, API, experimentation, and useful applications in scheduling.

---

[1] Pronounced as Ka-bek.

Part I of the third chapter by Salva *et al*, presents a simulation environment for cybersecurity attack analysis for network traffic. The chapter gives an overview of simulation, emulation, and virtualization. After this, a network case study focusing on network anomaly detection is presented.

In the second part, there are three critical reviews, each analyzing key literature related to modeling and simulation in the domain of modeling CACOONS of various kinds. The first survey by Akram and Niazi presents demand response management in the domain of smart grid. The chapter first gives an overview of smart grid and how this particular domain offers unique challenges to develop and understand by means of modeling and simulation. It next presents an overview of the problem domain of demand–response management in large-scaled CACOONS in the domain of smart grid. In terms of approaches, the chapter first presents learning-based approaches. Subsequently, it focuses on the complexity inherent to smart grid domain. Before concluding, the chapter then moves on to open research problems and directions in the domain.

The second chapter in the second part is by Akram and Niazi focusing on the use of agent-based computing, multiagent systems, and agent-based modeling in the domain of smart grid. It starts with an overview of the concepts and moves to applications ranging from learning to more. It concludes after giving an overview of key-open problems and issues in the domain.

The final chapter in second part is by Ventrella *et al*. This chapter focuses on the scale-free network topologies giving a detailed overview as well as literature review and more. The chapter starts with an overview of concepts pertaining to mapping the Internet. It then gives concepts of using traceroute to mapping. This is followed by IP options, subnet discovery, and router-level mapping. The chapter then presents internet models with a focus on graph theoretic approaches. It gives an overview of relevant concepts ranging from the basics to topological concepts such as scale-free, power-law, among others. An overview of network topology generation is also presented. Afterwards, the chapter moves on to the key topic of interest—namely shortest path models.

In the final part of the book, six modeling and simulation case studies are presented. The studies have been selected based on the criteria that first, these will be of interest not only to modelers and simulation experts but also to researchers and practitioners in the domain of complex communication and social networks.

The first chapter in the part is focused on the important topic of accurate modeling of VoIP traffic in modern communication networks by Toral-Cruz *et al*. The chapter starts by giving an overview of the importance and complexity in VoIP traffic in large-scale networks. It next presents the concepts of why modern networks have evolved from simple packet networks to multiservice networks. Subsequently, the chapter moves on to the importance of QoS in VoIP networks besides presenting VoIP frameworks such as H.323 and SIP. It then formally describes and models concepts related to QoS, one-way delay, jitter, self-similar processes, and more.

The second chapter in the last part presents implementation of two framework levels from the CABC framework in the domain of Internet of Things. The chapter starts with concepts related to the CABC framework and the simulator of choice.

It then presents research questions in the domain of 5G and the IoT. It then presents detailed results and discussion in the domain.

The third chapter in the part is by Attaullah *et al.* and focuses on the use of the DescRiptivE Agent-based Modeling (DREAM) from the CABC framework for the modeling and simulation of the Chord peer-to-peer (P2P) protocol. The chapter first introduces the Chord protocol describing its inherent complexities requiring the use of more advanced modeling and simulation techniques. After the description of chord protocols, the chapter presents the DREAM for the protocol allowing for a quantitative description using complex network centralities. The chapter also presents detailed results from both PeerSim as well as NetLogo-based simulations besides comparing DREAM with the previous approach—the so-called ODD approach originating from the domain of ecology and having been traditionally used in the past to model agent-based and individual-based models.

This is followed by another chapter employing the use of DREAM and ODD to model a P2P protocol commonly known as the Kademlia protocol. The chapter first gives an overview of the protocol and the challenges associated with the complexity of P2P protocols. This is followed by ODD and DREAM models, results, discussion, and a detailed comparison.

BitTorrent is a very commonly used P2P protocol in the real world. The next chapter in the part presents the use of the DREAM modeling level of the CABC framework for the modeling and simulation of the BitTorrent protocol. After presenting the background and overview of the torrent protocol, the chapter presents a BitTorrent case study for use in the simulation model. Next, ODD and DREAM are presented before a set of detailed discussion on the utility of the CABC framework in the modeling and simulation of CACOONS.

The final chapter in the book is by Khan and Niazi and presents the application of CABC level 1—complex network modeling level for the use of complex citation networks to analyze the domain of "Social networks." The chapter starts by introducing related concepts focused on measuring impact, citations, and scientometrics. It then presents the dataset retrieved for developing the complex citation networks. This is followed by a detailed network analysis demonstrating how this approach can be used to model, simulate, transform, and analyze various types of complex networks data.

The book presents first steps in the domain of consolidating material specifically focused on the modeling and simulation of complex communication networks—CACOONS. It presents a selection of key case studies as well as concepts with a primary focus on making the concepts accessible to a wide audience. However, like any text in such a large and vibrant domain, it is understandable that we were only able to present a sampling of key case studies and modeling paradigms in the domain. Readers are further recommended to follow Springer-Nature CASs Modeling journal for gaining access to more case studies and applications in the domain of modeling and simulation of complex communication networks—CACOONS.

While we have tried our level best to minimize errors, it is impossible to minimize all errors. If the book looks nice, it is all due to the efforts put in by the IET staff. And if there are any mistakes, I humbly accept them to be mine. As such, it is requested to kindly do keep sending your valuable and kind feedback and comments to the book editor at muaz.niazi@ieee.org.

Part I

Modeling and simulation

Chapter 1

Modeling and simulation: the essence and increasing importance

Tuncer Ören[1], Saurabh Mittal[2], and Umut Durak[3]

The technical aspects of the essence of simulation are elaborated based on the following definition: simulation is performing a goal-directed experimentation or gaining experience under controlled conditions by using dynamic models either to develop/enhance skills or for entertainment; where a dynamic model denotes a model for which behaviour and/or structure is variable over time. Hence, experimentation and experience aspects are explained. Several taxonomies, ontologies, and some ontology-based dictionaries are cited for a comprehensive and integrative perception of simulation. Finally, the evolution and increasing importance of simulation is explained.

1.1 Introduction

'Simulation as a discipline is like mathematics and logic. It can be studied per se to develop its own theories, methodologies, and tools, and it can be used in a multitude of problem areas in many disciplines. The uses of simulation involve this second aspect and make it a vital enabling technology for many disciplines'. The above is from the conclusion section of another publication [1]. A recent publication 'Guide to Simulation-Based Disciplines: Advancing our Computational Future' elaborates on the universality of simulation [2], and another publication 'The Profession of Modeling and Simulation' casts light on the professional aspect of simulation [3]. The clarifications given in this chapter on many aspects of simulation are relevant to the universality of simulation.

The term simulation has been in existence in English since fourteenth century. Its meaning is based on the concept of similarity. Depending on the goal of the similarity, the original non-technical use of the term simulation has positive and negative connotations. From a positive point of view, simulation implies imitation such as simulated leather or simulated pearl. From a negative point of view, simulation

[1]School of Electrical Engineering and Computer Science, University of Ottawa, Ontario, Canada
[2]The MITRE Corporation, United States
[3]German Aerospace Center (DLR), Institute of Flight Systems, Germany

implies disguised reality, e.g. counterfeit, feigning, false show, and hypocrisy. Later, the term simulation acquired technical meanings. However, still the term is also used with its original non-technical connotations. To denote its technical aspects, we use the following concise and comprehensive definition:

> Simulation is performing goal-directed experimentation, or gaining experience under controlled conditions by using dynamic models either to develop/enhance skills, or for entertainment; where a dynamic model denotes a model for which behaviour and/or structure is variable over time.

Due to its many aspects, there are many definitions of simulation. About 100 definitions of simulation were compiled and presented in nine categories by Ören [4], and a critical review of them was offered in a sequel publication [5]. As a testimony of the variety of simulation, Appendix A lists over 750 types of simulation. Appendix B, a list of 120 types of input variables, is yet another testimony of the richness of the field.

M&S is essentially composed of two separate activities: modelling and simulation. While modelling necessitates abstraction, simulation is purely an engineering activity that involves expertise from computer science and engineering discipline [2]. When we talk of simulation as a singular activity, it subsumes model building. From historical evidence, model building has been attempted by various non-technical means and a constant engagement with the problem-at-hand or the question under exploration. Model development has been undertaken in different disciplines in diverse manner. Some examples are as follows:

- *Engineering*: Model building is done for two purposes: *design* and *control*. The design aspect involves creating model(s) of a 'would be' system. The control aspect necessitates building the model of an 'existing system' that needs exploration of various control algorithms and mechanisms.
- *Science*: Model building is done to understand a natural phenomenon. New nomenclature, taxonomy, vocabulary and abstractions are developed. The critical part is the specification of assumptions that limit the complexity of real world in the model description.
- *Education*: Model building is done to explain, teach, understand, or learn a real-world phenomenon. The abstraction level is dependent upon the audience that is undergoing learning.
- *Training*: Model building is done to impart training or enhance skills (motor, decision-making and communication, and operation) of the trainee in a specific complex environment where it is cost prohibitive to involve real-world assets and systems.
- *Entertainment*: Model building is done to provide a fictional reality in real or staged environments for amusement purposes.
- *Decision support:* Model building is done to evaluate various courses of action of a real-world state of a system on an existing model. In such cases, it is cost prohibitive to perform real-world evaluation due to danger to life and property.

In all the disciplines mentioned above, model building incorporates the skill of developing abstractions. The determination of an abstraction level is contingent upon various factors such as the problem-at-hand, the desired goal, the available tools, and the available knowledge. For example, in each economic era, from the age of farming to Industrial Age and to the Information Era we currently are in, the problems, the desired goals, the tools, and the availability of knowledge have evolved, leading to new representation of models. Some models that describe the natural laws have withstood the test of time for example Newton Laws developed in eighteenth century, and sometimes, a completely known theory developed in twentieth century such as quantum mechanics fundamentally changes the perception of reality. Each economic era has led to the evolution of these four aspects and, consequently, model building has evolved accordingly. Model building takes its refuge in mathematics at the core level and involves constant subject–environment engagement to keep the developed abstractions attuned to the problem-at-hand. In times, today, much of the model development has moved to computerized workbenches, often called integrated development environments that bridge the gap between the model builder and the model representation.

The simulation activity builds upon the model-building activity and presents the challenge of running the model over time. In a computational environment, a simulator (a software entity) is tasked with managing the advancement of time. In a non-computational environment, the perception of movement of time becomes a critical factor in determining how effective the simulation is. For example, in a stage or theatre, if the modelled 'act' is executed in slower time or faster than real time, it would yield a completely different experience to the audience. Likewise, in a computational environment, the advancement of time delivers results that may or may not address the problem-at-hand. Handling time on an appropriate time base then becomes a paramount activity in simulation.

In the following sections, the following is done. Experimentation aspects of simulation are discussed in Section 1.2. In Section 1.3, experience aspects to develop/enhance three types of skills or for entertainment are discussed. Taxonomies and ontologies of simulation are mentioned in Section 1.4. Evolution and increasing importance of simulation is discussed in Section 1.5, and last section is for conclusion.

1.2   Experimentation aspects of simulation

From experimentation aspect, a concise definition of simulation is as follows: simulation is a goal-directed experimentation with dynamic models. Experimental conditions can be formally specified by experimental frames [6] and by experimentation scenarios.

Since Francis Bacon's Novum Organum (The New Organon) published in 1620, experimentation has been the essence of scientific approach. Simulation extends the scope of experimentation to many cases:

1. Real system *may not exist* (as in engineering problems where new systems are aimed to be built).
2. Real system may *not be reachable* for experimentation (e.g. testing lunar vehicles).

3. Experimentation on real system *may be dangerous* (e.g. testing nuclear detonation or simulation of forest fires).
4. Experimentation on real system *may not be convenient* (e.g. urban traffic simulation, instead of experimenting on the real system).
5. Experimentation on real system *may not be convenient time-wise* (economic systems would necessitate long time for experimentation, and some natural phenomena would be too fast to observe).
6. Experimentation on real system *may not be cost-effective*. Furthermore, testing a virtual prototype (via simulation) allows flexibility in finding optimal design.

Simulated experiments are used for description/explanation, decision support, exploration, and teaching/learning. For decision support alone, simulation can be used for the prediction of behaviour and/or performance, test of hypotheses about models or experimental conditions, sensitivity analysis of behaviour or performances, virtual prototyping, planning, acquisition, testing, and proof of concepts, as well as for evaluation of alternative models and experimental conditions. Since experimentation aspect of simulation is very widely used, several publications exist [7,8].

1.3   Experience aspects of simulation

Gaining experience under controlled conditions using dynamic models is one of the major motivations of simulation [9]. Experimentation aspect includes two major use cases: (a) training to develop or enhance skills and (b) entertainment. These use cases will be further elaborated in the following sections.

1.3.1   Simulation for training

Simulation is used in training to enhance or develop three types of skills, namely motor skills, decision-making skills, and operational skills. These application areas lead the community to a taxonomy of simulations: live simulations for operational skills, virtual simulations for motor skills, and constructive simulations for decision-making skills.

In live simulations, real people use real (or imitation) equipment in the real world. It puts the real equipment at work interconnected with computers where the assessment of trainee actions is conducted by computer algorithms [10]. They are used to train operational skills by using real life-like experience under controlled conditions. Live simulation has been a major method in defence for training warfighting techniques and tactics which inevitably require large number of assets and personal. With the advances in information and communication technologies, the trainings started to make use of virtual and constructive simulations to achieve a cost-effective mixture [11].

Following the definitions given by Ören in [7], in virtual simulations, real people use virtual equipment in virtual environments in order to enhance motor skills to gain proficiency in using the equipment. Typical examples of this type are flight simulators which have long been establishing training aids in aviation. Link trainer (sometimes called Blue Box) (Figure 1.1) is recognized as one of the first flight simulators. Ten

Figure 1.1   Link trainer in a US Army Air Force field

thousand link trainers were manufactured from 1934 to 1950 [12]. They provided means for training basic motor skill for pilots. The current flight simulator market is about USD 6 Billion and 2021 forecast is about USD 7.5 Billon [13].

In constructive simulation, simulated people use simulated equipment in virtual environment. The aim is to enhance decision-making and communication skills of trainees through interactions with the simulation systems. Air traffic control simulation systems are one of the typical examples [14]. Simulated pilots use simulated aircrafts in air traffic control simulation systems where the simulation provides the possibility to train controllers for decision-making and communication skills. One of the commercial-off-the-shelf products is MaxSim – air traffic control simulators from Adacel which can generate realistic air traffic based on defined scenarios and provides direct voice communication possibilities with virtual pilots via speech recognition features [15].

One of the key issues of simulation-based training is transfer of training which is defined as the degree to which trainees effectively apply the trained skills in real operation [16]. The research about transfer of training in flight simulators has quite a long history. Valverde has published a paper in 1973 that provides a review of flight simulator transfer of training studies since the 1950s [17]. In one of the recent studies, Pool and Zaal present a cybernetic approach to assess the transfer of training for manual control skills in flight simulators using multi-channel pilot models [18]. The fidelity, immersion, presence, and buy-in are defined as the four factors that drive the transfer of training [19]. Fidelity is defined as the extent to which the simulation matches the real world. While the immersion is the feeling of the individual to be absorbed by the experience, in the situated immersion, the presence is defined as the subjective experience of existence within the simulation [20]. Buy-in is eventually the user's acceptance of the experience as a useful training event.

1.3.2   Simulation for entertainment

Simulation is one of the key concepts of entertainment. It is not only used for interactive entertainment purposes such as computer games but also for computer-generated animation. With the advances in computer architectures, the interactive entertainment industry started to work on high quality motion synthesis, and physics-based simulation emerged as a field of study [21]. It was first referred as the study of the motion of virtual objects using the laws of motion. It evolved in years and now covers rigid body dynamics to impact and collision, deformable bodies to soft bodies, fluids and gasses to crowd simulation [22]. During the last decade, physics engines which can be defined as reusable simulation libraries for interactive entertainment became popular [23]. One of the well-employed open-source physics engines is Bullet Physics Library [24]. Its features include rigid-body and soft-body simulation and collision detection. Integrations plug-ins are available for bullet in order to use it with Autodesk® Maya [25] or Blender [26] 3D computer graphics software. Well-known computer games that utilize Bullet for physics simulations include Red Dead Redemption [27] from Rockstar Games and Toy Story 3 [28] from Disney Interactive Studios. PhysX [29] has been introduced as a proprietary real-time multi-platform game physics solution by NVIDIA. It also provides simulation-driven effects such as destruction or cloth simulation. Havok [30] is another major commercial multi-platform multi-threaded physics simulation library for interactive entertainment. It has been used in more than 400 games that have been published. In an ecosystem with various physical simulation, application programming interfaces exist; there is also an effort, namely Physics Abstraction Layer, which provide a high-level interface for various physics simulation libraries [31].

Physics-based simulation methods that are employed in interactive entertainment are also well-applicable in computer-generated animation to synthesize motion of objects. Non-real-time characteristics of physics-based simulations in animation applications allow further resolution in modelled behaviour. One of the recent examples is artistic simulation of curly hair by Disney/Pixar [32]. It has been used for animating Merida [33] in a famous animation film Brave [34]. Open source 3D animation software Blender also provides many physics-based simulation features such as fluids, smoke, cloth, or hair simulation. Reference [35] from Mullen which titled as 'Bounce, tumble, and splash!: simulating the physical world with Blender 3D' can be introduced as a reference book about physics-based simulation in Blender.

1.4   Taxonomies and ontologies of simulation

1.4.1   Background

Taxonomy, as the science of classification, is an indispensable aspect of scientific studies and is concerned with finding, describing, classifying, and naming of things. For example, taxonomies of plants and animals identify logical relationships of different species. In animal taxonomy, a living organism is assigned successively in a kingdom, a phylum, class, order, family, genus, and species. Another example is taxonomy of learning and Bloom's taxonomy of educational objectives [36]. Taxonomy of learning and Bloom's taxonomy of educational objectives are particularly important in simulation-based learning/teaching [37]. This importance is due to the fact that simulation-based disciplines are proliferating and teaching/learning them may satisfy different goals and hence should be targeted properly [2,9].

As a branch of metaphysics, *ontology* is concerned with the nature and relations of beings and deals with abstract entities.

Ontology-based dictionaries combine classification of entities and their definitions [38]. They show logical relationships of several related terms as well as their definitions; hence, they are helpful in teaching/understanding even the subtle differences of related terms.

1.4.2   Taxonomies of simulation

Perception of the big picture and richness of simulation [7,39,40] as well as discriminating the subtle differences in its many aspects would help appreciating many possibilities it offers. Due to the increasing importance, versatility, and high variety of simulation, several classification studies already exist. These taxonomies can be conceived under the following major groups: taxonomies of simulation, simulation languages, simulation software, simulation models, behaviour and processing of simulation models, simulation quality assurance, and elements of simulation models and simulation-based experimentations as well as simulation-based experiences. The taxonomies represent state of the art at the time of their preparations with rooms for future developments.

Taxonomies of simulation were developed by Ören [41] and Sulistio *et al.* [42]. The last one was reviewed by Ören [43]. Maier and Größler [44] developed a taxonomy of computer simulations to support learning about socio-economic systems. Roeder's Ph.D. thesis is on a taxonomy for discrete event simulations [45].

Taxonomies of simulation languages: An early taxonomy of simulation languages was developed by Ören [46]. Two updated versions of taxonomy of simulation languages are also by Ören [47,48].

Taxonomies of simulation software: A survey on taxonomies of discrete simulation software is prepared recently by Rachidi [49].

Taxonomies of simulation models: An early taxonomy of simulation models was published by Ören [50]. A taxonomy of symbolic processing of simulation models was published in 1987 [51]. Fishwick prepared a taxonomy of simulation modelling [52]. Hare and Deadman published a taxonomy of agent-based simulation models in environmental management [53]. A review of it was prepared by Ören [54]. Taxonomies of multimodels are prepared by Yilmaz and Ören [55] and by Ören *et al.* [56]. Lynch and Diallo prepared a taxonomy for classifying terminologies that describe simulations with multiple models [57]. Smith elaborates on the value of taxonomy in modelling [58].

Taxonomies of behaviour of simulation models: A taxonomy of types, as well as generation and processing techniques of model behaviour was prepared by Ören [59].

Taxonomy of more specific topics: As an example, an early taxonomy of input variables can be cited [60]. Wenzel *et al.* published a taxonomy of visualization techniques for simulation in production and logistics [61]. Le Digabel and Wild have a taxonomy of constraints in simulation-based optimization [62]. Goldstein and Khan prepared a taxonomy of event-time representations [63].

Due to the richness of modelling and simulation and its relationship with other relevant disciplines, several other taxonomies of specific topics will be useful. Even the most fundamental concepts have several terms to represent nuances. For example, there are over 150 terms related with 'variables', over 90 terms related with 'values', and over 1,000 terms related with or representing types of models (M&S Bok Index studies). To attest the richness of the field two appendices are given. Appendix A is a list of over 750 types of simulation and Appendix B is a list of 120 types of input.

1.4.3   Ontologies of simulation

Silver *et al.* prepared an ontology for discrete-event modelling and simulation [64]. The book edited by Tolk [65] is a very good source of information about simulation ontologies. From Tolk's book, the following are noteworthy contributions to simulation ontology: Partridge *et al.* [66]; Hofmann [67]; Heath and Jackson [68]; and Wang *et al.* [69]. An ontology for simulation systems engineering is developed by Durak and Ören [70].

An ontology-based dictionary of multimodels was prepared by Ören, Mittal, and Durak [9]. An ontology-based dictionary of machine understanding can be used for simulating systems with understanding abilities including systems able to understand emotions [38].

As a normative view, we think that development of new and updated as well as more diversified taxonomies, ontologies, and ontology-based dictionaries may be useful for learning several aspects of simulation, since it is progressing very rapidly and becoming infrastructure for many disciplines.

1.5   Evolution and increasing importance of simulation

A recent publication [9] clarifies nine phases of the evolution of simulation and documents the vital role of simulation for many disciplines:

1. In the pre-computer days, simulation started as non-computerized simulation such as thought experiments, role-playing, and sand-box simulation.
2. Computerized simulation was an important step in the evolution of simulation. At the beginning, the role of computers was limited to generation of model's behaviour (mostly trajectory and, sometimes, structural). With the advent of more powerful computers and maturity of simulation, computers were also used for the specification of simulation study as well as for other functions [71].
3. Contribution of system theories to simulation was the essence of formal simulation [72] especially in DEVS (Discrete Event System Specification) [73].
4. Contribution of machine intelligence (or artificial intelligence – AI) opens a new dimension for the full synergy of simulation and AI which consists of contribution of simulation to AI and contribution of AI to simulation. At the beginning, AI studies started as the simulation of natural intelligence [74,75]. Contribution of AI to simulation provided advanced (intelligent) knowledge-processing abilities to simulation systems [76].
5. Software agents are associated with the concept of autonomy (or quasi-autonomy). Similar to the synergy of simulation and AI, full synergy of simulation and software agents opened the possibility of agent-directed simulation which

consists of contribution of simulation to agents or agent-based simulation where intelligent and autonomous entities can be simulated and contributions of agents to simulation. The second aspect consists of contribution of agents during run-time, such as agent-monitored simulation, and contribution of agents to the simulation systems such as simulation systems with several types of understanding abilities [77].

6. When advances of computing reached the level of soft computing, fuzzy computation, evolutionary computation including genetic algorithms, Bayesian computation, and machine learning were part of the possibilities of computation. The contribution of soft computing to simulation is soft simulation.

7. Two types of complexity, namely complexity of simulation systems and especially complexity of simulands, laid the path to the synergy of simulation systems engineering [78]. Indeed, simulation systems engineering provides a proper paradigm for simulation to tackle complex problems.

8. Many advantages of experimentation and experience aspects of simulation are the essence of simulation-based activities which provide an indispensable and often a vital infrastructure for many disciplines. A recent volume, 'Guide to Simulation-Based Disciplines: Advancing our Computational Future' by Mittal *et al.* [2] covers in depth this phase of simulation.

9. Widespread availability and hence usage of simulation-as-a-service (SaaS) is a desirable phase for the maturity of the simulation discipline. This would also increase the usability and hence usefulness of simulation among many who are expert in their field of specialization but not in simulation techniques [79]. SaaS can be used by human users as well as by advanced agent-monitored systems.

The evolution of simulation at each phase of advancement brought to its users more and more advanced and powerful possibilities in simulation-based problem solving. Hence, the shift of paradigm from model-based approach [80] to simulation-based approach [9] appears to be very beneficial for many disciplines.

1.6   Conclusion

Experimentation and experience aspects of simulation have already made it an invaluable infrastructure for many disciplines and application areas. In this chapter, evolution and increasing importance of simulation are elaborated after clarifications of its experimentation and experience aspects. A comprehensive and integrative view of simulation would be helpful to appreciate many advantages it offers. For this reason, many taxonomies and ontologies and some ontology-based dictionaries are also presented.

Disclaimer

The author's affiliation with The MITRE Corporation is provided for identification purposes only, and is not intended to convey or imply MITRE's concurrence with, or support for, the positions, opinions or viewpoints expressed by the author(s). Approved for Public Release, Distribution Unlimited [Case Number: PR_17-3254-2].

Appendix A – A list of over 750 types of simulation

3-d simulation

A–
ab initio simulation
abstract simulation
academic simulation
accelerated simulation
accurate simulation
acoustic simulation
activity-based simulation
actor-based simulation
ad hoc distributed
  simulation
adaptive simulation
adaptive symbiotic
  multisimulation
adaptive-system
  simulation
adiabatic-system
  simulation
advanced distributed
  simulation
advanced numerical
  simulation
advanced simulation
agent-based
  multisimulation
agent-based participatory
  simulation
agent-based simulation
agent-controlled
  simulation
agent-coordinated
  simulation
agent-directed simulation
agent-initiated simulation
agent-mediated
  simulation
agent-monitored
  anticipatory
  multisimulation
agent-monitored
  multisimulation

agent-monitored
  simulation
agent simulation
agent-supported
  multisimulation
agent-supported
  simulation
agent-triggered
  simulation
aggregate-level
  simulation
agile simulation
AI-controlled simulation
AI-directed simulation
all-digital analog
  simulation
all-digital simulation
all-software simulation
allopoietic simulation
allotelic system
  simulation
ALSP-compliant
  simulation
alternative simulation
analog-computer
  simulation
analog multilevel
  simulation
analog simulation
analytic simulation
analytical simulation
ancestor simulation
anticipatory
  multisimulation
anticipatory perceptual
  simulation
anticipatory simulation
appropriate simulation
approximate simulation
approximate
  zero-variance simulation
array simulation
art-directed simulation

artistic simulation
as-fast-as-possible
  simulation
asymmetric simulation
asynchronous simulation
atomistic simulation
audio simulation
augmented live
  simulation
augmented physical
  simulation
augmented-reality
  simulation
augmented simulation
autopoietic simulation
autosimulation
autotelic system
  simulation

B–
backward simulation
base case simulation
baseline scenario
  simulation
baseline simulation
batch simulation
behaviourally adaptive
  simulation
behaviourally
  anticipatory
  multisimulation
behaviourally
  anticipatory simulation
big simulation
bio-inspired simulation
bio-nano simulation
biologically inspired
  simulation
biomimetic simulation
biosimulation
bisimulation
blended learning
  simulation

blended simulation
block-oriented simulation
bond graph simulation
boundary value
  simulation
branched simulation
built-in simulation

C–
case-based simulation
catastrophic simulation
cellular automaton
  simulation
centre-based simulation
chaotic simulation
classical simulation
closed-form simulation
closed-loop real-time
  simulation
closed-loop simulation
cloud-based simulation
cloud-hosted simulation
cloud simulation
cluster simulation
coercible simulation
coercion simulation
coercivity simulation
coersing simulation
cognitive simulation
cokriging simulation
collaborative
  component-based
  simulation
collaborative DEVS
  simulation
collaborative
  distributed-simulation
collaborative simulation
collaborative virtual
  simulation
collocated cokriging
  simulation
collocated cosimulation
collocated simulation

combined
  continuous/discrete
  simulation
combined simulation
combined-system
  simulation
common-use simulation
communal simulation
competition simulation
competitive simulation
complete simulation
component-based
  collaborated simulation
component-based
  distributed simulation
component-based
  simulation
component simulation
composable simulation
composite simulation
compressed-time
  simulation
computational simulation
computer-aided
  simulation
computer-assisted
  simulation
computer-based
  simulation
computer-mediated
  simulation
computer-network
  simulation
computer simulation
computerized simulation
conceptual simulation
concurrent simulation
condensed-time
  simulation
conditional Monte Carlo
  simulation
conditional simulation
conjoint simulation
conservative event
  simulation

conservative parallel
  simulation
conservative simulation
constrained simulation
constructive simulation
constructive training
  simulation
context-free simulation
continuous-change
  simulation
continuous simulation
continuous-system
  simulation
continuous-time
  simulation
continuous-time
  continuous simulation
conventional simulation
convergence simulation
convergent boundary
  simulation
convergent simulation
converging simulation
cooperation simulation
cooperative simulation
coopetition simulation
coopetitive simulation
cosimulation
coupled simulation
credible simulation
critical-event simulation
crowd simulation
customizable simulation
customized simulation
cyber-physical system
  simulation
cycle-based simulation

D–
data-driven simulation
data-intensive simulation
decision simulation
demon-controlled
  simulation

DES (Discrete Event
  Simulation)
descriptive simulation
detached eddy simulation
deterministic simulation
DEVS simulation
difference equation
  simulation
digital-analog simulation
digital-computer
  simulation
digital quantum
  simulation
digital simulation
direct numerical
  simulation
direct simulation
DIS (Distributed
  Interactive Simulation)
disconnected simulation
discontinuous simulation
discrete-arithmetic-based
  simulation
discrete-change
  simulation
discrete-event line
  simulation
discrete-event simulation
discrete simulation
discrete-system
  simulation
discrete-time continuous
  simulation
discrete-time simulation
discrete-variable
  simulation optimization
display-based simulation
dissimilar simulation
dissimulation
distributed agent
  simulation
distributed asymmetric
  simulation
distributed DEVS
  simulation

distributed event-driven
  simulation
distributed heterogeneous
  simulation
distributed interactive
  simulation
distributed lazy
  simulation
distributed-parameter
  system simulation
distributed real-time
  simulation
distributed simulation
distributed web-based
  simulation
DNA-based simulation
dynamic data-driven
  simulation
dynamic simulation
dynamic system
  simulation
dynamically composable
  simulation

E–
eddy simulation
edge simulation
emergence simulation
emergency egress
  simulation
emergent behaviour
  simulation
emergent simulation
emulation
endomorphic simulation
engineering simulation
entertainment simulation
entity-level simulation
equation-oriented
  simulation
error-controlled
  simulation
escapist simulation
ethical simulation

evaluative simulation
event-based agent
  simulation
event-based discrete
  simulation
event-based simulation
event-driven simulation
event-following
  simulation
event-oriented simulation
event-scheduling
  simulation
evolutionary simulation
evolutionary-system
  simulation
ex ante simulation
ex post simulation
ex situ simulation
exascale simulation
expanded-time simulation
explanatory simulation
exploration simulation
exploratory
  multisimulation
exploratory-
  multisimulation
  methodology
exploratory simulation
extensible simulation
extreme scale simulation
extrinsic simulation

F–
fast simulation
fair simulation
fault simulation
fault-tolerant simulation
faulty simulation
federated simulation
field simulation
finite-state machine
  simulation
first-degree simulation
fixed-topology simulation

flattened G-DEVS
  simulation
flattened simulation
fluid simulation
forward multisimulation
full cosimulation
full-immersive simulation
full-system simulation
fully coupled simulation
functional simulation
fuzzy simulation
fuzzy [system] simulation

G–
gate-level simulation
Gaussian copula
  simulation
Gaussian distribution
  simulation
Gaussian random
  function simulation
Gaussian sequential
  simulation
Gaussian simulation
gedanken simulation
general purpose
  distributed simulation
generalized mixed-mode
  simulation
generalized simulation
generative
  multisimulation
generative parallax
  simulation
generative simulation
generic simulation
genetic-algorithm
  simulation
goal-directed system
  simulation
goal-free simulation
goal-generating system
  simulation
goal-oriented system
  simulation

goal-processing system
  simulation
goal-setting system
  simulation
goal-determined
  simulation
goal-regression
  simulation
goal-seeking simulation
graphical simulation
grid-based simulation
grid simulation

H–
hand simulation
hands-on simulation
hard simulation
hardware-in-the-loop
  simulation
heterogeneous simulation
hierarchical simulation
high-fidelity simulation
high-level simulation
high-performance
  simulation
high-resolution
  simulation
high-speed simulation
historical simulation
HLA-based simulation
HLA-compliant
  simulation
holistic simulation
holographic simulation
holonic agent simulation
holonic simulation
holonic [system]
  simulation
human-centred
  simulation
human-in-the-loop
  simulation
human-initiated
  simulation

human-machine
  simulation
hybrid computer
  simulation
hybrid simulation

I–
ideal-seeking simulation
identity simulation
immersive simulation
imperative-driven
  simulation
importance-sampling-
  based
  simulation
in-basket simulation
in context simulation
in silico simulation
in situ simulation
in the large simulation
in the small simulation
in vitro simulation
in vivo simulation
inappropriate simulation
incremental simulation
indirect simulation
individual-based
  simulation
individual simulation
inductive simulation
industrial scale
  simulation
instructional simulation
integrated simulation
intelligent simulation
intelligent-system
  simulation
interaction-based
  simulation
interactive simulation
interactive graphical
  simulation
intermittent simulation
interoperable simulation

interpretational simulation

interpretive simulation

interval-oriented simulation

intractable simulation

intrinsic simulation

introspective simulation

inverse ontomimetic simulation

inverse simulation

J–

joint simulation

K–

knowledge-based simulation

kriging simulation

L–

L-system simulation

laboratory simulation

large eddy simulation

large-scale simulation

large simulation

lazy simulation

lean simulation

legacy simulation

library-driven simulation

Lindenmayer [system] simulation

line-of-sight simulation

linear [system] simulation

live instrumented simulation

live simulation

live system-enriching simulation

live system-supporting simulation

live training simulation

logic simulation

logical simulation

loosely coupled federated simulation

low-fidelity simulation

low-level simulation

ludic simulation

M–

machine-centred simulation

machine simulation

maintenance simulation

man-centred simulation

man-in-the-loop simulation

man-machine simulation

man-machine [system] simulation

manual simulation

Markov chain simulation

Markov simulation

massive-scale simulation

massively multiplayer simulation

mathematical simulation

mental simulation

mesh-based simulation

meshfree simulation

mesoscale simulation

metamodel-based simulation

metamorphic simulation

metasimulation

microanalytic simulation

microcomputer simulation

microgrid simulation

microsimulation

mission-level simulation

mission rehearsal simulation

mixed-mode simulation

mixed-signal simulation

mixed simulation

mobile-device activated simulation

mobile-device initiated simulation

mobile-device-triggered simulation

mobile simulation

mock simulation

model-based simulation

modular simulation

Monte Carlo simulation

moving-boundary simulation

multiagent-based simulation

multiagent participatory-simulation

multiagent-based simulation

multiagent simulation

multiagent-supported simulation

multiaspect multisimulation

multiaspect simulation

multibody simulation

multilevel simulation

multilingual simulation

multimedia-enriched simulation

multimedia simulation

multimethod simulation

multiparadigm simulation

multiperspective simulation

multiphysics simulation

multiplayer simulation

multiple-fidelity simulation

multiple-run simulation

multiprocessor simulation

multirate simulation

multiresolution multisimulation

multiresolution simulation

multiscale simulation

multisimulation
multistage
  multisimulation
multistage simulation
multi-user simulation
mutual simulation

N–
N-body simulation
nanoscale simulation
nano simulation
narrative simulation
nested multisimulation
nested simulation
netcentric simulation
network-oriented
  simulation
networked simulation
non-anticipatory
  simulation
non-convergent
  simulation
non-equation-oriented
  simulation
non-HLA-compliant
  simulation
non-line-of-sight
  simulation
non-stiff simulation
non-terminating
  simulation
non-zero-sum simulation
nonconvergent simulation
nondeterministic
  simulation
nondeterministic Turing
  machine simulation
nonlinear [system]
  simulation
nonnumerical simulation
normative simulation
null Turing machine
  simulation
numerical simulation

O–
object-oriented
  simulation
offline simulation
on-demand simulation
online role-play
  simulation
online simulation
ontology-based agent
  simulation
ontology-based
  multiagent simulation
ontology-based
  simulation
ontomimetic simulation
open form simulation
open loop simulation
open source simulation
optimistic parallel
  simulation
optimistic simulation
optimizing simulation
ordinary differential
  equation simulation
ordinary kriging
  simulation
outcome-driven
  simulation
outcome-oriented
  simulation

P–
parallax simulation
parallel discrete-event
  simulation
parallel distributed
  simulation
parallel simulation
parallelized simulation
partial differential
  equation simulation
partial equilibrium
  simulation
partial simulation
participative simulation

participatory agent
  simulation
participatory simulation
pathway simulation
peace simulation
pedagogical simulation
peer-to-peer simulation
pen and paper simulation
perceptual simulation
pervasive simulation
petascale simulation
Petri net simulation
physical simulation
physical [system]
  simulation
physics-based simulation
portable simulation
prediction simulation
predictive biosimulation
predictive simulation
prescriptive simulation
probabilistic
  bisimulation
process-based
  discrete-event
  simulation
process interaction
  simulation
process-oriented
  simulation
process simulation
program-based
  simulation
program-oriented
  simulation
proof-of-concept
  simulation
proxy simulation
pseudo-analytical
  simulation
pseudosimulation
public domain simulation
pure software simulation
purposeful simulation

Q–
qualitative simulation
qualitative mixed
   simulation
quantitative mixed
   simulation
quantitative simulation
quantum simulation
quasi-analytic simulation
quasi-continuous
   simulation
quasi-identity simulation
quasi-Monte Carlo
   simulation
quasi-identity simulation
queue simulation

R–
random simulation
rare-event simulation
rate-based simulation
ratio simulation
real-life simulation
real-system enriching
   simulation
real-system support
   simulation
real-time continuous
   simulation
real-time data-driven
   simulation
real-time
   decision-making
   simulation
real-time simulation
reasonable simulation
reasoning simulation
reconfigurable simulation
recursive simulation
reflective simulation
regenerative simulation
regular simulation
related simulation
reliable simulation
remote simulation

replicative simulation
reproducible simulation
resimulation
retrosimulation
retrospective simulation
reverse simulation
rigid-body simulation
risk simulation
role-play simulation
role playing simulation
rollback-based discrete
   simulation
rollback-based parallel
   discrete-event
   simulation
rollback-based simulation
rule-based simulation
rule-based system
   embedded simulation

S–
sandbox-style simulation
scalable simulation
scaled real-time
   simulation
scenario simulation
scientific simulation
second degree simulation
self-adaptive simulation
self-driven simulation
self-learning simulation
self-organizing
   simulation
self-organizing system
   simulation
self-regulating simulation
self-replicating system
   simulation
self-simulation
self-stabilizing system
   simulation
semiotic simulation
sensitivity simulation
sequential Gaussian
   simulation

sequential simulation
serial simulation
serious simulation
service-based simulation
shape simulation
simulation
simultaneous simulation
single-aspect simulation
single-component
   simulation
single-processor
   simulation
single-run simulation
single-user simulation
skeleton-driven
   simulation
smart phone activated
   simulation
smoothness simulation
soft body simulation
soft computing
   simulation
soft simulation
software-based
   continuous system
   simulation
software-based
   discrete-event
   simulation
software-based
   simulation
software-in-the-loop
   simulation
spatial simulation
spatiotemporal simulation
spreadsheet simulation
stand-alone simulation
state-maintaining
   simulation
static simulation
statistical simulation
steady-state simulation
stiff simulation
stochastic simulation

strategic-decision simulation
strategic simulation
strategy simulation
strong bisimulation
strong simulation
strong-strong simulation
strong two-way coupling simulation
structural simulation
structure simulation
successor simulation
suitable simulation
survivability simulation
sustainable simulation
swarm simulation
symbiotic simulation
symbolic simulation
symmetric simulation
system dynamics simulation
system-level simulation
system of systems simulation
system simulation
system-theory-based continuous system simulation
system-theory-based discrete-event simulation
system-theory-based simulation
systematic simulation

T–
t-simulation
tactical decision simulation
tactical simulation
tandem simulation
technical simulation
technology-enhanced simulation

teleological system simulation
teleonomic [system] simulation
teleozetic simulation
terascale simulation
terminating simulation
test and evaluation simulation
texture simulation
third degree simulation
thought controlled simulation
thought experiment simulation
thought simulation
throttled time-warp simulation
time-domain simulation
time-driven simulation
time-interval simulation
time-slice simulation
time-slicing simulation
time-stepping simulation
time-varying system simulation
time-warp simulation
timing simulation
trace-driven simulation
tractable simulation
training simulation
trajectory simulation
transfer function simulation
transparent reality simulation
trial simulation
trustworthy simulation
Turing machine simulation
turning bands conditional simulation
turning bands simulation
two-level simulation

U–
ubiquitous computing simulation
ubiquitous simulation
ultrascale simulation
uncertainty simulation
unconditional simulation
unconstrained simulation
uncoupled simulation
unified discrete and continuous simulation
unified simulation
uniformization simulation
universal Turing machine simulation
unsuitable simulation
utilitarian simulation

V–
value-free simulation
variable fidelity simulation
variable resolution simulation
variable-topology simulation
very large eddy simulation
very large simulation
virtual reality simulation
virtual simulation
virtual system simulation
virtual time simulation
virtual training simulation
virtualization simulation
visual interactive simulation
visual simulation

W–
war simulation
warfare simulation
weak bisimulation

weak classical simulation
weak simulation
weak-timed mutual
  simulation
wearable computer-based
  simulation
wearable simulation

web-based multi-user
  simulation
web-based simulation
web-centric simulation
web-enabled simulation
web-service-based
  simulation

Y–
yoked simulation

Z–
zero-sum simulation
zero-variance simulation

Appendix B – A list of 120 types of input

3–
3-D positional input

A–
acoustic input
actively perceived input
adequate stimulus
admissible input
alphabetical input
alphanumeric input
alternative input
ambiguous input
analog input
AND input
antenna input
anticipated fact
anticipated input
assumed goal
asynchronous input
audio input
autostimulant
aversive stimulus

B–
balanced input
batch input
bounded input
brain controlled input
brain signal
buffered input

C–
clock input
command-driven input

common-mode input
composite video input
conditioned stimulus
constant input
context-aware input
context-sensitive input
context-unaware input
conventional input
converted sensory data
counter input
credible input

D–
data input
delimiterless input
detected event
digital input
direct input
discriminative stimulus
distracting input

E–
emergent input
emotional input
endogenous input
equivalent input
evaluated input
excitation
exogenous input
external event
external excitation
external input
externally generated
  input

F–
fixed input
forced input

G–
geopositional input
gesture input
global position sensing
  input

H–
hand-gesture input
haptic input
high-level stimulus

I–
imposed input
impulse
inadequate stimulus
indirect input
inflow
input
intake
intermediate input
internal excitation
internal input
internally generated input
inverted input
inverting input
irrelevant input
irrelevant stimulus

K–
keyboard input

L–
limited input

M–
manual input
marginal input
microphone input
monotonous input
motive
multimodal input
multiple input
multisensory input

N–
neutral input
nociceptive input
nociceptive stimulus
noisy input
non-inverting input
non-speech audio input
numerical input

O–
OR input

P–
paltry input
passively accepted
 input
perceived endogenous
 input

perceived exogenous
 input
perceived external input
perceived input
perceived internal input
perceptual input
positional input
primary input

Q–
quantized input

R–
radar input
real-time input
reference input
relevant input
relevant stimulus

S–
self-excitation
self-test input
semantic input
sensed input
sensor input
sensory input
simulated input
single input
soft sensor input
sonar input
speech input
standard input

step input
stimulant
stimulator
stimulus
stylus input
subliminal stimulus
subthreshold stimulus
synchronized input
synchronous input
syntactic input

T–
tactile input
trigger input
two directional input

U–
unambiguous input
unconventional input
uniform input
user input

V–
variable input
video input
virtual sensor input
vision input
visual input
voice input

W–
wireless input

References

[1]   Ören T.I. 'Uses of simulation' in Sokolowski J.A., Banks C.M. (eds.). *Principles of Modeling and Simulation: A Multidisciplinary Approach*. New Jersey: John Wiley; 2009. pp. 153–179.

[2]   Mittal S., Durak U., Ören T. (eds.). *Guide to Simulation-Based Disciplines: Advancing our Computational Future*. Cham: Springer; 2007.

[3]   Tolk A., Ören T. (eds.). *The Profession of Modeling and Simulation: Discipline, Ethics, Education, Vocation, Societies, and Economics*. Hoboken, NJ: John Wiley & Sons; 2017.

[4]   Ören T.I. 'The many facets of simulation through a collection of about 100 definitions'. *SCS M&S Magazine*. 2011, vol. 2(2), pp. 82–92.

[5]   Ören T.I. 'A critical review of definitions and about 400 types of modeling and simulation'. *SCS M&S Magazine*. 2011, vol. 2(3), pp. 142–151.

[6]   Ören T.I., Zeigler B.P. 'Concepts for advanced simulation methodologies'. *Simulation*. 1979, vol. 32(3), pp. 69–82.

[7]   Ören T.I. 'Modeling and simulation: A comprehensive and integrative view' in Yilmaz L., Ören T.I. (eds.). *Agent-Directed Simulation and Systems Engineering*. Berlin: Wiley; 2009. pp. 3–36.

[8]   Ören T.I., Yilmaz L. 'Philosophical aspects of modeling and simulation' in Tolk A. (ed.). *Ontology, Epistemology, and Teleology of M&S: Philosophical Foundations for Intelligent M&S Applications*. Berlin, Heidelberg (Germany): Springer-Verlag; 2013. pp. 157–172.

[9]   Ören T., Mittal S., Durak U. 'The evolution of simulation and its contributions to many disciplines' in Mittal S., Durak U., Ören T. (eds.). *Guide to Simulation-Based Disciplines: Advancing our Computational Future*. Cham (Switzerland): Springer; 2017. pp. 3–24.

[10]   Bruzzone A.G., Massei M. 'Simulation-based military training' in Mittal S., Durak U., Ören T. (eds.). *Guide to Simulation-Based Disciplines: Advancing our Computational Future*. Cham (Switzerland): Springer; 2017. pp. 315–362.

[11]   Bezdek W.J., Maleport J., Olshon R. 'Live, virtual & constructive simulation for real time rapid prototyping, experimentation and testing using network centric operations'. *AIAA Modeling and Simulation Technologies Conference and Exhibit*, Honolulu, HI, 2008.

[12]   De Angelo J., George L.S., Moody J. *The Link Flight Trainer: An Historic Mechanical Engineering Landmark*. ASME International, History and Heritage Committee, & Roberson Museum & Science Center, Binghamton, New York, 2000.

[13]   MarketsandMarkets. *Flight Simulator Market by Application (Military, Commercial), by Type of Flight (Fixed Wing, Rotary Wing, Unmanned Aircraft), Military Component (FFS, FMS, FTD), Commercial Component (FFS, FBS, FTD), Geography – Global Forecast to 2021* [online]. Available from http://www.marketsandmarkets.com/Market-Reports/flight-simulator-market-22246197.html [Accessed 09 Sep 2017].

[14]   Hopkin V.D. *Human Factors in Air Traffic Control*. Bristol, PA: CRC Press; 1995.

[15]   Adacel. *ATC Simulation and Training* [online]. Available from http://www.adacel.com/solutions_services/downloads/brochures/2017_MaxSim_WEB.pdf [Accessed 11 Sep 2017].

[16]   Baldwin T.T., Ford J.K. 'Transfer of training: A review and directions for future research'. *Personnel Psychology*. 1988, vol. 41(1), pp. 63–105.

[17]   Valverde H.H. 'A review of flight simulator transfer of training studies'. *Human Factors*. 1973 vol. 15(6), pp. 510–522.

[18]   Pool D.M., Zaal P.M.T. 'A cybernetic approach to assess the training of manual control skills'. *IFAC-PapersOnLine*. 2016 vol. 49(19), pp. 343–348.

[19]   Alexander A.L., Brunyé T., Sidman J., Weil S.A. *From Gaming to Training: A Review of Studies on Fidelity, Immersion, Presence, and Buy-In and Their Effects on Transfer in PC-Based Simulations and Games.* DARWARS Training Impact Group, Woburn, MA: 2005.

[20]   Witmer B., Singer M. *Measuring presence in virtual environments.* U.S. Army Research Institute for the Behavioral and Social Sciences Tech. Report No. 1014, 1994.

[21]   Yeh T.Y., Faloutsos P., Reinman G. 'Enabling real-time physics simulation in future interactive entertainment'. *Proceedings of the 2006 ACM SIGGRAPH Symposium on Videogames*; Boston, MI; 2006.

[22]   Eberly D.H. *Game Physics.* 2nd edition, Boca Raton, FL: CRC Press; 2010.

[23]   Millington I. *Game Physics Engine Development.* San Francisco, CA: Morgan Kaufmann Publishers; 2007.

[24]   *Bullet Physics Library* [online] Available from http://bulletphysics.org/wordpress/ [Accessed 11 Sep 2017].

[25]   *Autodesk® Maya* [online] Available from https://www.autodesk.de/products/maya [Accessed 11 Sep 2017].

[26]   *Blender* [online] Available from https://www.blender.org/ [Accessed 11 Sep 2017].

[27]   *Red Dead Redemption* [online] Available from http://www.rockstargames.com/games/info/reddeadredemption [Accessed 11 Sep 2017].

[28]   *Toy Story 3* [online] Available from http://games.disney.com.au/toy-story-3-video-game [Accessed 11 Sep 2017].

[29]   *GameWorks PhysX Overview* [online] Available from https://developer.nvidia.com/gameworks-physx-overview [Accessed 11 Sep 2017].

[30]   *Havok* [online] Available from https://www.havok.com/ [Accessed 11 Sep 2017].

[31]   Boeing A., Bräunl T. 'Evaluation of real-time physics simulation systems'. *Proceedings of the 5th International Conference on Computer Graphics and Interactive Techniques in Australia and Southeast Asia*; Perth, Australia; 2007.

[32]   Iben H., Meyer M., Petrovic L., Soares O., Anderson J., Witkin A. *Artistic simulation of curly hair.* Pixar Animation Studios Technical Memo 12-03a, 2012.

[33]   *Merida* [online] Available from http://princess.disney.com/merida [Accessed 11 Sep 2017]

[34]   *Brave* [online] Available from http://movies.disney.com/brave [Accessed 11 Sep 2017].

[35]   Mullen T. *Bounce, Tumble, and Splash!: Simulating the Physical World with Blender 3D.* Indianapolis, IN: John Wiley & Sons; 2008.

[36]   Anderson L.W. *A Taxonomy for Learning, Teaching, and Assessing: Pearson New International Edition: A Revision of Bloom's Taxonomy of Educational Objectives, Abridged Edition.* London, England: Pearson Education Limited; 2013.

[37]   Ören T., Mittal S., Turnitsa C., Diallo S.Y. Simulation-based learning and education disciplines' in Mittal S., Durak U., Ören T. (eds.). *Guide to*

Simulation-Based Disciplines: Advancing Our Computational Future. Cham (Switzerland) Springer; 2017. pp. 293–314.

[38]   Ören T.I., Ghasem-Aghaee N., Yilmaz L. 'An ontology-based dictionary of understanding as a basis for software agents with understanding abilities'. *Proceedings of the Spring Simulation Multiconference (SpringSim'07)*. Norfolk, VA; 2007.

[39]   Ören T.I. 'Simulation and reality: The big picture'. *International Journal of Modeling, Simulation, and Scientific Computing*. 2010, vol. 1(1). pp. 1–25.

[40]   Ören T.I. 'The richness of modeling and simulation and an index of its body of knowledge' in: Obaidat M.S., Filipe J., Kacprzyk J., Pina N. (eds.). *Simulation and Modeling Methodologies, Technologies and Applications, Advances in Intelligent Systems and Computing*. Springer; 2014. pp. 3–24.

[41]   Ören T.I. 'Simulation: Taxonomy' in Singh M.G. (ed.). *Systems and Control Encyclopedia*. Oxford, England: Pergamon Press; 1987. pp. 4411–4414.

[42]   Sulistio A., Yeo C.S., Buyya R. 'A taxonomy of computer-based simulations and its mapping to parallel and distributed systems simulation tools'. *Software – Practice and Experience*. 2004, vol. 34(7), pp. 653–673.

[43]   Ören T.I. 'Review of the article: A taxonomy of computer-based simulations and its mapping to parallel and distributed systems simulation tools by Sulistio, O., Yeo, C., Buyya, R.'. *Software – Practice & Experience. ACM Computing Reviews*. 2005, vol. 34(7), pp. 653–673. February issue.

[44]   Maier F.H., Größler A. 'What are we talking about? A taxonomy of computer simulations to support learning about socio-economic systems'. *System Dynamics Review*. 2000, vol. 16(2), pp. 135–148.

[45]   Roeder T.M.K. *An information taxonomy for discrete event simulations*. Ph.D. Thesis, University of California, Berkeley, 2004.

[46]   Ören T.I. 'A basis for the taxonomy of simulation languages'. *Proceedings of the 1971 Summer Computer Simulation Conference*; Boston, MA, 1971.

[47]   Ören T.I. 'Simulation and model-oriented languages: Taxonomy' in Singh M.G. (ed.). *Systems and Control Encyclopedia*. Oxford, England: Pergamon Press; 1987. pp. 4303–4306.

[48]   Ören T.I. 'Simulation languages: Taxonomy' in Morris D., Tamm B. (eds.). *Concise Encyclopedia of Software Engineering*. Oxford, England: Pergamon Press; 1993. pp. 306–312.

[49]   Rachidi, H. 'Discrete simulation software: A survey on taxonomies'. *Journal of Simulation*. 2017, vol. 11(2), pp. 174–184.

[50]   Ören T.I. 'Simulation models: Taxonomy' in Singh M.G. (ed.). *Systems and Control Encyclopedia*. Oxford, England: Pergamon Press; 1987. pp. 4381–4388.

[51]   Ören T.I. 'Simulation models symbolic processing: Taxonomy' in Singh M.G. (ed.). *Systems and Control Encyclopedia*. Oxford, England: Pergamon Press; 1987. pp. 4377–4381.

[52]   Fishwick P.A. 'A taxonomy for simulation modeling based on programming language principles'. *IIE Transactions*. 1998, vol. 30(9), pp. 811–820.

[53]   Hare M., Deadman P. 'Further towards of a taxonomy of agent-based simulation models in environmental management'. *Mathematics and Computers in Simulation*. 2004, vol. 64(1), pp. 25–40.

[54]   Ören T.I. 'Review of the article: Further towards a taxonomy of agent-based simulation models in environmental management by Hare M. and Deadman P.'. *Mathematics and Computers in Simulation. ACM Computing Reviews*. 2004, vol. 64(1), pp. 25–40, May issue.

[55]   Yilmaz L. Ören T.I. 'Dynamic model updating in simulation with multimodels: A taxonomy and a generic agent-based architecture'. *Proceedings of SCSC 2004 – Summer Computer Simulation Conference*; 2004, San Jose, CA, pp. 3–8.

[56]   Ören T., Mittal S., Durak U. 'Induced emergence in social system engineering: Multimodels and dynamic couplings as methodological bases' Chapter 9 in: Mittal S., Diallo S., Tolk A. (eds.) (2018). *Emergent Behavior in Complex Systems Engineering: A Modeling and Simulation Approach*. Hoboken, NJ: Wiley; 2018, April.

[57]   Lynch, C.J., Diallo S.Y. 'A Taxonomy for classifying terminologies that described simulations with multiple models'. *Proceedings of the 2015 Winter Simulation Conference*, Huntington Beach, CA, 2015, pp. 1621–1632.

[58]   Smith R. 'On the value of taxonomy in modeling' in Tolk A. (ed.). *Ontology, Epistemology, and Teleology of M&S: Philosophical Foundations for Intelligent M&S Applications*. Berlin, Heidelberg (Germany): Springer-Verlag; 2013.

[59]   Ören T.I. 'Model behavior: Type, taxonomy, generation and processing techniques'. in Singh M.G. (ed.). *Systems and Control Encyclopedia*. Oxford, England: Pergamon Press; 1987. pp. 3030–3035.

[60]   Ören T.I. 'Software agents for experimental design in advanced simulation environments'. *Proc. of the 4th St. Petersburg Workshop on Simulation*; St. Petersburg, Russia, 2001. pp. 89–95.

[61]   Wenzel S., Bernhard J., Jessen U. 'A taxonomy of visualization techniques for simulation in production and logistics'. *Proceedings of the 2003 Winter Simulation Conference*; 2003, pp. 729–736.

[62]   Le Digabel S., Wild S.M. *A Taxonomy of Constraints in Simulation-based Optimization*. Argonne National Laboratory, Mathematics and Computer Science Division, Preprint ANL/MCS-P5350-0515, 2015.

[63]   Goldstein R., Khan A. 'A taxonomy of event time representations'. *Proceedings of the Symposium on Theory of Modeling and Simulation (TMS/DEVS'17)*; Virginia Beach, VA, 2017.

[64]   Miller J.A., Baramidze G.T., Sheth A.P., Fishwick P.A. 'DeMO: An ontology for discrete-event modeling and simulation'. *Simulation*. 2011, vol. 87 (9), pp. 747–773.

[65]   Tolk A. (ed.). *Ontology, Epistemology, and Teleology for Modeling and Simulation'? Philosophical Foundations for Intelligent M&S Applications*. Berlin, Heidelberg (Germany): Springer; 2013.

[66] Partridge C., Mitchell A., de Cesare S. 'Guidelines for developing ontological architectures in modelling and simulation' in Tolk A. (ed.). *Ontology, Epistemology, and Teleology of M&S: Philosophical Foundations for Intelligent M&S Applications*. Berlin, Heidelberg (Germany): Springer-Verlag; 2013, pp. 27–57.

[67] Hofmann, M. 'Ontologies in modeling and simulation: An epistemological perspective' in Tolk A. (ed.). *Ontology, Epistemology, and Teleology of M&S: Philosophical Foundations for Intelligent M&S Applications*. Berlin, Heidelberg (Germany): Springer-Verlag; 2013. pp. 59–87.

[68] Heath, B.L., Jackson R.A. 'Ontological implications of modeling and simulation in postmodernity' in Tolk A. (ed.). *Ontology, Epistemology, and Teleology of M&S: Philosophical Foundations for Intelligent M&S Applications*. Berlin, Heidelberg (Germany): Springer-Verlag; 2013. pp. 89–103.

[69] Wang W., Wang W., Li Q., Yang F. 'Ontological, epistemological, and teleological perspectives on service-oriented simulation frameworks' in Tolk A. (ed.). *Ontology, Epistemology, and Teleology of M&S: Philosophical Foundations for Intelligent M&S Applications*. Berlin, Heidelberg (Germany): Springer-Verlag; 2013, pp. 335–358.

[70] Durak U., Ören T. 'Towards an ontology for simulation systems engineering'. *Proceedings of the SpringSim'16*; Pasadena, CA, 2016.

[71] Ören T.I. 'Computer-aided modelling systems' in Cellier F.E. (ed.). *Progress in Modelling and Simulation*. London: Academic Press; 1982. pp. 189–203.

[72] Ören T.I., Zeigler B.P. 'System theoretic foundations of modeling and simulation: A historic perspective and the legacy of A. Wayne Wymore'. *Simulation*. 2012, vol. 88(9), pp. 1033–1046.

[73] Zeigler B.P. *Multifacetted Modeling and Discrete Event Simulation*. London: Academic Press; 1984.

[74] Simon, H.A., Newell A. 'Simulation of human thinking' in Greenberger M. (ed.). *Computers and the World of the Future*. Cambridge, MA: The MIT Press; 1962. pp. 94–131.

[75] Feigenbaum E.A., Feldman J. (eds.). *Computers and Thought*. McGraw-Hill Book Company; 1963

[76] Ören T.I. 'Artificial intelligence and simulation: A typology'. *Proceedings of the 3rd Conference on Computer Simulation*; Mexico City, 1995

[77] Yilmaz L., Ören T.I. (eds.). *Agent-Directed Simulation and Systems Engineering*. Berlin: Wiley-Berlin; 2009.

[78] Ören T.I., Yilmaz L. 'Synergy of systems engineering and modeling and simulation'. *Proceedings of the 2006 International Conference on Modeling and Simulation – Methodology, Tools, Software Applications (M&S MTSA)*; Calgary, AL, Canada, 2006.

[79] NATO-SaaS. *Modeling and Simulation as a service: New concepts and service-oriented architectures*. NATO STO Technical Report AC/323(MSG-131)TP/608, 2015.

[80] Ören T.I., Zeigler B.P., Elzas M.S. (eds.). *Simulation and model-based methodologies: An integrative view*. Berlin: Springer-Verlag; 1984.

Chapter 2

Flexible modeling with Simio

David T. Sturrock[1] and C. Dennis Pegden[1]

2.1 Overview

Simio is an object-oriented (OO), general-purpose modeling tool that can be applied in a broad set of applications including manufacturing, transportation, logistics, healthcare, and communication networks. Simio includes a comprehensive tool set for building models, providing 3D animation, verifying and validating the model, experimenting and optimizing results, and real-time planning and scheduling. Figure 2.1 illustrates a student model [1] of a rental car parking lot that has taken advantage of 3D animation to illustrate the solution. In this chapter, we will describe the general-purpose modeling features and capabilities of Simio but, where appropriate, describe these concepts as they relate to modeling of communication networks.

2.2 Simio object framework

Simio is a *sim*ulation modeling framework based on *i*ntelligent *o*bjects. The intelligent objects are built by modelers and then may be reused in multiple modeling projects.

Figure 2.1   3D animation of rental car operation

[1]Simio LLC, USA

Simio comes with pre-built libraries of objects. For example, the Standard Library is set of general purpose objects (source, server, path, sink, etc.) that is commonly used to model a wide range of discrete systems. Likewise, the Flow Library is a set of general purpose objects (e.g., tank, pipe, filler) that is used to model systems involving material flows such as liquids, sand, gravel, etc. Many other libraries are also available such as the Extras library that represents cranes, elevators, robots, and more.

In many cases, a modeling project is approached by first building a custom library of special purpose objects, and then those objects are used as building blocks for creating a model. For example, a complex communication network involving ships, tanks, airplanes, command centers, satellites, etc. can be modeled by first creating objects representing each of the physical components and then placing multiple instances of these objects into the final model. Objects can be stored in libraries and easily shared. A beginning modeler may prefer to use pre-built objects from libraries; however, the system is designed to make it easy for even beginning modelers to build their own intelligent objects.

As noted above, a Simio model is built by combining objects that represent the physical components of the system. A Simio model looks like the real system. The model logic and animation is built as a single step. An object is animated in 3D to reflect the physical object and its changing state. For example, a robot opens and closes its gripper, and a battle tank turns its turret. The animated model provides a moving picture of the system in operation. To simplify the effort of building animated 3D models, Simio can import 2D and 3D background objects as well as 2D and 3D object representations from the target domain. Simio also provides a direct link to Trimble 3D Warehouse, a free massive online library of 3D graphic symbols that contains high-quality 3D symbols from virtually every domain.

Objects are built using graphical processes and the concepts of object-orientation. There is no need to write programming code to create new custom objects. The activity of building an object in Simio is identical to the activity of building a model—in fact, there is no difference between an object and a model. This concept is referred to as the equivalence principle and is central to the design of Simio. Whenever you build a model, it is an object that can be instantiated into another model. For example, if you combine two satellite dishes and six missile launchers into a missile defense battery, the missile defense battery model is itself an object (see Figure 2.2) that can then be instantiated any number of times into other models. The missile defense battery model is an object just like the satellite dish and missile launchers are objects. In Simio, there is no way to separate the idea of building a model from the concept of building an object. Every model that is built in Simio is automatically a building block that can be used in building higher level models.

Composite objects: The previous example in which we defined a new object definition (missile defense battery) by combining other objects (satellite dish and missile launcher) is one example of how we can create object definitions in Simio. This type of object is called a *composed object* because we create this object by combining two or more component objects. This object-building approach is fully

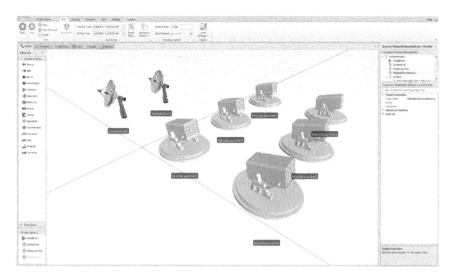

Figure 2.2   Placing a missile defense battery object in a model

hierarchical, i.e., a composed object can be used as a component object in building higher level objects. This is only one way of building objects in Simio—there are two other important methods.

Base objects: The most basic method for creating objects in Simio is by defining the logical processes that alter their state in response to events. For example, a router object might be built by defining the processes that alter the router state as events occur such as packet arrivals, breakdowns, etc. This type of modeling is like the process modeling done in traditional modeling systems in use today such as Arena™ or GPSS™. A base object can in turn be used as a component object for building higher level objects.

Derived objects: The final method for building objects in Simio is based on the concept of inheritance. In this case, we create an object from an existing object by overriding (i.e., replacing) one or more processes within the object, or adding additional processes to extend its behavior. In other words, we start with an object that is almost what we want, and then we modify and extend it as necessary to make it serve our own purpose. Whereas in a programming language, we extend or override behavior by writing methods in a programming language; in Simio, we extend or override behavior by adding and overriding graphically defined process models. With Simio, the skills required to define and add new objects to the system are modeling skills, not programming skills.

As an example of the idea of inheritance, we might build a specialized satellite dish from a generalized satellite dish object by adding additional processes to handle the failure and replacement of one of its critical components. An object that is built in this way is referred to as a derived object because it is subclassed from an existing object.

Regardless of which method is used to create an object, once created it is used in the same way. An object can be instantiated any number of times into a model. You simply select the object of interest and place it (instantiate it) into your model.

Entity is a class of objects that represents physical items in the system that can be created and destroyed and moved through 3D space, such as ships, satellites, tanks, etc., or packets of information that move between physical items. When modeling systems comprise the IoT, the objects can represent the "things" or the "messages" that are sent back and forth between the things.

Objects in Simio have static inputs referred to as *properties*, and dynamic variables referred to as *states*. For example, a missile might have a property that specifies its maximum travel speed, and states that specify its current 3D position, direction, and speed. The number and types of properties and states are defined by the creator of the object. Both properties and states are strongly typed and can be defined as numeric, Boolean, list, etc. There are two basic types of states: discrete and continuous. A discrete state is a value that only changes at event times (packet arrival, machine breakdown, etc.). A continuous state (e.g., degrees of tank turret rotation, position of a satellite, etc.) has a value that may change continuously over time by specifying its rate of change.

2.3 Simio object classes

There are six basic classes of objects in Simio. These six classes of objects provide a starting point for creating intelligent objects within Simio. By default, all six of these classes of objects (Figure 2.3) have very little native intelligence, but all can gain intelligence. You build intelligent versions of these objects by modeling their behavior as a collection of event-driven processes. The Standard Library included with Simio provides a rich and customizable set of objects that are derived from these six basic classes of objects.

The first class is *Fixed Objects*. This object has a fixed location in the model and is used to represent the things in your system that do not move from one location to another. Fixed objects are used to represent stationary equipment such as routers, servers, etc.

Links and *Nodes* are objects that are used to build networks over which entities may flow. Note that links and nodes can model networks for both communication movements as well as physical movements. A link defines a pathway for entity

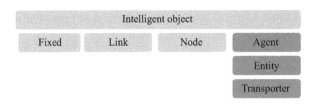

Figure 2.3 Basic Simio object classes

movement between objects. A node defines a starting or ending point for a link. Links and nodes can be combined into complex communication and physical networks. Although the base link has little intelligence, we can add behavior to allow it to model unconstrained flow, congested traffic flow, or complex material handling systems such as accumulating conveyors or power and free conveying systems.

Agents are objects that can freely move through three-dimensional space. Agents are also typically used for developing agent-based models. This modeling view is useful for studying systems that are composed of many independently acting intelligent objects that interact with each other and in so doing create the overall system behavior. Examples of applications include market acceptance of a new product or service, or population growth of competing species within an environment. Note that in Simio, all objects graphically defined processes provide intelligence to control their behavior rather than requiring Java or other programming code as in most other products.

Entities are objects that can freely move through three-dimensional space. Entities can move through the system from object to object over a network of links and nodes or move directly between objects through free space. Examples of entities include communications such as information packets, or physical items such as tanks, satellites, ships, etc. Note that in traditional modeling systems, the entities are typically passive and are acted upon by the model processes. However, in Simio, the entities can have intelligence and control their own behavior.

The final class of object is a *Transporter* and is subclassed from the entity class. A transporter is an entity that has the added capability to pick up, carry, and drop-off one or more other entities. By default, transporters have none of this behavior, but by adding model logic to this class, we can create a wide range of transporter behaviors. A transporter can model an airplane, ship, subway car, automated guided vehicle (AGV), or any other object that can carry other entities from one location to another. The Standard Library contains a vehicle object and a worker object, both of which are derived from a transporter object.

A key feature of Simio is the ability to create a wide range of object behaviors from these six basic classes. The Simio modeling framework is application domain neutral—i.e., these basic classes are not specific to communications, manufacturing, service systems, healthcare, military, etc. However, it is easy to build application-focused libraries comprising intelligent objects from these classes designed for specific application. For example, it is relatively simple to build an object (in this case a link) that represents a complex accumulating conveyor for use in manufacturing applications. The design philosophy of Simio directs that this type of domain-specific logic belongs to the objects that are built by users, and not programmed into the core system.

2.4   Modeling movements

A major focus in modeling typical communication networks is representing the movement of both physical items and information packets through the system. In Simio,

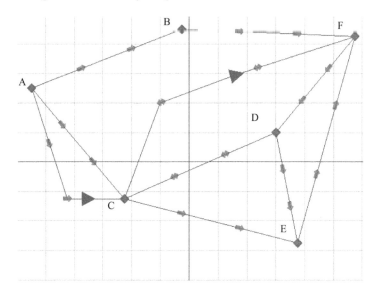

Figure 2.4   Example network

both items are modeled with entities, which move through the 3D model in one of two ways. The first is to simply move in free space with no constraints in movement. In this case, the entity can set its own direction, speed, and acceleration. In free space, the entity is in complete control of its own movement. The second method is to move over a network of nodes and links, where the network may control and limit the movements of the entities. Networks are very useful for modeling complex movements.

Networks comprise one or more links, where each link starts and ends at a node. A node can have any number of incoming and outgoing links. Links can be unidirectional or bidirectional, have a capacity that limits traffic on the link, and can have a maximum speed to limit traffic speed. Links also have a selection weight that can be used in decision rules for routing entities through the network. The example network in Figure 2.4 has six nodes (labeled A–F) and ten links connecting the nodes, where the triangles are entities moving through the network.

The complete set of all links in a model is referred to as the global network. However, links can also belong to one or more subnetworks. For example, the communication links between a set of satellite dishes might be represented by a subnetwork that is limited for use by signals traveling between satellite dishes, and pathways where ships travel may be specified by a separate network.

The Standard Library contains four link objects and two node objects. The connector, path, time path, and conveyor are derived from the link object and the basic node and the transfer node are derived from the node object. The connector moves entities across the link in zero-time. This type of link is used to model movements such as signals that travel at the speed of light, for which the travel time is negligible and can be ignored. The path is a type of link used to model entity movements where each entity can travel at its own speed and either pass or not pass other entities based on a property that is specified on the path. The time path is used to model situations where the travel time on the link is specified by an expression (perhaps involving random variables and other system status variables). The conveyor is a type of link that is used to model both accumulating and non-accumulating conveyors that are found in typical manufacturing and warehousing applications. Although the links and nodes provided by the Standard Library work for many applications, users can also create their own custom nodes and links.

Each entity in the network can have a single destination where it is headed, or it can follow a specified travel sequence through the network. A travel sequence is an ordered list of nodes (e.g., A, C, D, F in Figure 2.4) that must be visited in the specified order on the way to its last node in the sequence. In either case, an entity may have more than one possible routing to its next destination. For example, an entity traveling from C to E could either take the direct path from C to E or travel from C to D and then D to E. This might be advantageous, for example, if the link to C to E was congested, and the travel speed on the alternate route through D was faster and warranted the extra travel distance. The decision for which route to take when moving to its next intermediate or final destination is based on properties that are specified on the transfer node. The *Outbound Link Rule* property specifies that the link should be selected based the shortest path or on decision weights that can be assigned to each link. The *Link Preference Property* specifies if all links are to be considered, only links that are currently available or a specific link is desired.

2.5 Modeling physical components

When modeling communication networks, it is also important to be able to model the physical components of the system, such as vehicles, machines, workers, etc. This can be done using objects in the Standard Library, Flow Library, or other custom libraries. Although custom libraries are the most flexible, many applications can be done using the Standard Library. The Standard Library consists of pre-built objects to model a wide range of systems. We have already briefly discussed the four link objects and two node objects that are included in this library. Table 2.1 summarizes the 15 objects in the Standard Library.

The Flow Library is designed to represent situations where flow is continuous (e.g., fluid in a pipe or ore on a conveyor) or so fast that it can best be modeled as continuous (e.g., pills on a conveyor). The Flow Library also handles conversion between discrete and flow (e.g., a filling machine that converts continuous fluid flow into filled bottles). Table 2.2 summarizes the ten objects in the Flow Library.

Every object has properties that control its behavior. For example, the source object properties view shown in Figure 2.5 has an arrival mode property that specifies the mechanism for creating entities (either based on an interarrival time, data in a table, an event, or based on a time-varying arrival rate). Each property value may in turn switch on additional properties; e.g., if the arrival mode is specified as interarrival time, then a property is switched on that specifies the expression for computing the

Table 2.1   Objects in Simio Standard Library

Name	Class	Description
Source	Fixed	Creates entities that arrive to the system
Sink	Fixed	Destroys entities and records statistics
Server	Fixed	Models a multichannel service process with input/output queues
Resource	Fixed	Models a resource that can be used by other objects
Combiner	Fixed	Combines entities in batches
Separator	Fixed	Separates entities from batches
Workstation	Fixed	Models a three-phase workstation with setup, processing, and teardown
Vehicle	Transporter	Carries entities between objects and serves entities at a fixed location
Worker	Transporter	Carries entities between objects and serves entities at a fixed location
Basic node	Node	A simple intersection of links
Transfer node	Node	An intersection where entities set destination and wait on transporters
Connector	Link	A zero-time connection between two nodes
Path	Link	A pathway between two nodes where entities travel based on speed
Time path	Link	A pathway with a specified travel time
Conveyor	Link	An accumulating/non-accumulating conveyor device

Table 2.2   Objects in Simio Flow Library

Name	Class	Description
Flow source	Fixed	Generates a flow of fluid or other mass of a specified entity type
Flow sink	Fixed	Destroys flow entities representing quantities of fluids or other mass that have finished processing in the model
Tank	Fixed	Models a volume or weight capacity–constrained location for holding entities representing quantities of fluids or other mass
Container entity	Entity	Models a type of simple moveable container (e.g., barrels or totes) for carrying flow entities representing quantities of fluids or other mass
Filler	Fixed	Fills containers with flow entities representing quantities of fluids or other mass
Emptier	Fixed	Empties the flow contents of container entities
Item to flow converter	Fixed	Converts entities representing discrete items into flow entities representing quantities of fluids or other mass
Flow to item converter	Fixed	Converts flow entities representing quantities of fluids or other mass into entities representing discrete items
Flow node	Node	Regulates the flow of entities representing quantities of fluid or other mass
Flow connector	Link	A zero-time connection between two flow nodes

Figure 2.5   Selected properties of source object

interarrival time (typically a random variable). All properties can be used for either deterministic or stochastic arrivals. Other properties provide flexibility in terminating the arrival stream, for example, after a specified time or specified number of arrivals. Many objects also use events to customize their behavior. *Events* let objects easily communicate with other objects. For example, an event triggered elsewhere in the model might cause a source object to create an arrival or entirely stop creating new arrivals.

Figure 2.6 shows a simple model built using the source, server, and sink objects, along with path links to define the movements between these objects. In this example, entities are created at the source, travel to the server where they queue up and wait for processing, and then travel to the sink where they depart the model.

The objects shown in Figure 2.6 all have their default generic graphics; in a typical model, these would be replaced by more appropriate graphics. For example, if the server represented an ATM machine at a bank, we would typically replace the rectangular server symbol with a graphic symbol of an ATM machine from 3D Warehouse. We could also replace the triangles representing entities with animated

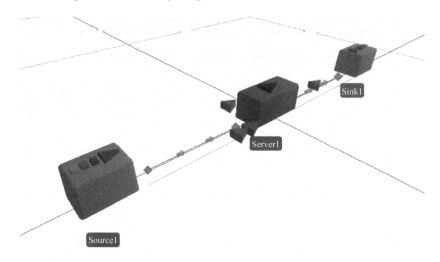

Figure 2.6   Example model using source, server, and sink objects

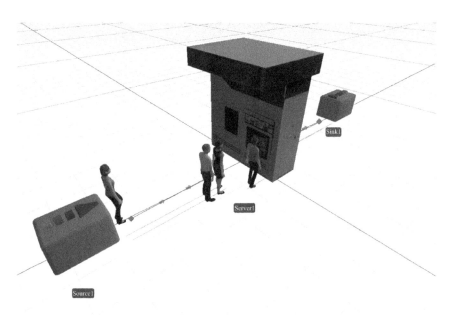

Figure 2.7   Example model with domain-specific 3D animation

walking people. Figure 2.7 enhances that same model with the default graphics for the server and entity replaced, which required only a few minutes to create.

The *server* object that is used in this simple model is one of the most powerful and commonly used objects in the Standard Library. It can model a wide range of

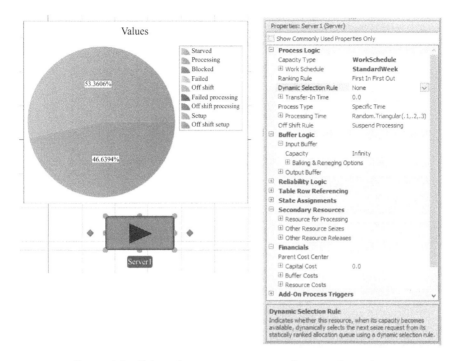

Figure 2.8   Selected server properties and optional graphics

physical elements of a system that constrain the movement of entities based on one or more activities that must take place, secondary resources that may be required, and material that may be consumed. Figure 2.8 illustrates many of the common properties of the server object as well as an optional attached pie chart in the facility view that indicates possible resource states.

The server can model multichannel processors, follow complex work schedules, and incorporate failure/repair patterns. The server can also model complex operations composed of a network of tasks that follow precedence relationships and operate parallelly and/or sequentially. For example, Figure 2.9 illustrates a generic six-step task sequence where each task has prerequisite tasks. Not only is the number and relationship of tasks unlimited, but each individual task could require resources or materials, or even be defined to execute one or more other objects, which themselves might have networks of tasks.

The *worker* and *vehicle* are two other objects that are commonly used in Simio models. The worker object is used to model operators or crew members that move around the system and perform tasks. For example, a server may request that a worker must come to the server to set it up before processing an entity. The vehicle object is used to model ships, trucks, AGVs, etc., that travel through the model, picking up and dropping off entities. Vehicles have flexible work selection and allocation logic,

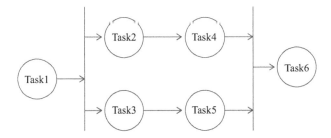

Figure 2.9   Generic parallel/serial task sequence

reliability logic, and many options to control both behavior and animation such as load and unload time, dwell logic, and automatic parking and homing options.

While each object has object-specific properties as mentioned above, each object also has categories of properties that are found across many different objects. For example, objects that incorporate buffers or queues typically have a *Buffer Logic* category that contains properties to describe the capacity of those buffers, as well as the logic that governs balking (bypass queue entry) and reneging (abort queue waiting). Most objects have a *Financials* category that specifies the properties to support comprehensive activity–based costing and supporting all world currencies.

Objects that typically represent some types of machine or equipment have a *Reliability* category where failure-related properties such as downtime mode, period between downtimes, and time to repair are specified. Downtime modes include calendar or processing time between failures, processing count between failures, and event-based failures.

Many objects also have categories to provide higher level interaction with other objects such as state assignments, statistics, customized animation, and data logging. Two broad interaction mechanisms—processes and data tables—are discussed in Sections 2.6 and 2.7, respectively.

2.6   Processes

The use of library objects permits fast, highly productive modeling. But unless the library is designed to closely match your application, you will often have to customize objects in order to model accurately enough to meet project objectives. In most OO simulation products, this customization can only be accomplished by modifying the object definition using programming code like Java, C++, or a proprietary language. Doing so takes a level of expertise often not readily available. Simio provides two alternatives, both based on the patented concept of processes.

A *process* is a graphical way of defining the logic behind an object. Processes can be used to make decisions, seize or release resources, search collections of objects or data, wait for or trigger communications events, assign state variables, record custom

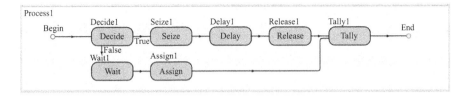

Figure 2.10   Example process logic

Table 2.3   Commonly used process steps

Assign step is used to assign a new value to a state variable

Create step is used to generate objects into the system

Decide step may be used to determine the flow of a token through process logic

Delay step delays the arriving token in the step for the specified time duration

Destroy step destroys either the parent object or the executing token's associated object

EndTransfer step may be used to indicate that the entity object associated with the executing token has completed transfer into an object or station

Execute step may be used to execute a specified process

Find step may be used to search the value of an expression over a specified range of one or more index variables. The expression will typically involve array variables (vectors or multidimensional arrays) or indexing-related functions

Fire step may be used to fire an object event

Move step may be used to request a move from one or more moveable resources that have been seized by either the parent object or object associated with the executing token. The executing token will be held in the Move step until the resources have arrived to the requested locations

Release step releases capacity of one or more objects on behalf of the parent object or the object associated with the executing token

Scan step may be used to hold a process token at the step until a specified condition is true

Search step may be used to search a collection of objects

Seize step may be used to seize capacity of one or more objects on behalf of the parent object or the object associated with the executing token

SetNode step may be used to set the destination node of any entity object

Tally step tallies an observation for each token arriving to this step

Transfer step may be used to transfer the entity object associated with the executing token between objects and between free space and objects

Wait step may be used to hold the arriving token in the step until a specified event occurs

---

statistics, and much more. Figure 2.10 illustrates a process that makes a decision, then either seizes, delays, and releases a resource or waits for an event, assigns a state, and then records an observational statistic. Although over 60 steps are available, the most commonly used steps are described in Table 2.3.

A process can be used in an object definition to define the logic in a new object or customize the logic in a subclassed object. But in many cases an even simpler alternative is available. Most library objects have "hooks" called *add-on process triggers* that can be used to supplement the logic in a specific object instance. Figure 2.11 illustrates the add-on process triggers available in the server object. These support the

Figure 2.11   Server add-on process triggers

application of custom logic based on triggers related to starting/ending the process-ing, starting/ending a failure, going on/off shift, and much more. This allows users to graphically specify custom logic—for example, examining the current system state before deciding what to do with any entities in process when a server is about to transition off shift.

In addition to the flexibility in easily customizing logic, processes provide two other important benefits: spanning time and allowing parallel processing. In most programming solutions to customized object, the logic cannot span time, so the seize–delay–release sequence and the alternate wait sequence illustrated in Figure 2.9 would not be possible in a single program function. Another advantage is that while entities or agents are the most primitive construct in most simulation products, Simio has the concept of *Tokens*, which are delegates of an object—tokens execute processes. Since an object can have many tokens, this allows for an entity to take multiple actions at once by using multiple tokens. For example, a server object might execute a setup process while at the same time summoning a mechanic to do a concurrent repair operation.

2.7   Data tables

Simio contains a set of process steps that provide comprehensive input/output capabil-ities including the ability to read/write CSV files, Excel™ files, and a wide variety of

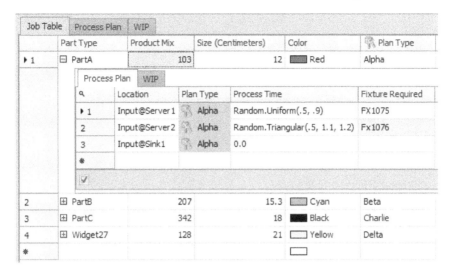

Figure 2.12   Relational tables with partial master-detail view expansion

database files. But processing such files incrementally during a model run can often be slow and inconvenient. So Simio extends this commonly available capability to also create in-memory data repositories called data tables. In-memory tables execute extremely fast.

The schema or design of data tables is under user control—you can have any number of columns of different data types, in any order you want. A table can be designed to be most convenient to the modeler or could be designed to perfectly match an external data source to avoid transforming the data on each use. Data tables can be simple tables, like a spreadsheet, or can be comprehensive sets of hierarchical relational tables linked by keys and foreign keys. Figure 2.12 illustrates a set of three tables, Job Table, Process Plan, and WIP, that are related by a key field. The master-detail view is expanded on the first part type to show the relationships.

Tables can be built and used entirely within Simio, but it is more common to import the table data from an external source. Simio incorporates sophisticated table-input mechanisms. In addition to CSV, Excel, and databases, Simio directly supports reading data tables in the Business to Manufacturing Markup Language (B2MML). Since B2MML is used to integrate business systems such as ERP and software such as SAP [2] with manufacturing systems such as manufacturing execution systems (MESs), it is a rich source of predefined information for use in simulations. Simio can also generate tables directly from Wonderware™, a leading MES software. Tables can be configured to import on demand, or tables with frequently changing data can be configured to automatically import with each run. Not only does Simio have extensive built-in support of data import (see Figure 2.13), but also it provides the capability, including sample source code, to customize data import with programming in any of over 60 .NET languages.

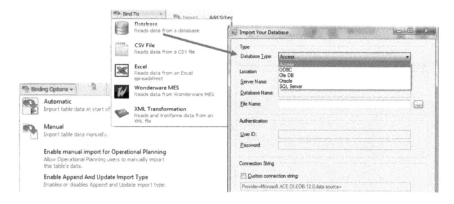

Figure 2.13   Data table binding and import options

Simio tables are key to implementing two important modeling strategies which are having major impact on the simulation industry.

Data-driven modeling is a way of structuring a model, so much of the model data is in data tables rather than disseminated throughout the model and allowing the configuration of the model to take place in data tables (or associated external files). The combination of these features makes models easier for the modeler to understand, maintain, and share with others, and makes it easier (e.g., "lowers the bar") for stakeholders to use and update the model without comprehensive knowledge of simulation.

Data-generated modeling is a mechanism for building all or most of a model entirely from external data. For example, a fairly complete model can be built directly from data using the B2MML, ISA 95, or Wonderware™ import mechanisms mentioned above. Alternatively, the Simio application programming interface (API) can be used to import model data from virtually any database, spreadsheet, or other data source. Importing major parts of system configuration and descriptive information can dramatically lower the time and expertise required to create a model-based solution to a pressing problem.

2.8   Experimentation with the model

When you build a model in the facility view, you can run it, view real-time 3D animation, interact with the model, use sophisticating debugging techniques, like step, break, watch, and trace, and view sample output results. But in stochastic models, it is vitally important to run multiple replications for statistical analysis and validity. Simio has a built-in experiment window to very efficiently and quickly run multiple replications, configure controls (what changes) and responses (KPI's), configure and compare multiple scenarios, and even run automatic optimization.

While the experiment window offers standard textual reports, most people prefer the built-in *Pivot Grid* reports. Like the pivot tables featured in many top data analysis packages, Simio allows you to filter, sort, and recategorize the data. This allows you to generate concise, custom reports in literally just a few clicks, then you can save those reports and reuse them anytime. The details of all scenarios are shown along with statistical measures like mean, minimum, maximum, and half-width. Both the summary and the detailed results can also be exported for additional analysis in external programs.

Simio's experiment window will automatically run any number of replications, using all your available processors (defaults up to 16). If you have the common configuration of a dual-threaded quad core processor, you can run eight replications in about the same time it would take to run one. With higher versions of Simio, you can also take full advantage of other computers in your workgroup, and you can even extend the limit of 16 to take full use of a server farm or network of workgroup computers. Another approach to running replications is to use the *Simio Portal*. This Azure™-based software as a service offering allows you to bring the processing power of the cloud and scale up to run massively parallel replications and instantly distribute the results across the internet.

Running many replications quickly is most important when you want to compare multiple scenarios. Simio allows you to define referenced properties in your model, which are displayed as *controls* in your experiment. Controls describe how one scenario differs from another, for example, number of workers or number of servers. Simio also allows you to define *Responses* in the experiment. A response is like a key performance indicator (KPI) that is a quick measure of the performance of each scenario. Additional statistical information is also recorded on responses to support the *response results* view. The response results view is an enhancement of the measure of risk and error (MORE) analysis technique described by Nelson [3] that makes it easier for people without a strong statistical background to gain important insight into their data.

When you have many possible scenarios to evaluate, manually generating them can be tedious. And it is important to minimize or completely avoid the execution of poor scenarios. Simio is tightly integrated with *OptQuest*®, the leading simulation-based optimization product. OptQuest uses metaheuristics to guide the search algorithm to quickly find better solutions. OptQuest combines Tabu search, scatter search, integer programming, and neural networks into a single composite search algorithm that is orders of magnitude faster than other approaches [4].

2.9   Application programming interface

The Simio libraries are comprehensive, especially when supplemented with add-on processes. Many users also choose to create custom objects, either by modifying the open-source Simio libraries or using processes to graphically create entirely new objects. But in rare instances, users desire even more customization.

Simio has an extensive API that allows customization of virtually all aspects of Simio using the API and any of over 60 .NET languages. Users with programming background can create new tools and customize the menus to display those tools to your stakeholders. You can create new steps and elements—the fundamentals of Simio processes. You can create design-time add-ins that support building models from external data. You can build in new import/export capabilities to support a unique or proprietary data source for importing individual items, entire data tables, or even generating models directly from external data. With the API, you can even add custom experimentation. For example, both OptQuest and the Select Best Scenario tools were implemented as experiment add-ins.

As an example of highly customized menu items, a customer who was scheduling weather-sensitive operations used the API to add a "Get Weather" menu item which would log on to a weather subscription service and download regional weather forecasts into a Simio data table that was directly accessed during planning. This was combined with other application-specific items to create a custom ribbon using customer terminology.

While Simio already includes a comprehensive set of scheduling rules (see Section 2.10) you can customize these or create your own rules. On the analysis end, although Simio already includes extensive support for experimentation and optimization, the API supports creation of custom design of experiments and add-ins such as a custom optimization algorithm.

In addition to providing comprehensive documentation of the API, Simio supplies extensive sets of sample code for all the items mentioned above.

2.10   Applications in scheduling

Although simulation tools in the past have been primarily used in the design of complex networks, Simio is specifically designed to also support applications for scheduling these same systems. In scheduling applications, the focus is on simulating the actual flow of entities through the network, given an initial starting state for the system. The simulation of the entity flow then produces an operational schedule of the system, for a given starting state and a given set of entities. The purpose of the simulation is to forecast the operational performance of the network in an actual real-time setting.

An example of a complex network scheduling application is the scheduling of computational tasks to be executed on a network of processors. This application can be represented as a directed graph, where each node on the graph is processor that can perform one or more independent computational tasks. However, because of data interdependencies, the tasks have precedence relationships that define the permissible sequences for execution for these tasks. This is a complex scheduling application; however, it is relatively simple to model the basic network system using Simio (using routing sequences or task sequences).

Traditional simulation tools lack the necessary features to use them effectively in a real-time scheduling environment. However, Simio has been designed from the ground up to support these applications by incorporating key features to support critical scheduling functions. These features include the ability to

1. Drive the model from relational data tables.
2. Initialize the model state—particularly with entities in process.
3. Incorporate complex decision logic—particularly involving dynamic rules that must be executed each time a decision must be made.
4. Log-detailed transactional data for creating reports, dashboards, and Gantt charts that depict the planned schedule.
5. Evaluate schedule robustness and the associated risk.

In the following sections, we will describe each of these key features.

As discussed in Section 2.7, Simio has extensive capabilities in representing, storing, and importing flat and relational data files. This is particularly useful in scheduling applications. Resource data (e.g., machines, machine groupings, fixtures, manpower) must be mapped to model constructs. Dynamic data (e.g., order information, material arrivals, man power schedules) must drive system arrivals. Simio's data-generated modeling features are designed to "reach into" existing data repositories and configure or entirely build a simulation model from that data.

Up-to-date WIP information must be used to initialize the scheduling system before each schedule is generated. This data is usually tracked in a MES or other real-time data system. Simio data tables have user-configurable schemas and comprehensive importing capabilities that support flexible mapping to existing data. It also features the capability to configure data import, so sensitive data is automatically refreshed. Figure 2.14 illustrates a high-end deployment with ERP and MES integration. While typically such deployments include a human scheduler in the loop, some deployments are done with the MES and Simio automatically detecting problems and taking corrective actions.

Most planning and scheduling tools are highly limited in the complexity of the system that can be represented [5]. As a general-purpose simulation language, Simio provides the capability to model the devices like electronic switches, AGVs, cranes, tanks, ovens, and conveyors that are often critical to facility operation. Most planning and scheduling tools are also limited in the decision logic controlling device interaction. Figure 2.15 illustrates the dynamic ranking and selection rules that form a part of Simio's decision logic. Simio also includes a comprehensive set of standard dispatching rules (Table 2.4) that can be used to provide both local and global system optimization through routing and resource allocation. In addition to these built-in rules, the Simio API allows creation of custom rules and optimization algorithms unique to a facility.

While design-focused simulation is often concerned mainly with summary statistics like cost and utilization, scheduling models need to record and display transactional data. *Logs* of the start and end times of every significant transaction (like a resource, entity, or material state change) must be maintained along with important transactional details. These logs are then used to generate custom reports like Resource Dispatch Reports and Workflow Constraints Analysis. Those same logs are used to create interactive displays like the resource plan Gantt (Figure 2.16) and the

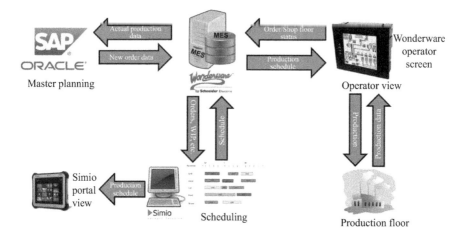

Figure 2.14   Example scheduling deployment integrated with ERP and MES

Ranking Rule	First In First Out
⊟ Dynamic Selection Rule	**Standard Dispatching Rule**
Dispatching Rule	**CriticalRatio**
Tie Breaker Rule	FirstInQueue
Filter Expression	
Look Ahead Window (Days)	**2**

Figure 2.15   Simio ranking and selection

entity plan Gantt. These two forms of Gantt charts graphically display activity from a resource or entity perspective. The Simio implementations provide extra tracking options such as graphical material inventory, resource states, downtime, schedules, constraints, and detailed background information on any Gantt item. Data logs are also the basis for user-designed Dashboard Reports (Figure 2.17). In addition to extensive tracking and diagnostics, using a drag and drop interface, the Gantt charts can be used to interact with the plan by such actions as specifying overtime or downtime periods, or selecting alternate process flows. While these Gantt and log-related features are primarily designed for planning and scheduling applications, many users have found them to also be extremely valuable in providing debugging and clear communication for traditional design applications.

A common problem with most scheduling systems is that the schedule must be created deterministically. There is no good way to generate a schedule that accurately predicts system downtime, material delays, extended processing times, and other commonly encountered variability. One approach is to ignore such variability entirely, which results in an optimistic schedule that becomes infeasible the first time something goes wrong. Another approach is to build-in extra processing time or idle

Table 2.4   Simio standard dispatching rules

Dispatching rule	Selection criteria
FirstInQueue	The entity ranked nearest the front of the queue
LargestPriorityValue	The entity with the largest priority state value
SmallestPriorityValue	The entity with the smallest priority state value
EarliestDueDate	The entity with the earliest due date
CriticalRatio	The entity with the smallest critical ratio. Critical ratio is the time remaining until the entity's due date divided by the total operation time remaining
LeastSetupTime	The entity with the least setup time
LongestProcessingTime	The entity with the longest operation time
ShortestProcessingTime	The entity with the shortest operation time
LeastSlackTime	The entity with the least slack time. Slack time is the time remaining until the entity's due date minus the total operation time remaining
LeastSlackTimePer Operation	The entity with the least average slack time per its remaining operations
LeastWorkRemaining	The entity with the least total operation time remaining to complete its assigned sequence
FewestOperations Remaining	The entity with the fewest number of operations remaining to complete its assigned sequence
LongestTimeWaiting	The entity that has been waiting longest in the queue
ShortestTimeWaiting	The entity that has been waiting the least time in the queue
LargestAttributeValue	The entity with the largest value of the specified expression
SmallestAttributeValue	The entity with the smallest value of the specified expression
CampaignSequenceUp	The entity that has a campaign value equal to or next largest compared to the value of the last processed entity
CampaignSequenceDown	The entity that has a campaign value equal to or next smallest compared to the value of the last processed entity
CampaignSequenceCycle	Alternates back and forth between a campaign sequence up and a campaign sequence down

time to allow for when things go wrong. But unfortunately, this is time that is wasted when things go well.

Simio initially creates a deterministic plan based on no variability, then it makes additional stochastic replications that consider all the potential problems and calculates the risk of key milestones or targets being missed. The colored markers on each order in Figure 2.18 indicate the risk associated with each order. For example, even though Order-01 and Order-02 have similar slack time, the lower likelihood of Order-02 achieving its release date target might be due to utilizing an unreliable machine or consuming materials that are often late. With the knowledge of this risk before deploying the schedule, the scheduler can use Simio to objectively evaluate the most cost-effective ways to reduce the risk. This in-turn makes the schedule more robust, e.g., it stays useful for a longer time. A related benefit is the ability to quickly replan. When a major event (e.g., an equipment failure) invalidates the plan, in Simio, the replan time is typically a few minutes versus the hours required in most other approaches.

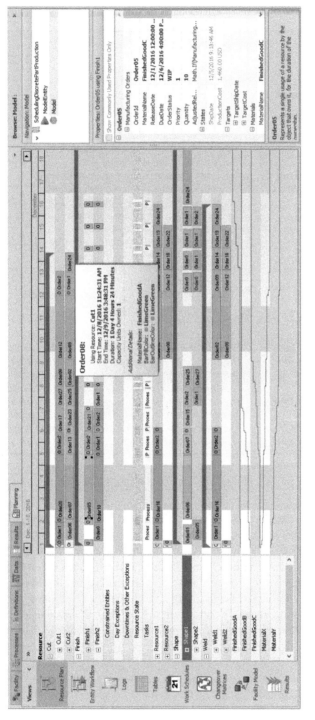

Figure 2.16 Resource plan Gantt

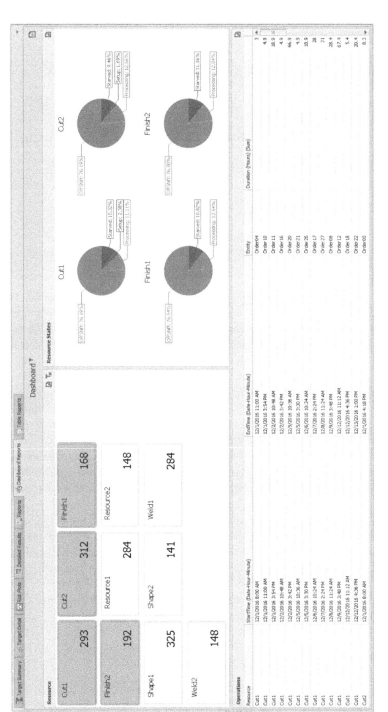

Figure 2.17 Interactive custom dashboard

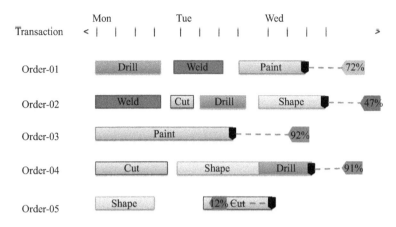

Figure 2.18   Entity Gantt with target risk analysis

2.11   Summary

Library-based modeling has long provided a faster way to build models, but unless the library was closely matched to your application, it was often necessary to make model approximations which made your solutions less accurate. The flexibility promised through OO technology has the potential for dramatic improvements but often has problems scaling to large models, and the customization of objects and libraries still required high programming expertise.

Simio was invented with two primary goals in mind. The first was to bring new technology to the OO simulation field to allow users to more effectively build objects, libraries, and models without programming. The second goal was to extend the field of discrete-event simulation beyond the traditional system design applications into planning and scheduling. Rather than "bolting on" features as needed, Simio was designed from the ground up to incorporate all the features needed to solve problems in design, planning, and scheduling, in a single tool using a single model.

The creation of Simio with its data-driven and data-generated modeling features was timely—just as the concepts of the smart factory promise a new way of operating our production systems. The smart factory [5], also referred to as the fourth industrial revolution or Industry 4.0 (Figure 2.19), represents the concept of physical systems where the components are monitored and connected to a virtual system model to predict and improve system performance. The virtual factory model provided by Simio is a key component of the smart factory of the future.

Glossary

B2MML: Business to Manufacturing Markup Language as defined in [6] is a set of XML schemas implementing the ISA-95, Enterprise-Control System Integration family of standards, known internationally as IEC/ISO 62264.

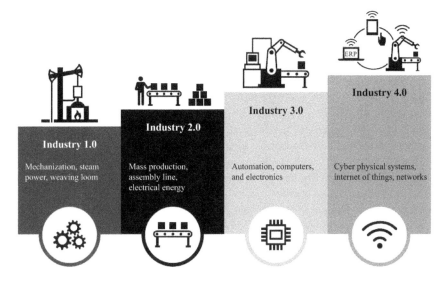

Figure 2.19   Simio is a key technology to implement smart factory [5]

Dispatching rule: An algorithm for deciding which job to process next in a production facility, such as which job has the earliest due date or which requires a minimum changeover.

Enterprise resource planning (ERP): Enhancements of the original material requirements planning (MRP) functions to bring together accounting, human resources, and other functions into a fully integrated IT system. ERP also incorporated supply chain management (SCM) to extend inventory control over a broader scope, including distribution.

Entity: Part of an object model and can have its own intelligent behavior. They can make decisions, reject requests, decide to take a rest, etc. Entities have object definitions just like the other objects in the model. Entity objects can be dynamically created and destroyed, moved across a network of links and nodes, move through 3D space, and move into and out of fixed objects. Examples of entity objects include customers, parts, or workpieces.

Event: A notification that can be given by one object and responded to by several. It alerts other objects that an action has occurred.

Experiment: Part of the project that is used for output analysis. The user defines one or more sets of inputs/outputs (scenarios) and runs multiple replications to get statistically valid results from which to draw conclusions.

Finite capacity scheduling (FCS): A scheduling approach that accounts for the limited production capacity of the system. This contrasts with the enterprise resource planning system that typically assumes an infinite capacity.

Gantt chart: A chart used in scheduling applications for showing activities over a timeline. A resource Gantt and an entity Gantt show the same information, but from two different perspectives.

Manufacturing execution system (MES): A computerized system used to track and document the transformation or raw materials into finished goods, including the status of resources and the flow of work.

Model: A representation of real-world object or collection of objects that interact with each other. Models are usually used to make decisions and are defined by their properties, states, events, external view, and logic. A model is an object that is executable.

Optimization: The process of defining and evaluating experiment scenarios to determine the overall "best" scenario. OptQuest is a tool that is highly integrated with Simio that will automatically optimize against a single objective, multi-objective weighted, or pattern frontier.

Planning: The process of creating a high-level production plan that identifies the that work needs to be done, the materials that are required to perform that work, and where the work will take place.

Risk-based planning and scheduling (RPS): The use of a custom-built stochastic simulation model for both system design and scheduling. RPS incorporates risk measures for assessing the robustness of a production schedule.

Scheduling: The process of turning a master production plan into a detailed, actionable schedule that can be followed to produce required items while meeting key objectives. Scheduling requires a detailed model of all the critical constraints in a system, and a good schedule is dependent upon good planning.

SimBit: A brief, well documented, example of solving a specific modeling issue. Basic and advanced SimBit search engines are provided via the Support Ribbon or descriptive information is in Help.

Smart factory (Industry 4.0): A fully connected and automated production system, based on digital part data, interconnected devices, and a virtual factory model to plan and project the future of products and production facilities. The Industry 4.0 initiative is focused on creating the smart factory of the future.

Table: A table is a set of rows and columns to hold data during a run. It may be relational and may contain special columns like times or destinations in addition to general model data. Tables may be imported and exported. An object may be associated with a specific table and row.

Transporter: A transporter object is a special type of entity that can pick up entity objects at a location, carry those entities through a network of links or free space, and then drop the entities off at a destination.

References

[1]   Student Models, *Student Simulation Competition* [online]. Available from https://www.simio.com/academics/student-competition.php.   Accessed   Nov 2018.

[2]   Junot Systems, Inc. *Advanced SAP MES Integration* [online]. Available from http://mes-to-sap.com/. Accessed Nov 2018.

[3]   Nelson, B. L. 2008. "The MORE Plot: Displaying Measures of Risk & Error From Simulation Output." In *Proceedings of the 2008 Winter Simulation Conference*, edited by S. J. Mason, R. R. Hill, L. Mönch, O. Rose, T. Jefferson, J. W. Fowler, Piscataway, New Jersey: Institute of Electrical and Electronics Engineers, Inc.

[4]   OptTek Systems, *OptQuest* [online]. Available from http://www.opttek.com/products/optquest/. Accessed Nov 2018.

[5]   Pegden, C. D. Deliver on Your Promise, How Simulation-Based Scheduling Will Change Your Business. Pittsburgh, Simio LLC, 2017.

[6]   MESA International, *Business to Manufacturing Markup Language (B2MML)* [online]. Available from http://www.mesa.org/en/B2MML.asp. Accessed Nov 2018.

Chapter 3

A simulation environment for cybersecurity attack analysis based on network traffic logs

Salva Daneshgadeh[1], Mehmet Uğur Öney[2], Thomas Kemmerich[3], and Nazife Baykal[1]

The continued and rapid progress of network technology has revolutionized all modern critical infrastructures and business models. Technologies today are firmly relying on network and communication facilities which in turn make them dependent on network security. Network-security investments do not always guarantee the security of organizations. However, the evaluation of security solutions requires designing, testing and developing sophisticated security tools which are often very expensive. Simulation and virtualization techniques empower researchers to adapt all experimental scenarios of network security in a more cost and time-effective manner before deciding about the final security solution. This study presents a detailed guideline to model and develop a simultaneous virtualized and simulated environment for computer networks to practice different network attack scenarios. The preliminary object of this study is to create a test bed for network anomaly detection research. The required dataset for anomaly or attack detection studies can be prepared based on the proposed environment in this study. We used open source GNS3 emulation tool, Docker containers, pfSense firewall, NTOPNG network traffic–monitoring tool, BoNeSi DDoS botnet simulator, Ostinato network workload generation tool and MYSQL database to collect simulated network traffic data. This simulation environment can also be utilized in a variety of cybersecurity studies such as vulnerability analysis, attack detection, penetration testing and monitoring by minor changes.

3.1 Introduction

A computer network is a set of connected network devices at the edge of the network which are used in personal and professional lives such as PCs, tablets, iPads and

[1] Department of Information Systems, Informatics Institute, Middle East Technical University, Turkey
[2] Department of Computer Engineering, Atılım University, Turkey
[3] Department of Information Security and Communication Technology, Norwegian University of Science and Technology, Norway

smartphones. Furthermore, it encompasses the network cores such as network switching and routing. Additionally, computer networks offer dedicated network services to the connected devices as well as to applications. These network services can be DHCP, NTP and security services like encryption or packet filtering [1]. Recently, network researchers have also started to use simulation tools to test and evaluate communication protocols and network behavior. Every network simulation requires network topologies, traffic models between senders and receivers, background noises introduced by other devices and possible dynamic events. Moreover, researchers require virtualization tools to understand, verify and analyze the complicated behavior in network simulation [2].

3.1.1   Network simulation

Living in the age of information and telecommunication technologies, our personal and professional lives are extremely dependent on Internet-based technologies. A thriving, innovative and secure network topology and networking protocols can be seen as a key for successfulness of the businesses operations. Therefore, an increasing trend is observed for designing, developing and managing novel, high-performance and security enabled–computer networks. However, designing and validating each new network technologies requires huge investments. Subsequently, network designers and researchers rely on network simulation. Generally, there are two types of network simulation: analytical modeling and computer simulation. Analytical modeling uses mathematical analyses to characterize a network, but it is mostly a too simple model to emulate the dynamic features of the network. Computer simulation models the behavior of real events in a real life scenario in association with time [3]. Simulator tools also are able to model interaction between different network entities (e.g., routers, switches, links) and their related events such as link changes, route changes, link failures and link overloading [4]. Network simulation empowers network designers to investigate different design options before coming to an agreement on a final network design.

Evaluation of network performance is one of the preliminary aims of network designers to simulate the network topology before realizing it. Other motivating factors of network simulation include failure analysis, network design and network resource planning [5]. The network simulation not only saves a lot of time and money but also increases efficiency by allowing testing of multiple behaviors of the network in a controlled and reproducible manner.

Network simulations have been initialized in the early 1990s by the advent of the Network Simulation Testbed (NEST) tool. NEST can be seen as a backbone of the many modern simulation tools [6]. It had a graphical environment for simulation and rapid prototyping of distributed networked systems and protocols. In the 1990s, it was used by designers of the distributed networked systems to measure the performance of the systems under a different situation such as failure of the links or switches. NEST simulation tool was composed of a network server and monitors as clients. The client/server architecture of it allowed multiple remote accesses to a shared test bed. NEST was based on the standard Unix and its server and clients were UNIX libraries and functions. Therefore, users could easily modify its functions based on their own needs [7].

REAL (realistic and large) network simulation tool was based on a modified version of the NEST 2.5 simulation test bed. Its initial developing motivation was to compare the "fair queuing" gateway algorithm with first-come-first-served scheduling and with competing proposals from Digital Equipment Corporation. REAL was composited of two parts: a simulated server and a display client. The Berkeley UNIX socket was used to connect the server to the client. It supported packet switched, store and forward networks similar to the existing Xerox corporate net and the DARPA Internet. REAL was able to model many details of the flow in the network and transport layers [8].

In general, each network simulation or emulation study requires a simulation scenario which defines the input configuration. According to Bajaj *et al.* [9], each simulation scenario is usually made up of four components:

1. Network topology: which defines the physical interconnects between nodes and the static characteristics of links and nodes.
2. Traffic model: which defines the network usage patterns and locations of unicast and multicast senders.
3. Test generation: which creates events such as flooding traffic toward specific node.
4. Network dynamics: such as node and link failures.

Additionally, NS2, NS3, OMNeT++, SSFNet, J-Sim, OPNET and QualNet are some other examples of the well-known network simulation tools [10]. According to Wehrle *et al.* [11], simulation tools have to model different network elements as following:

- Network nodes: which illustrate end nodes such as PCs, laptops, servers, tablets and network devices such as routers, hubs and switches.
- Network devices: which illustrate the physical devices that connect nodes to Ethernet network interface card, a wireless IEEE 802.11 device, etc.
- Communication channels: which illustrate the medium for sharing information among network devices such as fiber-optic-point-to-point links, shared broadcast media, wireless spectrum, etc.
- Communication protocols: which model the implementation of standardized and experimental network protocols such as User Datagram Protocol (UDP), Domain Name System (DNS), etc.
- Protocol headers: which illustrate the special data related to the specific protocol in the network packets.
- Network packets: which are the main parts of the information exchange in computer networks. Network packets consist of protocol header and payload data.

Conjointly, Wehrle *et al.* [11] emphasize on the importance of the realism rather than abstraction in network simulation, as the high level of the abstraction might result in abundant divergence from the experimental results.

3.1.2   Network emulation

Network emulation is an integration of simulated networks with real end-systems such as computers, routers, switches, etc. The connections between real world and simulated environments are done in a seamless manner as the connection among real network objects.

3.1.3   The application of network simulation and emulation in network security

Conducting real experiments such as attack scenarios in an operational network environment causes high risks. Therefore, simulation environments become more popular in network security research. Network anomaly and intrusion detection is one of the interesting research subjects in the area of network security. Most of the existing studies in the field of network anomaly detection validate their methods using simulated datasets, because practicing attack scenarios on real and live networks may causes a network crash [12].

3.1.4   Virtualization

Computer virtualization techniques were first developed in 1960s by IBM [13]. Virtualization techniques enable users to divide the physical computer to multiple isolated environments called virtual machines or guest machines. Virtual machines also can be seen as an emulation of physical machines. Virtual machines are another solution which is used to model networks. There are two types of virtualization: virtual machines which are powered by hypervisors and container-based virtual machines (Docker).

3.1.5   Virtualization using hypervisor

All of virtual machines working on a physical computer (host) share their physical resources such as the memory, disk and network devices by means of a software which is called hypervisor or a control program. VMware, KVM, Xen, Xbox and Hyper-V are the most well-known hypervisors. The hardware resources are allocated by the virtual machines on request. Resource sharing and isolation are two prominent advantages of the virtual machines. Virtualization empowers system administrators to create a simplified abstract view of the system as a working space of the software application. Additionally, most of the cybersecurity phenomena cannot be studied experimentally due to the potential risk of collapsing or infecting the experimental environment, high cost and legal issues. Therefore, virtual machines are absolutely essential for cybersecurity studies. Virtual machines empower researchers to develop intrusion prevention and intrusion detection systems (IDSs) by safely testing suspicious activities in a virtual environment.

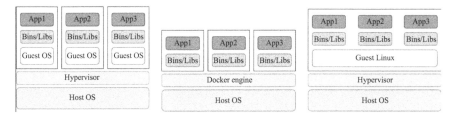

Figure 3.1   Architecture of Docker and hypervisor-based virtual machines

3.1.6   Virtualization using container

Linux containerization is a virtualization method which enables running of multiple processes on multiple isolated environments or operating systems only by means of a single kernel. It is an operating system level virtualization environment which takes advantages of Linux Cgroups[1] and namespace[2] to allow different containers to run on a single host [14]. Docker provides very light vitalization, because it does not require hardware level virtualization. It wraps software and its dependencies such as shared applications and services into a standardized container [15]. In other words, a Docker enables a transparent and an independent workspace for each application running on it by dividing operating system resources [16]. Figure 3.1, displays the difference of traditional virtual machines and Docker.

The advantages of using a Docker image rather than a traditional virtual machine image are agility, portability and controllability of the application environments. It enables users easy to maintain customized execution environments, in the shape of lightweight Docker images instead of bulky virtual machine images. This paves the way for the micro-services architectural pattern to rise.

3.1.7   Virtual machines and simulation

Virtual machines are suitable for modeling mid-scale networks, but the representation of huge networks with thousands of network objects are not practical using virtual machines [17]. In this chapter, we took advantage of both simulation and virtualization techniques to create a safe virtual lab for imitating different attack scenarios and collect traffic logs. These network traffic logs can be used to develop novel detection and defense methods in the field of network security.

The remainder of the chapter is organized as follows: Section 3.2 describes the background of network simulation and virtualization in the field of anomaly and attack detection; Section 3.3 provides the methodology used; Section 3.4 describes the network topology of our test bed; Section 3.5 provides detailed information about the way that different objects of the network topology were configured; Section 3.6

---

[1] Control group is a Linux kernel feature which provides isolated workstation with limited resources called container.
[2] Namespace is a Linux feature that prevents observation of resources used by different groups.

presents discussion and results; and finally, Section 3.7 summarizes the study and presents a road map for the future work.

3.2 Literature review

This section presents a short overview of network anomalies and their detection techniques, network workload generators and some simulation practices in the field of network security.

3.2.1 Network anomalies and detection methods

In a large dataset, nonconforming patterns are often called anomalies, outliers, exceptions, aberrations, surprises, peculiarities or discordant observations in various domains. However, outliers and anomalies are the most frequently used terms in the context of network intrusion detection [18]. Network events which are far from normal or expected normal behavior are suspicious from the perspective of the security. Not certainly all anomalies reflect a malicious activity in the network. Anomalies also can be observed when the meaning or scope of the normality changes. Therefore, the number of anomalies is always equal or higher than the number of malicious points in a given dataset. On the other hand, anomalies which are the result of the unauthorized attempt to access information, unauthorized information manipulation and attempt to make a system unreliable or unusable are categorized as malicious activities. There are bunch of studies in the literature which investigate network anomalies using different methods including statistical methods, classification, clustering, information theory. References [19–22] are examples of many survey studies which present the state of the art in the field. All network anomaly detection researches require dataset to validate their detection methods. Nevertheless, there are only few real datasets which are publicly available such as FIFA World Cup Dataset 1998, DARPA Intrusion Detection Data Sets 1998, KDD cup Dataset 1999, UCLA Dataset 2001, CAIDA DDoS Attack Dataset 2007 and TUIDS DDoS Dataset 2012. The KDDCUP'99 dataset is the most widely used benchmark dataset in network anomaly and attack detection studies. On the other hand, some researchers have blamed it for its inherited problems as following [23]:

- Both the background and the attack data were synthesized for the privacy issues.
- Data's false alarm characteristics were neglected; therefore, it is difficult to claim that the available dataset is similar to the observed data.
- The workload of the synthesized data does not seem to be similar to the traffic in real networks.
- More probably, the TCPdump data collector tool was overwhelmed during the heavy traffic load and drop packets.
- There is no exact definition of the attacks for some cases such as probing or buffer overflow.

Gogoi *et al.* [24] emphasize on the nature of the input data as the key aspect of any anomaly detection system. Input data is defined as a collection of data with some attributes of same or different types such as binary, categorical or continuous. As the nature of attributes determines the applicability of an anomaly detection technique, it is so prominent to employ the dataset with desired attributes. It is not likely to find any publicly available real dataset which perfectly matches attribute requirements of all anomaly detection studies. In a nutshell, the combination of real data and realistic synthetic dataset which represents the real environment could be seen as a coherent choice to validate novel anomaly detection engines in the rapidly growing computer and information technology area.

3.2.2 Network workload generators

There are only a few studies in the literature regarding the realistic network workload generation in contrast to the huge amount of studies on the characterization, modeling and simulation of computer networks. In general, a synthetic network workload generator should be able to appropriately capture the complexity of real workload in different scenarios, modify the properties of workload based on specific demands of the scenario and finally measure indicators of the performance for the workload at network level [25]. OSTINATO [26], SEAGULL [27], Tmix [28], RUDE/CRUDE [29], MGEN [30], KUTE [31] and BRUTE [32] are some examples of the network workload generation tools. Network workload generation approaches can be classified into two groups such as following:

- Trace-based generation: In this approach, the content and the timings of traffic traces are mimicked based on previously collected data in the real scenarios.
- Analytical model-based generation: In this approach, flows and packets are generated based on statistical models.

A comprehensive network workload generation tool should employ both approaches depending on the characteristics of the various scenarios[25].

3.2.3 Network simulation for security studies

In recent years, simulation and virtualization have gained popularity in network security research as well. They empower researchers to run vulnerability-related programs against IT-systems and IT-applications and then develop solutions to detect and mitigate vulnerabilities. Network anomaly detection is one of the well-known research focus in the area of network security. Enormous research efforts have been spent on anomaly-based network intrusion detection using mathematical, machine learning, artificial neural network, fuzzy set, knowledge-based and combination learning techniques [33]. In general, all these techniques have a common point of intersection: they need network traffic datasets to validate their anomaly detection approaches.

Real and simulated datasets are two major types of datasets which are used in network anomaly detection studies. However, each of them has some advantages and disadvantages. Most of the real datasets in the field are outdated and anonymized for privacy concerns. Furthermore, there is no existing dataset that can meet all requirements of researchers for various attack detection methods [34]. Therefore, researchers struggle to validate their approaches using real datasets in experiments or simulations. On the other hand, real system–based experiments are very complicated, time consuming and expensive. As a result, researchers have launched to prepare their own datasets in simulation environment due to the controllability, reproducibility and scalability of simulated data [35]. In this section, we presented some recent examples of the studies which have benefited the simulation tools to investigate network anomalies.

Kuhl *et al.* [36] used ARENA simulation software for modeling a network setup, for modeling cyberattacks, for simulating cyberattacks and generating IDS data. Their simulation environment composited of three main components: machines, connectors and subnets. Machines were individual client computers or servers which were employed as attacker or target machines. Target machines were equipped with IDS sensors to detect cyberattacks and create alarms. In this study, attack scenarios were defined in 5 major groups and 23 subgroups. The simulated environment enabled users to choose different types of attacks to be occurred over a period of time along with a specified quantity of network noise. The main goal of this study was to generate automated attack and produce IDS alert files which included alerts from both attack actions and noise. Correspondingly, these alert files were used to test and evaluate cybersecurity systems.

Elejla *et al.* [37] proposed a flow-based IDS for detecting ICMPv6-based DDoS attacks. Since there was no existing dataset that could meet their criteria for detecting ICMPv6-based DDoS attacks, researchers created their own dataset in virtual environment using GNS3 emulation tool. The GNS3 allowed them to collect a normal traffic from the real-life network of a university and generate attack data in simulated environment. Their simulated environment consisted of one victim and two attacker machines: Kali OS with THCtoolkit and an Ubuntu OS machine with SI6 attacking tool. In this study, different ICMPv6 DDoS attack scenarios were performed, and the related network traffic was captured using Wireshark[3] tool in PCAP file format.

Balyk *et al.* [38] used the GNS3 tool to simulate DDoS attacks. The experimental network topology consisted of three virtual PCs of regular users, one Fedora core 22 64 bit Linux system running apache 2.4.12 web server and one attacker host. The network was realized by Cisco Ethernet switches and Cisco routers. For DDoS attack simulation, a simple Perl script was used to create multiple parallel connections to destination port 80 of the web server. Wireshark was used to capture the traffic flows on the closest switch to the web server. This study only simulated a single DDoS attack scenario, while the concentration of the study was on web server parameters settings and defense modules settings in GNS3 simulations.

Al Kaabi *et al.* [39] developed a virtual lab called DoS_VLab to allow students to practice five types of DDoS attacks including ARP cache poisoning, Switch CAM table corruption, TCP SYN flood, Land and ARP storm attacks in secure academic environment. The lab was based on virtualization and GNS3 network simulation for

---

[3]https://www.wireshark.org/.

building virtual networks. It consisted of two Windows XP Virtual Machines (VMs): one of them acts as the attacker host and the second as the victim host. The study only mentioned that appropriate tools were installed on the attacker machine and different types of DDoS attacks were performed, but the names of the tools were not implicitly indicated.

For developing a traffic analyzer to detect DDoS attacks, Ojeniyi *et al.* [40] proposed a simulated environment using GNS3 tool. The network topology consisted of an attacker and a victim machine both running Kali Linux. In order to simulate the DDoS attack, the built-in Hping3 tool of the Kali Linux on the attacker's machine was used to perform DDoS attack toward the victim.

For proposing a method to discriminate DDoS attacks from the flash events (FE), Behal and Kumar [41] developed a test bed to generate dataset of low-rate and high-rate DDoS attacks and of FE. Their test bed was a combination of real and emulated systems. Their test bed consisted of 75 physical nodes that run Ubuntu and Windows OS, Ethernet switches, routers and Linux server. The Core emulator tool was used to increase the number of the identical virtual nodes. The httperf and D-ITG traffic generator tools were used to generate legitimate HTTP traffic and BoNeSi botnet simulator tool was used to generate DDoS attack traffic.

Zhao *et al.* [42] evaluated different methodologies to detect network anomalies. In order to prepare datasets for their experiments, they defined the topology of the simulation network in NS2 format, then they used Malicious traffic Composition Environment (MACE) tool[4] as the malicious traffic generator and the LTProf tool[5] as the legitimate traffic generator. Moreover, Scalable URL Reference Generator (SURGE) tool[6] was used to produce extra workload on the web servers to test stress tolerance of the network. NetFlow was generated for all passing network traffics in both directions of the link (in the form of compressed nfdump files) and passed to anomaly detection engine for further analyses.

Sieklik *et al.* [34] developed a simulated environment to investigate the amplification DoS attack based on the Trivial File Transfer Protocol (TFTP). Their network topology was consisted of three routers, a computer and two servers. GNS3 simulation software was used to simulate the network topology in this study. VMware virtualization software was utilized to implement more flexible simulation systems. The three virtual machines (attacker, target and amplifier) were connected to the router running in GNS3. The attacker server ran Kali Linux 5 R3 OS with various penetration testing tools. Both amplifier and target were running Windows XP SP3. The attacker computer ran several TFTP service, and the target machine ran default Windows XP UDP services whereas any other UDP service could be employed. They investigated a realistic amplification scenario by creating spoofed packets using Scapy,[7] then the attacking machine using a loop command to send these multiple crafted packets to the amplifying server.

---

[4]MACE is a toolkit to generate divers set of attacks [43].
[5]TheLTProf collects legitimate traffic samples from public traces [44].
[6]SURGE is a web workload generation tool which mimics a set of real users accessing a server [45].
[7]It is a special network analysis tool written in Python to create network packets [46].

3.3   Methodology

The most challenging aspect of simulation based anomaly detection research is proving the reliability and dependability of simulated datasets in comparison to real-life datasets. On the other hand, well-designed simulation environment offers repeatability, programmability and extensibility of the validation instrument [12].

The main purpose of this chapter is to introduce a simulation environment using VMware virtualization software to design a flexible and reliable simulation environment. In order to realize the simulated environment, some software and hardware were required such as VMware workstation, GNS3 software and Ubuntu Docker image. We also used open-source pfSense firewall, NTOPNG and MYSQL to apply network rules, collect network flow data and store network flow data, respectively. Moreover, we utilized botnet simulator and network traffic–generator tools for creating DDoS attack and normal traffic data samples. We used GNS3 simulator to develop our experimental environment; as mention in [38], the results of the GNS3 simulation tool matched the results obtained from the Cisco network. Additionally, GNS3 is a well-tested and established network-simulation tool, which is also used by many other companies like Exxon, Walmart, AT&T, NASA, etc. [38].

3.4   Defining a simulated and virtualized test bed for network anomaly detection researches

We have implemented the virtual lab named Cyber Security Simulated Lab (CSSL) in order to create an isolated platform to simulate, test and analyze different types of security threats. Our infrastructure was built by means of a VMware virtualization software on one physical machine. In order to connect the virtual machine to the network, we mapped the external Internet connection of our host machine to the internal VM network. The CSSL allows us to configure different network topologies for simulating different attack scenarios. Our virtual test bed is an isolated environment to mainly fabricate and collect simulated DDoS attack data. As network technologies are growing rapidly, we primarily employed open platforms to include different efforts and different packages whenever there is a need [10]. Moreover, using open source tools and applications facilitates the repeatability of the study. We initially defined the network topology as shown in Figure 3.2 using GNS3. Attacker and target machines are Ubuntu Docker appliance for GNS3. pfSense is an appliance of the GNS3. VMnet8 is our exit point to the Internet. We disabled all incoming and outgoing traffic to/from the VMnet8 using firewall rules during the experimental phase for security concerns. (For more information refer to Section 3.5.3.)

3.4.1   GNS3

GNS3 is a graphical network emulation tool which can provide simulation/emulation of entire networks and many network devices such as links, switches, routers

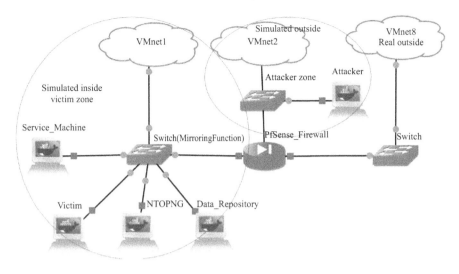

Figure 3.2 Network topology of the test bed

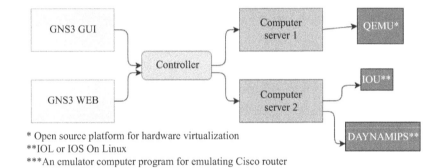

Figure 3.3 GNS3 architecture

firewalls, etc. As it can be seen in Figure 3.3, GNS3 has a similar architecture to Linux computers based on internal interfaces (network to device driver) and application interfaces (sockets) [11]. All the communications in GNS3 tool are done over HTTP using JSON; therefore, HTTP basic authentication can be used to securely access to the application programming interfaces [47]. Additionally, GNS3 enables packet filtering and raw-packet capturing in the network using its direct interface to Wireshark application [48]. We installed both GNS3 windows application and GNS3 virtual machine image [49]. As it can be seen in Figure 3.4, we also connected them to each other by setting the remote main server address of the GNS3 windows application to the IP address of the GNS3 virtual machine.

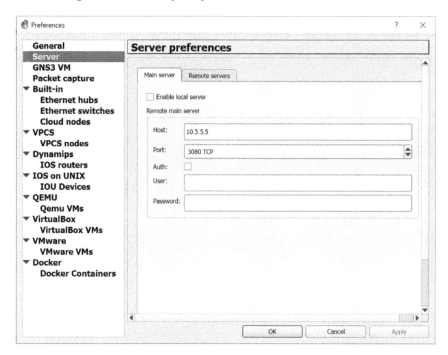

Figure 3.4   *Connecting GNS3 application to the remote server*

3.4.2   Ubuntu

We used the following command to pull GNS3 Ubuntu Docker container on GNS3 VM from Docker registry.

```
docker pull gns3/Ubuntu: xenial
```

The Ubuntu Docker container encompassed networking tools such as net-tools, iproute2, ping and traceroute, curl (data transfer utility), host (DNS lookup utility), iperf3, mtr (full screen traceroute), socat (utility for reading/writing from/to network connections), ssh client, tcpdump and telnet [50].

3.4.3   Network interfaces

Virtual machine allows defining and setting of different network adapters. As it can be seen in Figure 3.5, we created three network interfaces as following [51]:

● NAT (network address translation): The virtual machine does not have an IP address on the external network. Therefore, it translates the addresses of virtual machines in a private VMnet network to that of the host machine. Subsequently,

Name	Type	External Connection	Host Connection	DHCP	Subnet Address
VMnet1	Host-only	-	Connected	-	10.5.6.0
VMnet2	Host-only	-	Connected	-	10.5.7.0
VMnet3	Custom	-	-	-	10.5.8.0
VMnet4	Host-only	-	Connected	-	192.168.253.0
VMnet8	NAT	NAT	Connected	Enabled	10.5.5.0

Figure 3.5   Host-only, NAT and custom virtual network interfaces

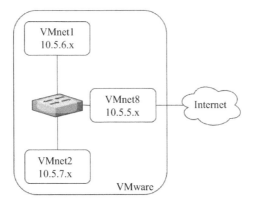

Figure 3.6   Virtual machine–adapter setting

it uses the host computer network connection in order to connect to the Internet. VMware virtual DHCP server assigns an address to the virtual machine. It provides a transparent and easy-to-configure method to access to network resources.

- Host-only: It provides a network connection between the virtual machine and the host computer. The virtual machine is connected to the host-operating system using a virtual Ethernet adapter that is visible to the host-operating system on a virtual private network. It is not visible to the outside host.
- Custom: It is a more complicated networking configuration option which provides customized setup for virtual network adapters. After selecting "Custom" option, the user should choose a virtual switch to connect the virtual machine's adapter to that switch.

Accordingly, we created three corresponding network adapters for the GNS3 virtual machine like Figure 3.6. We also assigned IP addresses for each of the interfaces of GNS3 virtual machine. Figure 3.7 demonstrates the assignment of IP addresses to interfaces of VMnet8.

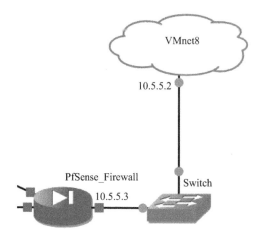

Figure 3.7   IP address assignment for VMnet8

3.5   Simulated environment for network anomaly detection researches

As a simplest topology, we required an attacker, a target, a firewall, normal and attack traffic generators, a log collector and a data repository.

3.5.1   Victim machine

As a victim machine, we installed the Apache2 server on an Ubuntu Docker container and made a new container called Victim_Machine_Template.

3.5.2   Attacker machine

As the attacker machine, we installed the attack tool BoNeSi [52] on an Ubuntu Docker container and made a new container called Attacker_Machine_Template.

1. We were able to add any penetration testing or attack simulation tool on the attacker machine based on our needs. We needed to simulate a DDoS attack; therefore, we used BoNeSi botnet simulator tool with 50k bots to simulate DDoS attack.
2. BoNeSi is able to generate different types of flooding attacks using ICMP, UDP and Transmission Control Protocol (TCP) protocols using different 50k IP addresses. BoNeSi empowers users to configure rates, data volume, source IP addresses, URLs, target port, time to live, etc. It also supports the simulation of the HTTP-GET floods attacks. As cited in [52], BoNeSi can generate up to 150,000 packets per second on an AMD Opteron with 2 GHz. Its rate can be duplicated using recent AMD Phenom II X6 1100T with 3.3 GHz. We simulated the following DDoS attacks:
    ```
    Bonesi -r 3000 -p tcp -d eth0 10.5.5.50:80
    ```
    (Sends 3,000 packets per second to port 80 of victim machines)

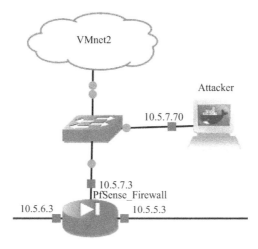

Figure 3.8   IP address assignment for WAN and LAN interfaces of pfSense

```
Bonesi -r 3000 -s 320 -p tcp -d eth0 10.5.5.50:80
```
(Sends 3,000 packets with size of 320 byes per second to port 80 of victim machines)
```
Bonesi -r 3000 -i ip_list -p tcp -d eth0 10.5.5.50:80
```
(Sends 3,000 packets per second to port 80 of victim machine using source IPs in the ip_list text[8] file)
```
Bonesi -r 3000 -i ip_list -p tcp d eth0 10.5.5.80.80
```
(Sends 3,000 packets per second to port 80 of the victim machines using source IPs in the ip_list text file)
```
Bonesi -i ip_list -p tcp -d eth0 10.5.5.80.80
```
(Floods as much as packets it can to port 80 of the victim machines using source IPs in the ip_list text file)

3.5.3   pfSense firewall

We added the pfSense appliance of GNS3 to this project. Consequently, we assigned the WAN and LAN interfaces of pfSense and their correspondent IP addresses such as shown in Figure 3.8.

3.5.3.1   Firewall configuration

We accessed the web configurator of the firewall using the http://10.5.6.3 URL as it has been applied in the LAN interface setting of the firewall. Subsequently, we defined rules for floating, WLAN, LAN and attacker zone. For instance, we allowed and logged all of traffic between victim zone and attacker zone. We also blocked all traffic from/to real-outside, because the virtualized environment

---

[8]The list of source IP addresses to participate in the DDoS attack can be provided by a text file and then pass to BoNeSi using '-i' parameter.

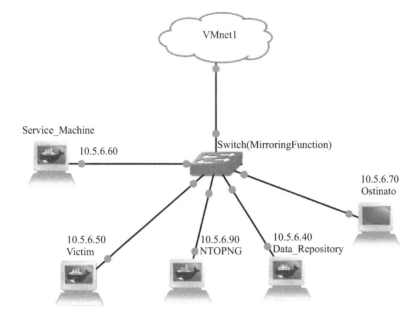

Figure 3.9   IP address assignment for victim zone

is vulnerable to all the traditional attacks and exploits worse than normal environments [52]. Moreover, we wanted an isolated environment to create and test attacks without affecting the real systems.

3.5.4   NAT and VMware host-only networks

The network interfaces of the real outside, attacker zone and victim zone in our project were mapped to corresponding network interfaces adapters. Figure 3.9 demonstrates the IP address assignment of interfaces in the victim zone.

3.5.5   Traffic generator machine

Ostinato is a packet crafter, network traffic generator and analyzer which is supported with a user friendly graphical user interface (GUI) [26]. Ostinato allows users to create desired data streams by manually configuring each packet at different layers of the Open Systems Interconnection (OSI) model. Ostinato uses a client/server architecture. The GUI is used to create desired packets and then send them out of the traffic interfaces using Ostinato's server (drone) [26]. The Ostinato GUI can be used to configure IP and MAC addresses of the interfaces (eth1, eth2, eth3, etc.). By default, the interface eth0 is used to connect to the local machine. Figure 3.10 demonstrates the main user interface of Ostinato tool.

We added the Ostinato appliance of GNS3 to this project in order to fabricate and send packets of several streams. Ostinato generates stateless streams which are

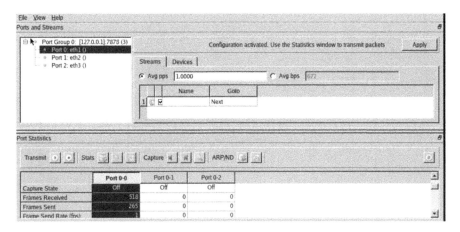

Figure 3.10   GUI of Ostinato

dropped by pfSense firewall. Therefore, we placed it behind the firewall. In order to use the GUI of the Ostinato, we set the console type to virtual network computing (VNC) in the node configuration window of GNS3. GNS3 supports ThightVNC [53] viewer, so we installed ThightVNC[9] viewer on our local machine. The IP address of the remote host in ThightVNC was set to the IP address of GNS3 virtual machine (10.5.5.5). We set the several parameters at different OSI layers when a new stream was created as following:

1. Configuring protocols in different layers (MAC, Ethernet, IPv4, TCP and Text for physical, data link, network, transport and application layer correspondingly).
2. Configuring MAC address of the sender and receiver to MAC address of attacker and victim machines, respectively.
3. Configuring source and destination IP addresses to 10.5.7.70 (attacker machine) and 10.5.6.50 (victim machine).
4. Configuring destination port number to 80.
5. Writing "Network Traffic Generation" in payload of the data.
6. Configuring the number of packets to 10,000 and the transmission rate to 200 pps.

Ostinato also allows to set checksum and different packet flags including URG, ACK, PSH, RST, SYN and FIN. Figure 3.11 demonstrates the properties of the crafted stream in detail.

3.5.6   NTOPNG tool

We added the Docker image of the NTOPNG as an appliance to our GNS3 project. We utilized NTOPNG tool to collect the processed network traffic data in our GNS3

---

[9]VNC is a graphical desktop sharing system based on the Remote Frame Buffer protocol to remotely access and control another computer [54].

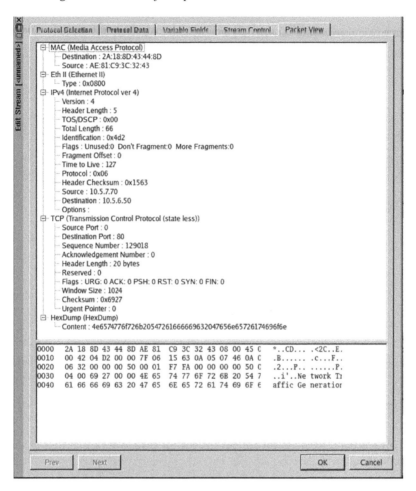

Figure 3.11   Detailed packet view in Ostinato

environment. NTOPNG is a network traffic probe that monitors network usage. NTOPNG is based on `libpcap` and has a capability to virtually run on every Unix platform, MacOSX and on Windows as well.

Tools like TCPdump and Wireshark collect raw data. It means that when a packet is received and sent, data is captured and logged. On the other hand, NTOPNG is able to collect data in a connection-based format. It enabled us to capture main features of network traffic including source IP address, destination IP address, source port, destination port, duration, number of bytes and packets which are sent and received. It also provides an extra feature to geographical information of the IP addresses [55].

Figure 3.12 show the summary of active hosts in non-attack period, and Figure 3.13 shows the time-based graph of traffic follows in the NTOPNG interface (eth0) which is connected to the victim host during the simulated DDoS attack period.

IP Address	Location	Alerts	Name	Seen Since	ASN	Breakdown	Throughput	Traffic
172.217.19.67	Remote	0	172.217.19.67	1 min, 21 sec	Google Inc.	Sent	0 bps —	3.46 KB
144.122.145.153	Remote	0	144.122.145.153	56 sec	Middle East Technical University	Sent	0 bps —	6.43 KB
10.6.2.101	Remote	0	10.6.2.101	1 min, 32 sec		Sent Rcvd	0 bps —	1.45 KB
10.5.6.50	Local	4	10.5.6.50 ⚠	24 min, 48 sec		Sent Rcvd	0 bps —	49.1 MB

Figure 3.12   Sent/Received traffics in interface of NTOPNG (eth0) during attack

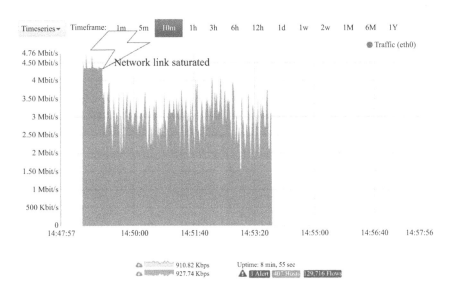

Figure 3.13   Active network hosts during non-attack time period

Figure 3.14 shows the summary of a time-based graph of the packets' arriving rate during simulated DDoS attack in NTOPNG.

3.5.6.1   NTOPNG configuration

As Figure 3.12 shows, we attached NTOPNG to the same switch which the victim machine was also connected. In addition, we configured port mirroring and copied all traffic on the switch port which was connected to the victim to the mirror port. We used this technique in order to send all network traffic to the victim machine to NTOPNG [56].

3.5.6.2   NTOPNG configuration to dump logs to Mysql machine

We set a start command of the NTOPNG machine using its configuration interface by typing the following connection string.

```
(mysql;<host>;<dbname>;<table name>; <user>; <pw>)
```

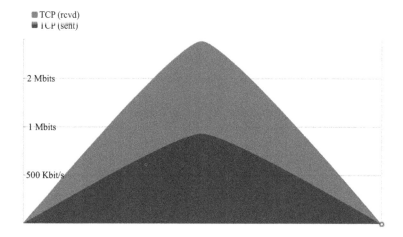

Figure 3.14   Sent/Received traffics in victim host during attack

Table 3.1   Fields of NTOPNG logs and their corresponding data types

Field	Type	Key
idx	Int	Yes[a]
VLAN_ID	Small Int	No
L7_Protocol	Small Int	No
Source IP address	Int	No
Source port number	Small Int	No
Destination IP address	Int	No
Destination port number	Small Int	No
Protocol	Tiny Int	No
Bytes	Int	No
Packets	Int	No
First_Switched	Int	No
Last_Switched	Int	No
Info	Int	No
Json	Blob	No
Profile	Varchar	No
NTOPNG_Instance_Name	Varchar	No
Interface_ID	Small Int	No

[a]Auto incremental.

Consequently, NTOPNG created a database and table on the Data_Repository machine as Table 3.1 displays.

3.5.7   Repository machine

As NTOPNG supports the exportation of the monitor data to MySQL, Elasticsearch and Logstash, we installed MYSQL on an Ubuntu Docker container and made a new container called Data_Repository_Template. We connected it to the NTOPNG machine to collect network traffic from NTOPNG. As Figure 3.9 demonstrates, the Data_Repository machine has been placed in the same network with NTOPNG and the victim machine (10.5.6.0/24).

3.5.7.1   Repository machine configuration

We altered the configuration of Data_Repository machine for the listening port. By default, MYSQL listening port is port 3306 from the local host. If the "dump traffic data" option of the NTOPNG is set, it automatically will send the log data to MYSQL. In this situation, MYSQL should receive data (listen) out of local host. We set bind_address variable to the IP address of the NTOPNG machine, then stop and restart the Data_Repository machine.

3.5.7.2   Give a remote root access to Data_Repository machine

By default, the remote root access to MYSQL is disabled for security reasons. We locally enabled it using the following SQL query in order to authorize NTOPNG for writing network follow logs in data repository machine.

```
GRANT ALL ON *.* TO root@'10.5.6.90' IDENTIFIED BY
'test'; FLUSH PRIVILEGES;
```

3.6   Discussion and results

To best of our knowledge, there is no comprehensive study in the literature which merely concentrates on step-by-step implementation of the simulated lab, based on virtual machines and Docker containers to produce directional (sent/received) DDoS attack data and collect these data according to many criteria including IP address, port, L7 protocol, throughput, RTT, TCP statistics (retransmissions, out of order packets, packet lost), bytes/packets transmitted, etc. We think this chapter fills in the gap by providing guidelines for cybersecurity researchers and practitioners to develop an isolated simulation environment. Usage of open-source tools ensure the repeatability and comparability of results among studies which will apply the proposed simulated environment in their research. The simulate environment is suitable to create normal and malicious network traffics and collect flow-based traffic logs for network anomaly detection analyses. Minor changes might be required based on different attack scenarios. For example, different attack tools can be installed on attacker machine to launch variety of exploits toward victim machine.

3.7   Summary

In this chapter, we provided information on some key concepts of network simulation and virtualization and their differences. Subsequently, we have presented a network simulation environment with a focus on network attack and anomaly detection scenarios. We took advantages of both hypervisor and Docker-based virtual environments. We utilized preconfigured GNS3, NTOPNG, Docker containers, Ostinato and pfSense virtual image to realize our network topology. On the other hand, we made our own containers including Data_Repository, Victim_template and Attacker_template which can be used in future industry and academic studies. Virtualization and simulation play a significant role not only in reducing the operational costs but also in keeping operational environments safe from threats. Therefore, they provide an intrinsic potential to be used in the field of network security, attack and anomaly detection. In this study, we almost opted open source tools to encourage adoption, modification and improvement of this simulated environment by future researchers. This is an example of using GNS3 and computer virtualization for network simulation for the investigation of network security and attack scenarios. Based on this example, several different simulation scenarios can be realized.

The simulation results depend strongly on the performance of the used hardware.[10,11] The same simulation with slower/faster hardware could deliver divergent results. This is also depending on bus speed, NIC speed, etc. This has to be taken into account for performance testing purposes. But the fundamental results concerning security investigation using this simulation method are not infected.

References

[1]   Kurose JF, Ross KW. Computer networking: A top-down approach. Addison-Wesley, Reading; 2010.

[2]   Breslau L, Estrin D, Fall K, *et al.* Advances in network simulation. Computer. 2000;33(5):59–67.

[3]   Sarkar NI, Halim SA. A review of simulation of telecommunication networks: Simulators, classification, comparison, methodologies, and recommendations. Cyber Journals: Multidisciplinary Journals in Science and Technology, Journal of Selected Areas in Telecommunications (JSAT). 2011;2(3):10–17.

[4]   Chang X. Network simulations with OPNET. In: Proceedings of the 31st Conference on Winter Simulation: Simulation—A Bridge to the Future-Volume 1. ACM; 1999. p. 307–314.

[5]   Gyires T. Network simulation. In: Iványi A, editor. Algorithms of informatics. vol. 2. Budapest: MondAt Kiadó. 2007.

[6]   Chan KFP, De Souza P. Transforming network simulation data to semantic data for network attack planning. In: ICMLG 2017 5th International Conference on Management Leadership and Governance. Academic Conferences and Publishing Limited; 2017. p. 74.

[7]   Dupuy A, Schwartz J, Yemini Y, Bacon D. NEST: A network simulation and prototyping testbed. Communications of the ACM. 1990;33(10):63–74.

[8]   Keshav S. REAL: A network simulator. University of California Berkeley, Berkeley, CA, USA; 1988.

[10] Host: Intel(R) Core(TM) i7-7700HQ CPU @ 2.80GHz, 16.0GB RAM, 64-bit OS.
[11] Virtual Appliance: 1×4 Core processor, 6.1GB RAM.

[9] Bajaj S, Breslau L, Estrin D, *et al.* Improving simulation for network research. University of Southern California, Tech. Rep; 1999.

[10] Pan J, Jain R. A survey of network simulation tools: Current status and future developments. Washington University in St. Louis, Tech. Rep; 2008.

[11] Wehrle K, Günes M, Gross J. Modeling and tools for network simulation. Aachen: Springer Science & Business Media; 2010.

[12] Behal S, Kumar K. Trends in validation of DDoS research. Procedia Computer Science. 2016;85:7–15.

[13] Bitner B, Greenlee S. z/VM a brief review of its 40 year history. Dosegljivo: http://www vm ibm com/vm40hist pdf (pridobljeno: 26 4 2016). 2012.

[14] Preeth E, Mulerickal FJP, Paul B, Sastri Y. Evaluation of Docker containers based on hardware utilization. In: Control Communication & Computing India (ICCC), 2015 International Conference on. IEEE; 2015. p. 697–700.

[15] Morris D, Voutsinas S, Hambly N, Mann R. Use of Docker for deployment and testing of astronomy software. Astronomy and Computing. 2017;20: 105–119.

[16] Geng X, Zeng X, Hu L, Guo Z. An novel architecture and inter-process communication scheme to adapt chromium based on Docker container. Procedia Computer Science. 2017;107:691–696.

[17] Grunewald D, Lützenberger M, Chinnow J, Bye R, Bsufka K, Albayrak S. Agent-based network security simulation. In: The 10th International Conference on Autonomous Agents and Multiagent Systems-Volume 3. International Foundation for Autonomous Agents and Multiagent Systems; 2011. p. 1325–1326.

[18] Bhattacharyya DK, Kalita JK. Network anomaly detection: A machine learning perspective. New York: Chapman and Hall/CRC; 2013.

[19] Gogoi P, Bhattacharyya D, Borah B, Kalita JK. A survey of outlier detection methods in network anomaly identification. The Computer Journal. 2011;54(4):570–588.

[20] Jyothsna V, Prasad VR, Prasad KM. A review of anomaly based intrusion detection systems. International Journal of Computer Applications. 2011;28(7):26–35.

[21] Chandola V, Banerjee A, Kumar V. Anomaly detection: A survey. ACM Computing Surveys. 2009;41(3), Article 15:1–58.

[22] Ahmed M, Mahmood AN, Hu J. A survey of network anomaly detection techniques. Journal of Network and Computer Applications. 2016;60: 19–31.

[23] Tavallaee M, Bagheri E, Lu W, Ghorbani AA. A detailed analysis of the KDD CUP 99 data set. In: Computational Intelligence for Security and Defense Applications, 2009. CISDA 2009. IEEE Symposium on. IEEE; 2009. p. 1–6.

[24] Gogoi P, Bhuyan MH, Bhattacharyya D, Kalita JK. Packet and flow based network intrusion dataset. In: International Conference on Contemporary Computing. Springer; 2012. p. 322–334.

[25] Botta A, Dainotti A, Pescapé A. A tool for the generation of realistic network workload for emerging networking scenarios. Computer Networks. 2012;56(15):3531–3547.

[26]   Ostinato. Github; 2018. Available from: https://ostinato.org/.

[27]   Seagull: an Open Source Multi-protocol traffic generator. sourceforge; 2018. Available from: http://gull.sourceforge.net/.

[28]   Weigle MC, Adurthi P, Hernández-Campos F, Jeffay K, Smith FD. Tmix: A tool for generating realistic TCP application workloads in ns-2. ACM SIGCOMM Computer Communication Review. 2006;36(3):65–76.

[29]   RUDE & CRUDE. SourceForge; 2018. Available from: http://rude.source forge.net/.

[30]   Multi-Generator (MGEN). U.S. Naval Research Laboratory; 2018. Available from: https://www.nrl.navy.mil/itd/ncs/products/mgen.

[31]   Kernel-based Traffic Engine (KUTE). Swinburne University of Technology; 2018. Available from: http://caia.swin.edu.au/genius/tools/kute/.

[32]   Brawny and RobUst Traffic Engine (BRUTE). MIUR project EURO "University Experiment of an Open Router"; 2018. Available from: https://code.google.com/archive/p/brute/.

[33]   Bhuyan MH, Bhattacharyya DK, Kalita JK. Network anomaly detection: Methods, systems and tools. IEEE Communications Surveys & Tutorials. 2014;16(1):303–336.

[34]   Sieklik B, Macfarlane R, Buchanan WJ. Evaluation of TFTP DDoS amplification attack. Computers & Security. 2016;57:67–92.

[35]   Ringberg H, Roughan M, Rexford J. The need for simulation in evaluating anomaly detectors. ACM SIGCOMM Computer Communication Review. 2008;38(1):55–59.

[36]   Kuhl ME, Kistner J, Costantini K, Sudit M. Cyber attack modeling and simulation for network security analysis. In: Proceedings of the 39th Conference on Winter Simulation: 40 years! The Best is Yet to Come. IEEE Press; 2007. p. 1180–1188.

[37]   Elejla OE, Anbar M, Belaton B, Alijla BO. Flow-based IDS for ICMPv6-based DDoS attacks detection. Arabian Journal for Science and Engineering. 2018; 43(12):1–19.

[38]   Balyk A, Karpinski M, Naglik A, Shangytbayeva G, Romanets I. Using graphic network simulator 3 for DDoS attacks simulation. International Journal of Computing. 2017;16(4):219–225.

[39]   Al Kaabi S, Al Kindi N, Al Fazari S, Trabelsi Z. Virtualization based ethical educational platform for hands-on lab activities on DoS attacks. In: Global Engineering Education Conference (EDUCON), 2016 IEEE. IEEE; 2016. p. 273–280.

[40]   Ojeniyi JA, Balogun MO, Sanjo F, Ugochukwu O. Development of a traffic analyzer for the detection of DDoS attack source. 2016. In: International Conference on Information and Communication Technology and Its Applications (ICTA 2016). 2016. p.111–117.

[41]   Behal S, Kumar K. Detection of DDoS attacks and flash events using information theory metrics—An empirical investigation. Computer Communications. 2017;103:18–28.

[42] Zhao X, Qian YK, Wang CS. A framework of evaluation methodologies for network anomaly detectors. In: Advanced Materials Research. vol. 756. Trans Tech Publ; 2013. p. 3005–3010.

[43] Sommers J, Yegneswaran V, Barford P. A framework for malicious workload generation. In: Proceedings of the 4th ACM SIGCOMM Conference on Internet Measurement. ACM; 2004. p. 82–87.

[44] Mirkovic J. D-WARD: Source-end defense against distributed denial-of-service attacks. University of California, Los Angeles, CA; 2003.

[45] Barford P, Crovella M. Generating representative web workloads for network and server performance evaluation. In: ACM SIGMETRICS Performance Evaluation Review. vol. 26. ACM; 1998. p. 151–160.

[46] Scapy's documentation; 2018. Available from: https://scapy.readthedocs.io.

[47] GNS3 Architecture. GNS3 Academy.; 2018. Available from: http://api.gns3.net/en/latest/general.html#architecture.

[48] Neumann JC. The book of GNS3: Build virtual network labs using Cisco, Juniper, and More. San Francisco: No Starch Press; 2015.

[49] GNS3 Software. GNS3 Inc.; 2018. Available from: https://www.gns3.com/software.

[50] Goldstein M. BoNeSi – The DDoS Botnet Simulator; 2016. Available from: https://github.com/Markus-Go/bonesi.

[51] Ali I, Meghanathan N. Virtual machines and networks-installation, performance study, advantages and virtualization options. arXiv preprint arXiv: 11050061. 2011.

[52] Reuben JS. A survey on virtual machine security. Helsinki University of Technology. Tech. Rep; 2007.

[53] TightVNC Software. TightVNC Group; 2018. Available from: https://www.tightvnc.com.

[54] Virtual Network Computing. AT&T Laboratories Cambridge; 2018. Available from: http://www.hep.phy.cam.ac.uk/vnc_docs/protocol.html.

[55] NTOP Software. NTOP Inc.; 2018. Available from: https://www.ntop.org/products/traffic-analysis/ntop/.

[56] DrayTek. Vigor3300V user guide V3.0; 2009. https://www.draytek.com/en/products/products-a-z/router.all/2016/03/30/vigor3300v/.

Part II

Surveys and reviews

Chapter 4

Demand–response management in smart grid: a survey and future directions

Waseem Akram[1] and Muaz A. Niazi[1]

Nowadays, one of the key areas of research in smart grid (SG) is demand–response management (DRM). DRM assists in simplifying interactions between the customers and the utility-service providers. It also helps in the improvement of energy efficiency as well as effects on load balancing. Studies on DRM have brought a number of interesting, technical discussions and research contributions. Many of these studies work toward making energy-efficient systems. However, there is a need to work in the domain of customer satisfaction; this area needs considerable new advances. From past few decades, a number of studies have been carried out in SG regarding DRM. However, there is no such work that presents a comprehensive analysis of these works. There is a need to investigate different techniques, their advantages, as well as limitations. By focusing on DRM from a customer satisfaction perspective, in this chapter, we present a detailed overview of different solutions for developing DRM. We also group existing solutions and identify trends and challenges in an SG domain from DRM perspective.

4.1 Overview

We first start by giving an introduction of SG. Then background and basic concepts are given. Next, we present a detailed review of different literature from DRM perspective in SG. Then open-research problems are given. Finally, we present conclusion at the end of this chapter.

4.2 Introduction

The traditional power system provides one-way power flow to the consumers. On the other side, the energy demands are continuously growing from consumer sides. This makes the traditional power system difficult to respond to the ever-changing and

[1]Computer Science Department, COMSATS Institute of Information Technology, Pakistan

rising energy demand of consumers. Due to this issue, the energy sector has started working for efficient and sustainable energy system. This effort introduced the SG concept in the energy domain.

The SG introduced a two-way dialog where electricity and information can be exchanged between utility and consumers. It integrates advanced information and communication technology (ICT), smart meters, smart appliances, and other sensing mechanisms [1]. It is a developing network of distributed nodes, where all operations of the system are controlled by an intelligent and autonomous system [2]. The SG involves the transmission of energy to the consumers in a controlled and smart way, which benefices both utility and end users [3].

DRM plays an important role in SG environment. It enables the dynamic adjustment of energy demand from consumers in response to the price signals and incentives. This process shifts higher demand to lower demand, thus reducing energy cost [4,5]. It assists in the interaction between end users, appliances, and utility service provider which minimizes end-user effort in controlling power usage devices [6,7]. It also helps in fault detection and prevention in the system, thus improves system reliability and sustainability [2].

There are several research challenges related to the DRM. The deployment of ICT, smart meters, and renewable energy resources is a challenging task [1]. Renewable energy resources have unpredictable fluctuation in power generation. It is difficult to predict energy for the day ahead [8]. Another big challenge is decision-making for demand and consumption at consumer side. Consumers are making a decision about how much energy is required for a certain type of appliance in a particular time period. This makes the consumer decision more complex. The users' demand for energy changes with time (variable demand), this needs an adaptive strategy of grid unit that can modify their capacity according to the user demand [9]. Reliability is another issue in an SG environment [10]. Some naturally accruing events lead to the cascading failure of SG [6,11–14], where supervisory control and data-acquisition system is used to detect and prevent a fault in the system [3]. The SG presents heterogeneous structures composed of distributed nodes. All operations are controlled through a communication network. The current communication techniques are inefficient due to the large and complex systems.

The deployment of renewable energy resources needs more coordination and controlling techniques to achieve reliable and efficient system. A multi-agent system (MAS) is a useful tool for coordination and controlling all operations within the SG, due to its distributed and autonomous property. MAS is widely used in SG applications. In articles [13,15–17], MAS is adopted for DRM [18], fault handling [14], and voltage and storage control [17], [19]. In the last couple of decades, researchers have made a number of contributions to the DRM and have made the efficient system in the SG environment. However, there is still a need for improvement in consumer satisfaction domain.

From past few decades, a number of studies have been carried out in SG regarding DRM. However, there is no such work that presents a comprehensive analysis of these works. There is a need to investigate different techniques, their advantages, and limitations. So here in this part, we present a comprehensive and detailed review of

DRM techniques. We review different scientific publications and investigate their features as well as open-research problems.

Our expected contributions are listed as follows:

1. To review large number of literature in the domain of SG from DRM perspective.
2. To propose a classification of DRM techniques used in previous literature.
3. To highlight their key features as well as open-research problems.

4.3 Backgrounds

In this section, we are going to present basic background and concepts for understanding DRM in the SG.

4.3.1 Smart grid

The traditional power system is responsible for generation and transmission of energy to end users. However, the user demand changes with time (variable demand), so the static approach cannot deal with variable demand. This problem gained the attention of researchers and introduced SG technology. SG is a complex system that is being formed from the traditional power system [20]. This integrates advanced communication and control technology that enables the system to perform the automated operation. It also consists of other various technologies like smart meters, smart homes, generators, storage devices, appliances, load, etc. This presents a network composed of distributed nodes; all operations of the system are controlled intelligently and autonomously. The key benefit of an SG is to achieve an efficient energy system [2].

NIST [21] presented a conceptual model for SG domain called NIST SG framework 1.0 in the National Institute of Standards and Technology, US Department of Commerce. This model represents seven different actors/applications that are interacting with each other. The conceptual model for SG has been shown in Figure 4.1. Each actor in this model is described below:

1. Customer: Represent end users that consume and store energy. They may be residential, industrial or commercial.
2. Market: Operators in the electricity market.
3. Operation: They manage all energy transmissions.
4. Service provider: They provide services and facility to the utility and customer.
5. Generation: They generate and store energy.
6. Transmission: They carry energy over large distances.
7. Distribution: They carry energy to and from customers.

The model components:

1. Social components: Electricity consumers, producers, grid operators.
2. Technical components: Loads (consuming devices), generators, power lines, buses.

The interaction and behavior of these actors will influence and be influenced by the technical system. The changes in the configuration of the technical system will

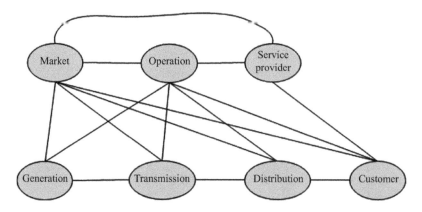

Figure 4.1   SG conceptual model adapted from [21]

affect the actors' behavior, and changes in actors' behaviors will affect the technical system configuration. Therefore, there is a need to consider coupled social-technical system in order to achieve reliable, sustainable, and resilient power system.

4.3.2 Demand–response management

DRM plays an important role in the SG environment. This refers to the process in which end users reduce energy consumption in response to the incentives [22]. This results in scalability and efficiency of the system. It also provides different strategies such as time-of-use (TOU) price, real-time pricing (RTP), critical peak pricing, incentives based, pricing and capacity market [23]. The authors of articles [24,25] defined these strategies as follows:

1. TOU: It uses different unit prices for different time periods.
2. RTP: It involves changing the energy cost on hourly bases.
3. Critical peak price: Combines both TOU and RTP strategy.
4. Incentive-based: This strategy involves three different pricing schemes. Direct load control, in which different power usage devices are directly controlled (on/off). Interruptible service, in which service provider offers a discount rate to the end users. Biding strategy, in which end users bid for consumption pattern.
5. Capacity market: In which end users have to send back energy to the grid unit.

DRM is an important resource for the SG system. It integrates modern technology like sensor, ICT, smart meter, appliances, etc. These technologies enable end users to interact with the power system. They also allow users to reduce energy consumption, which has a considerable impact on energy-cost reduction [26].

The integration of modern technology and renewable energy resources faces great challenges in SG system. Recently, DRM has gained a great deal of attention from the research community and brought a number of research contributions and technical discussions on DRM. During our literature study, we found two types of literature; learning-based techniques and complex system to address DRM in the SG.

4.3.3 Complex systems

The complex system represents an object with many interconnected elements or agents. There exist relatively many relations among each element or agent. The behavior of each element depends on the behavior of others. Another term, emergence, is also used to describe complex system [27]. The complex adaptive system is a type of complex system that concerns with agent's behaviors in complex systems. These agents are capable of learning and adapting in response to interaction with other agents. They have nonlinear and hierarchy properties and behaviors. The complex adaptive system represents a set of natural and artificial complex systems like market environment [28], cell organism, the internet with user and servers, a power system with consumers and grid unit, etc.

Recently, a number of studies have been carried out on the topic "DRM in power system environment from a complex system perspective." In articles [4,16,29,30], complex system approach was adopted and developed complex models by integrating consumers feedback, agents behavior, generation unit, a distribution unit, storage, and consumption devices.

4.3.4 Learning-based approaches

Learning-based approach is one of the major domain of artificial intelligence (AI) in which agents learn from past experience [31]. Agents can act according to the situation, hence they are able to have dynamic behavior against the dynamic situation. Experience gained through previous action state pair for existing event and training data. Learning-based approach is used in agent-based modeling for optimization and predictive analysis [32].

Reinforcement learning (RL) is a type of machine-learning approach and also a branch of AI. It concerns with agents' behaviors and actions in the environment. These agents learn the states and available actions as well as other agent's action and state. It perceives the input from the environment and performs a specific action and gets a reward corresponding to that action for any state. In articles [32–34], RL approach is adopted to train agents to get an optimal policy to make demand–supply balance in the power-system environment.

4.4 A review of demand–response management in SG

In this section, we are going to present a detailed overview of surveyed work about DRM in the SG. We have grouped DRM into three main categories which are learning-based approaches, complex system, and other techniques. Figure 4.2 shows the classification of the reviewed work on the SG from DRM perspective. In learning-based approach, different learning techniques are used like artificial neural network (ANN) and RL. While complex system consists of collaborative, adaptive, particle

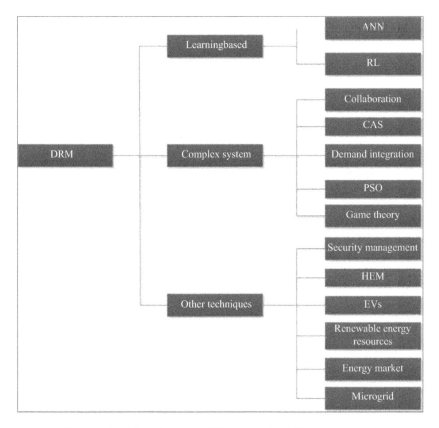

Figure 4.2   Classification of literature for DRM in smart grid

swarm optimization (PSO) and game-theory approaches, other techniques comprise communication management, home-energy management (HEM), renewable energy sources, electric vehicles (EVs), energy market, and microgrid.

4.4.1   Learning-based approaches

In learning-based approach, agents learn and adapt their behavior against the dynamic environment. The learning-based approach is categorized into ANN and RL techniques.

4.4.1.1   Artificial neural network

Hernandez *et al.* [35] proposed a MAS for the virtual power plant. The virtual power plant consists of small elements or a single unit. ANN is applied for efficient control and management of operations in a virtual power plant. The experimental results showed 1.5% error rate. This model works on a small level, and it needs enough information for predicting the future state of the system.

4.4.1.2 Reinforcement learning approach

There needs an intelligent and accurate model for prediction of energy consumption in an SG environment. In [36], Mocanu *et al.* presented a model for energy prediction based on RL without using any historical data. This model integrates the RL with the deep belief network. It estimates the state space and then finds optimal policy by using the RL technique. Experimental results showed 91.42% accuracy of the system. However, this model is not implemented for the different level in the SG.

In [10], Lakić *et al.* presented an agent-based model using SA-Q learning technique. It learns how much the system reserves to offer power at the different time. It also increases the ratio between economic cost and its benefits. The results showed an improvement in performance and economic outcome to users. However, this method is not applied to the multi-agent framework.

In [37], Dusparic *et al.* proposed a multi-agent scheme based on RL for demand–response problem in the SG. This method uses current load information and predicts load for the next one day by using load forecasting approach. The agents learn how to fulfill user demands from available energy. The results showed peak usage reduction is 33% and off-peak increased by 50%. However, in this method, there is no collaboration and communication among agents.

Wen *et al.* [38] addressed demand–response problem in SG and proposed framework base on RL technique. In this method, demand–response problem is decomposed over each device. This technique performs the self-initiation job and handles many flexible requests. The complexity of the proposed algorithm is linear. The results showed that for a broad range of trade-off parameters, it outperforms. However, this technique focuses on the demand–response problem of a single unit or building.

In [33] by Ruelens *et al.*, research work is carried out on demand–response problem using a batch RL technique. The batch RL technique covers the inefficient information problem of RL method. This method uses a batch of experiences to find out optimal policy. In this work, two agents, water and residential building agent, are used. For dynamic pricing, to minimize cost, a closed loop policy and for the day ahead scheduling, open-loop policy is followed. The results showed that energy cost is reduced by 19%, and consumption rate is increased by 4%.

When economic dispatch and demand response are treated as separate and sequential operation, energy efficiency decreases. Zhang *et al.* [39] presented optimal energy-management strategy in order to maximize social welfare. This method operates through coordination of demand response and economic dispatch. Economic dispatch is provided by generator and demand response by the customers. This method is also used for the discovery of power demand–supply mismatch. The simulation results showed convergence rate of 40 iterations.

Another approach for demand response is studied by O'Neill *et al.* [40] and proposed consumer-automated energy system. This technique reduces residential energy cost and usage. This method uses online energy-cost estimation and user decision policy. This is the independent approach to energy price and system behavior. In this method, users decide which device will use energy and how much. The results showed 40% cost reduction by using price unaware of energy scheduling.

4.4.2   Complex system

In this section, we present related work carried out in SG domain from a complex system perspective. We grouped the existing techniques into collaborative, complex adaptive system, demand-side integration, PSO, and game-theory approach.

4.4.2.1   Collaborative approach

MASs are widely used for controlling and managing in an SG. In [41], Manick-avasagam proposed and developed intelligent energy control center (ECC) mechanism for the SG. This technique consists of two layers, the one is DER which serves as a client and the other is ECC as a server. ECC is controlled and monitored by a fuzzy logic controller (FLC). Communication and negotiation between client servers take place through internet protocol. The simulation results are stored in an excel database acting as a monitoring agent. ECC used these results for decision-making in DERs. However, communication between results and FLC is not taken into account.

The mismatch between supply and demand reduces system performance. Parallel Monte Carlo tree search (P-MCTS) can produce an optimal solution for power balancing, but it has no coordination support. In [18], Golpayegani *et al.* extended the P-MCTS work by introducing collaborative and coordination concept. Agents negotiate with each other and present their proposal. This method resolves problems of agent's conflict, load-shifting, and charging capacity. The results showed that charge capacity increased from 33% to 50%. However, this model does not deal with prediction of data.

In [42], Le Cadre and Bedo worked on uncertainty in an SG environment and present decentralized hierarchy based on the learning-game approach. It is composed of supplier, generator, and consumer agents. Agents forecast demand and production of the grid in a collaborative manner. It determines the price that balance power and demand. The results showed that in a shared information network, faster convergence rate is achieved using cooperative learning as compared with an individual learning.

In [43], Huang *et al.* presented a novel model for the demand–response problem with the conjunction of the elastic dispatch process. In the study system, the elastic economic dispatch process is used as a feedback controller and flexible load cost as a control signal. The control signal balances demand and response. In this method, max-min and interval mathematic technique is used for boundary calculation. This estimates the uncertainty which is the difference between the present and the target value. Simulation results showed that the interval mathematic technique is efficient as compared to Monte Carlo approach. Its convergence time is also 1% less than Monte Carlo approach. However, this technique does not handle probability distribution.

4.4.2.2   Complex adaptive system

In [44], Kremers *et al.* presented a bottom-up approach for the SG. It consists of two layers: physical layer for electrical power transmission and logical layer for communication. This model has the ability to integrate new devices in an SG environment. It provides dynamic load management, power, and communication controlling and monitoring. Experiment results showed 40% reduction factor in energy consumption. However, this model is not capable of handling high-load management.

In [4], Thimmapuram and Kim proposed an agent-based model using elasticity market complex adoption system to an SG domain. This technique handles user elastic demand and lower cost. This method reduced peak load in the range of 8%–5%. However, the cost of energy for some user is increased.

4.4.2.3 Demand-side integration

Demand-side integration in SG results in security, quality, efficiency, and reduction in cost. In [45], Mocci *et al.* proposed a MAS for integration of demand and EVs. The load agents calculate power demand and act as master agents. The master agents with cooperative agents send power load and global data to the demand side. Its demand–response rate is 85%, and it also reduced the flow of data. However, this technique is not able to calculate the state of batteries of different storage at the different time.

In [22], Nunna *et al.* proposed a priority banking scheme, its concerns with user's demands. This method gives some share to the users from available resources; it monitors user's demand and updates their priority. This method reduced network loss by 50% and also reduced dependency on overall grid. However, this technique provides fewer shares to users.

4.4.2.4 Particle swarm optimization

Advance power-grid shifting from vertical to a horizontal structure which requires efficient management system. In [46], Hurtado *et al.* proposed biding energy management system. This technique provides interaction between different environments. It used PSO approach for maximizing comfort level and energy efficiency. In this hierarchical infrastructure, lower level agents abstract information and provide to higher level agents. Performance is described by weight factor; fair scenario showed 0.5 weight while bias scenario results in 0.3 weight. However, this technique generates some unbalance situation.

4.4.2.5 Game-theory approach

The retailer and market-price management in SG gained great attention in research work. In [47], Wei *et al.* focused on energy price and dispatch problem in an SG environment. This study proposed a two-stage two-level model. Customer demand and price is considered as the first stage using the Stackelberg game approach, while the operation of storage devices is considered as the second stage using linear max–min problem. Then the model is translated into mixed integer linear problem (MILP). The results showed 5% improvement in system performance, and it also increased retailer profit. However, this method is very sensitive and knowledge gathering process is very difficult.

In [48], Chai *et al.* addressed demand–response issue and present two-level game approach. This technique handles multiple utility company and users. The utility company is modeled as noncooperative and communication between users as evolutionary. The results showed that the proposed technique reduced cost payment from 3,197.7

to 2,425.6 and the energy demand is increased from 1,224.9 to 1,478.4. However, this work does not handle constraint on power consumption.

In [49] by Song *et al.*, another framework for optimal nonstationary demand-side management in an SG environment is proposed. In this method, the user selects their energy usage pattern according to their priority and needs. They used a repeated game approach which provides interaction among foresighted price anticipating users. This method showed 50% reduction in energy cost and robustness in error. However, higher threshold value results in a trade-off between cost and peak average ratio.

In [50], Nunna and Doolla carried out research work on management of demand response in multiple microgrid networks. In this work, customers participate in demand–response strategy. This study proposed a priority index approach through which customers participate in the market. This method reduced peak demand. It is found that customers with high priority index get power at low cost.

In [51], O'Brien *et al.* focused on DRM in SG application. In this work, demand response is modeled as the game-theoretic environment, and Shapley-value (SV) is used for payment distribution process. RL technique is used to estimate SV. Simulation results showed that for random sampling, 1,000,000 samples take 58.2 s execution time, while for sigmoid sample, 51,129 samples take 6.5 s. The results also showed that uniform sample balances demand and response. However, this method is not suitable for distribution scheme and its direct estimation is difficult. The literature summary of DRM has been shown in Table 4.1.

4.4.3   Other techniques

Apart from learning-based and complex-system approaches, there are also several other studies that have carried out in the SG domain. We grouped these studies into security management, HEM, EVs charging, energy market, and microgrid.

4.4.3.1   Security management

In demand–response (DR) process, communication occurs between consumers and suppliers in order to balance demand and supply. During communication process, sensitive information is exchanged between consumers and suppliers. This communication process needs to be secure from unauthorized users. To address this issue, Rahman *et al.* [52] have proposed a private secure bidding protocol for DR using incentive-based strategy. This proposed technique uses cryptographic primitives. It comprises three entities that are a supplier as a registration manager, DR automation server as bidding manager and bidder. The proposed protocol is implemented in Java by using primitive operations. The empirical results showed the feasibility of the protocol in terms of computation cost and a number of primitives. However, in the proposed scheme, RM and BM are assumed as secure.

There are some issues with existing communication framework in SG, i.e., there is no currently available ideal communication architecture and no such work which analyze and investigate bit error rate during the communication process. In this regard, in [53], Moghaddam *et al.* proposed a cloud-based DR in an SG environment. In this work, a communication model is developed for the smart distributed system. The evaluation criteria are considered as cloud DR and distributed DR. The simulation results

Table 4.1   Literature summary of demand–response management

Ref	Technique	Strength	Limitations
Hernandez *et al.* [35]	Energy Forecasting in VPP, ANN	1.5% Error rate	Works on single unit
Mocanu [36]	Energy prediction for building, RL + belief neural network	91.42% accuracy	Not implemented on different level
Lakić [10]	Learning how much system reserve to offer at different time, SA-Q learning	Ratio between cost and benefit increased	Not implemented on MAS framework
Dusparic [37]	Energy prediction of next day-ahead for EVs, RL+ load forecasting	33% usage reduced	No communication
Wen [38]	Segmentation of each device, RL	Linear complexity	Single unit
Ruelens [33]	Thermostatic load controlling, batch RL	19% cost reduction	Single unit
Zhang [39]	Finding mismatch between demand and supply, economic dispatch	Convergence rate 40 iteration	Only local communication
Golpayegani [18]	Conflict management between Evs, CP-MCTS	Battery capacity increased 17%	No prediction
Kremers [44]	Agent-based model of simple smart grid, CAS	40% consumption reduced, and peak load lies in the range 5%–8%	Not handle high load
Mocci [45]	Controlling of integrated demand and EVs, DSI	Response rate achieved 85%, network loss rate reduced 50%	No calculation of battery state
Hurtado [46]	Controlling of interoperation of smart building, PSO	Performance in the form of weight factor, which is achieved 0.5	Unbalance situation
Wei [47]	Management of energy price and dispatch problem, MILP	5% performance improved	Difficult to gather information
Chai[48]	Controlling multiple utility centers and end users, two-level game	Cost reduced from 3,197.0 to 2,425.6 RS	No constraints on power consumption
O'Brien [51]	Payment distribution process, Shapley value distribution	Converge in 58.2 s	Not suitable for energy distribution

showed that on clusters creation, the bit error is very large. It has been shown that by using UDP protocol for communication, the broadcast showed no optimal solution, while the TCP protocol showed a high bandwidth capacity. However, the convergence rate is increased. With high DR usage, effective communication is achieved at a high

Table 4.2   Literature summary of DR security management

Ref	Technique	Strength	Limitations
Rahman [52]	Secure communication process, incentive-based DR	Computation cost is reduced	Some entities are assumed pre-secure
Moghaddam [53]	Investigation of bit error rate during communication, cloud-based DR	TCP protocol showed high bandwidth capacity	No local communication
Tsai [54]	Large-scale communication network, randomized alternative technique	50% balance state is achieved	Not handle corrupt data
Wada [55]	Privacy management in distributed system, RTP scheme	Balance state of the system is achieved	Not tested on other DR scheme

cost. Another disadvantage of this study is that the distributed DR is applied on huge distance; there is no local communication among neighbor's channels.

In [54], Tsai *et al.* have worked on distributed DR for large-scale consumers load with the conjunction of renewable energy resources. In this work, a neighbor-communication strategy is applied. This results in low communication cost. They used a randomized alternative direct technique of multipliers for distributed DR. In this method, there is no need for communication synchronization. With few messages, the balance state can be achieved by the system. The results showed 50% balance state is achieved. It showed outperformance by using a RTP scheme over the existing distributed DR. However, they assumed that all consumers involved in communication process are trustable. The proposed scheme cannot handle wrong data transmission.

Although DR creates an energy-efficient system by reducing energy demand from peak-hour to off-peak. In this process, consumers and utility service providers always communicate with each other. Consumers transmit their energy demand profile to the grid unit, while from grid side, the energy-cost information is routed to the consumers. In this communication process, this information can be accessible to unauthorized users. So there is a need to make the communication system secure. In this regard, Wada *et al.* [55] worked on privacy management and proposed masking method to secure the privacy of each individual in a smart distributed energy system. In this scheme, every agent uses a mask signal along with their states. Then, during the communication process, each agent exchanges their mask with other agents. To obtain the correct signal, agents subtract the obtained signal from their own state. The RTP scheme of DR is applied. The results showed that this method can protect information of each agent along with a balanced state of the system. The literature summary of security management in SG has been shown in Table 4.2.

4.4.3.2   Home-energy management system

Nowadays, DR is getting more importance in-HEM systems (HEMSs). However, dynamic pricing scheme is not useful without a combination of DR and HEM. In [56],

Ghazvini *et al.* have proposed a new HEM algorithm. The proposed scheme schedule appliances, EVs, and electric water heater (EWH) with a combination of energy storage. EV is considered as a dispatchable energy source. In this work, renewable energy resource such as photovoltaic voltage (PV) is also used. They used simple rule-based algorithm under different pricing scheme which schedules EV charging and EWH heating process. The simulation results showed 29.5%–31.5% energy-cost reduction.

In [57], Luo *et al.* have worked on large scale ice-thermal storage system with the investigation to find out how to use it for fast voltage-control strategy with the conjunction of renewable energy resources. The work presents a modified version of the conventional system for thermal load management. In this work, a refrigerator is used for the ice-thermal load. This work showed that the proposed technique can effectively reduce the ratio of power imbalance in smart homes. The proposed technique is implemented on computer-simulation tool. The results showed the total fluctuation in voltage frequency reduced. The possible extension of this work can be the use of the proposed scheme on the large-scale distributed power system.

In smart homes, a smart meter is used that monitors the user load and demand profiles. However, the load forecasting of individuals at large scale is a challenging task due to stochastic nature of the individual demand. In this regard, in [58], Yu *et al.* have worked on this issue and proposed the use of the sparse coding technique to model individual loads at a large scale of the distributed power system. In this work, data of 5,000 homes based on a project with the collaboration of electrical power board in Chattanooga (2011–13) was used. The objective function was to forecast and predict next-day and next-week total load. The results showed that 10% accuracy of the system improved. However, the proposed scheme needs to be tested on others sparse methods like change point detection in a distributed system for getting a more accurate system.

In most previous studies, the game theoretic approach has been used for DRM. However, their computation cost is very large for finding Nash equilibrium. In [59], Li *et al.* have proposed a sparse load-shifting based DRM that schedule different smart home appliances. In this work, bidirectional communication is used that improve the searching process for Nash equilibrium. The objective function to minimize peak to average ratio (PAR) was used. The proposed algorithm showed the linear cost for finding Nash equilibrium. The results showed convergence rate of 500 iterations. The deployment of DR needs appropriate policy design and new technology. In this regard [60], a MAS is developed for residential DR in a distributed energy network. In this work, two agents, i.e., home agent and retailer agent, are used. The home agent predicts the load profile of consumers. The RTP scheme of DR is used in this work. The convex programming is used to model the consumption pattern of consumers. They used two objective functions, i.e., energy-cost minimization and users' waiting time. In this work, two case studies were considered. The simulation results showed that in Case 1, PAR and cost are reduced by 2.32$ and 62.7$, respectively. For case 2, PAR and cost are reduced by 1.54$ and 51.82$, respectively.

In [61], Huang *et al.* have introduced the use of the smart-gateway network in the SG. In this work, a single home with multiple rooms is considered along with a

Table 4.3   Literature summary of HEM

Ref	Technique	Strength	Limitations
Ghazvini [56]	HEM with EWH scheduling, rule-based algorithm	Cost reduced 29.5%	Single home
Luo [57]	Voltage management of large-scale, ice-thermal energy scheduling	Reduced voltage fluctuation	Not implemented for distributed system
Yu [58]	Load forecasting of appliances on large scale, sparse coding technique	Improved 10% accuracy	Not used other sparse methods like change point detection and clustering making for distributed system
Li [59]	Scheduling load of different smart homes, sparse load shifting technique	Linear computation cost, convergence rate 500 iterations	Energy cost was ignored
Wang [60]	Designing appropriate policy for DR, RTP scheme	PAR reduced 2.32, cost reduced 62.7$	Single home is considered
Huang [61]	Appliances scheduling of multiple rooms, minority-game based DR	Peak load reduced 38.5%	Single home is considered

single power grid and one PV. They used the multi-agent framework. First, energy-demand pattern of each room is extracted with some uncertainties assumptions. Each room is considered an agent. In this work, a dataset of a single building is used. A minority-game-based DR is used for peak demand reduction. The simulation results showed that peak load is reduced 38.5% in summer and 5.8% in winter. The literature summary of HEM in SG has been shown in Table 4.3.

4.4.3.3   Electric vehicles charging

In SG, EVs are used to store energy and then to use this energy later. In this regard, in [62], Yao *et al.* have proposed a real-time charging scheme for EVs with the conjunction of DR. The scenario of parking station is considered. The charging and discharging of EVs is modeled as a binary optimization problem. The binary (on, off) optimization fasts the charging and discharging process, while the exhaustive search is expensive in terms of computational cost. Therefore, a convex relaxation technique is applied that searches and schedules the charging and discharging periods. The simulation results demonstrate satisfactory results for charging EVs. With maximum numbers of EV, minimization of total energy cost is achieved. The computation time is noted as 0.19 s.

Regarding EVs charging, in [63], Le Floch *et al.* worked on two types of EV load management. The one is fixed power load that can be changed, while the second is flexible load but always remain fixed. They used a hierarchal control scheme with the objective function of minimizing cost. This technique comprises two steps. First of

Table 4.4   Literature summary of EVs

Ref	Technique	Strength	Limitations
Yao [62]	EVs charging, binary optimization	Computation time 0.19 s is achieved	Energy cost is ignored
Le Floch [63]	Two type of EV load management, price-based DR	PAR reduced 40%	Only feasible under limited threshold voltage
Jannati [64]	Optimal management of EV with parking plots, time-of-use DR	Operational cost reduced 4.30%	Not tested on other DR strategies

all, it computes voltage capacity of the system. Then customizes load profile of the consumers by using a price-based DR. The proposed scheme showed feasibility for some specific cases like if the voltage remains within fixed limits, the flexible load is achieved. The work is implemented on IEEE 55-bus radial distribution network. The results showed 40% reduction in PAR.

The main benefit of renewable energy resources is to reduce air pollution produced by fuel consumption in power grid. In this context, a number of EVs, as well as their parking plots, also increase to reduce the burden on the power grid. However, there needs an optimal operation of these EVs with the conjunction of parking plots. In this regard, in [64], Jannati and Nazarpour proposed an optimal management system for EVs and their parking plots. They integrated the model with the wind, PV, and local generators. They also used hydrogen and fuel cell storage system. The TOU DR strategy is applied to schedule charging and discharging process of EVs. The objective function to minimize operational cost along with charging and discharging of EVs cost is considered. Then mixed integer linear programming is used in four case studies. In Case 1, hydrogen storage and DR was not applied. In Case 2, hydrogen storage was integrated with the model. In Case 3, DR strategy was integrated, and in Case 4, both hydrogen storage and DR strategy were used. The simulation results of Case 2, Case 3, and Case 4 were compared with Case 1. The obtained results showed that Case 2, Case 3, and Case 4 reduced operation cost by 1.79%, 4.07%, and 4.30%, respectively. The literature summary of electric vehicles has been shown in Table 4.4.

4.4.3.4   Renewable energy sources

Nowadays, DERs such as PV play an important role in the SG. Therefore, there is a need to promote the use of PV in urban as well as ruler areas. In [65], Wang *et al.* have worked on PV promotion by using a game theoretic approach. The proposed scheme was analyzed from a different level of DR with RTP along with a different number of PV and batteries. In this work, 5 levels of consumers and 32 PV were considered. The optimization problem was modeled as a mixed compliment problem. The simulation results showed that consumers using high response get larger PV; they need less battery, meet energy demand at the real time, and also with less cost.

With the advent of smart homes, electrical sector encourages users to use renewable energy which benefits both users and grid as the total energy cost can be reduced. A buy-back strategy encourages users to generate more power from renewable energy resources that reduce the load on the main power grid. In [66], Chiu *et al.* worked on buy-back scheme with dynamic pricing technique. Dynamic pricing is modeled as a convex optimization dual problem. In this work, a day-ahead time-dependent pricing scheme is used. It also integrates wind, PV, and battery storage in the system. The objective function to achieve maximum user and company benefit was used. The simulation results showed that 1.28 PAR was achieved and peak load was reduced from 881.11 to 754.18/kW h.

Nowadays, the uses of wind energy resource are increasing. However, due to the stochastic nature of it, the mismatch of energy demand and power generation is also increasing. This introduced micro-combined heat power (CHP)—a hybrid energy system. However, there is a need to analyze the impact of DR with the conjunction of CHP at large scale with wind-energy resources. In [67], Jiang *et al.* addressed this issue and proposed an operation model representing the residential hybrid energy system. The proposed scheme uses price response, micro-CHP, smart appliances, and also load aggregator. The load aggregator is used to centralize different consumers load. The scheme is implemented on IEEE 118-bus. The simulation results showed that wind power curtailment is reduced 78% in 6-buses. It also reduced energy cost 10.7% and operation cost 11.7% on 118 buses. HEMS use DR to schedule home appliances. However, currently, there is no accurate method that predicts load consumption of appliances within a residential building. In [68], Hu and Xiao have worked on load prediction within the residential sector. In this work, the air conditioner appliance is used to train the thermal model. The historical data of indoor and outdoor temperature was used. The optimization algorithms like trust region algorithm, genetic algorithm (GA), and PSO were used to schedule the load of the air conditioner. They also used two strategies for the temperature which are set point and precooling. The simulation results showed power reduction 26%. However, they used single speed compressor. The proposed methodology should be tested on the inverter-driven air conditioner.

In [69], Amrollahi and Bathaee have worked on modeling a stand-alone microgrid that is far from the main power grid. This work investigates DR in the component size of optimization of the microgrid. They considered only wind and solar energy system. In this work, component size optimization and cost reduction are done by time-shift and load scheduling. The simulation results showed that a number of batteries, inverters, PV capacity, and energy cost reduced to 35.6%, 35%, 1.8%, and 17.1%, respectively.

In SG, a thermostatic load such as heat and air conditioner also help in reducing energy cost. In [82], Behboodi *et al.* have worked on thermostatic load control with controlling the real-time energy market by using transactive control paradigm. In this work, an ABM is developed that models DR for thermostatic loads. The proposed scheme can control thermostatic load under heating and cooling condition. The simulation results showed 10% energy cost reduction. However, this work ignored other appliances and just focused on the heater and air conditioner. The proposed work can be extended to integrate others appliances as well as renewable energy resources.

Table 4.5 Literature summary of EVs

Ref	Technique	Strength	Limitations
Wang [65]	Promotion the use of PV, game-theoretic approach, RTP scheme of DR	Energy cost is reduced, fewer batteries, consumers with high response get larger PV	Not implemented on other DR strategies
Chiu [66]	Modeling renewable energy as buy-back scheme, dynamic pricing technique of DR	PAR achieved 1.28	Maintain centralize communication infrastructure, not useful in the case of blackout
Jiang [67]	Modeling large-scale CHP with DR. Price response DR	Reduced energy cost 10.7%	PAR is ignored
Hu [68]	Load profile prediction trust region algorithm, GA, PSO	Power reduction 26%	Only modeled single speed compressor
Amrollahi [69]	Modeled stand-alone microgrid time-shift, DR scheme	Energy cost reduced 35.6%	Only wind and solar energy is considered
Behboodi [82]	Optimization of thermostatic load, transactive control scheme, agent-based model	Energy cost reduced 10%	Only focused on heater and air conditioner
Shakeri [70]	Control scheme for thermal energy storage novel optimization algorithm	Energy cost reduced 20%	Demand and PV capacity prediction was ignored

Regarding HEMS, in [70], Shakeri *et al.* have proposed a new control strategy for thermal and storage-management system. The working of the proposed algorithm is that it receives price information in advance and purchases energy at an off-peak hour. This work also integrated batteries and PV in a residential home. Total 26 appliances are used. Results showed 20% energy cost reduction. However, the proposed scheme was not tested for demand forecasting and prediction of PV capacity. The literature summary of renewable energy sources has been shown in Table 4.5.

4.4.3.5 Energy market

Samimi *et al.* [71] have proposed a stochastic framework for coupling active and reactive market in SG application. Active and reactive power is provided by distributed energy resources (DERs). A distributor company buys active and reactive power. The whole seller sells this power via the market environment. Demand buyback program (DBP) is used in which aggregators participate in the market. The scheduling process is modeled by optimization problem using mixed integer linear programming. In this work, evaluation criteria are set as the cost of energy, reactive power from DER, CO_2 emission, cost of DBP, and minimization of total cost. The simulation results showed the effectiveness of the DBP scheme, and also it reduced energy cost. However, aggregators only offer load reduction at a price they want to be reduced.

Dynamic pricing strategy is used to implement DRM for optimization of energy consumption pattern. It also helps in reduction of peak load. In [72], Srinivasan *et al.* proposed a game theoretic approach for dynamic pricing scheme using a special case of the Singapore energy market. They focused on the residential and industrial sector. In this work, five different loads along with price dataset were used. The pricing strategies half-hourly, RTP, TOU, and day–night pricing were used. The simulation results showed the RTP scheme that demonstrated high reduction of peak load. The peak load of residential and commercial was noted as 10% and 5%, respectively. However, in this work, the dataset used was not sufficient.

In some cases, there may be power transmission from generation unit to consumers at large scale. For this type of scenarios, there needs a dynamic behavior of transmission system. So whenever, a new consumer enters in the system, the system must be compatible with this new addition. However, in this area, the research focus is very less. DR strategy can be applied to control the whole system from short to the large-scale power system. Regarding this issue, in [73] have proposed a nonlinear economic model of consumers load. In this method, the price elasticity of demand and customer benefit is used. A multi-objective function consists of transmission planning, DR program is used. The power model is integrated with wind energy resource. The proposed scheme is implemented on IEEE reliability test system and Iran 400-kV. The simulation results showed that the total energy cost has successfully reduced with the proposed scheme. The future direction of this work can be the testing of other DR programs.

Although in previous literature a number of studies have been carried out in SG regarding DR, there still exist some challenges like the selection of optimal buses in power system using DR strategy. Regarding this issue, in [74], Dehnavi and Abdi have worked on selection and searching of optimal locations for the distrusted power system. In this work, IEEE-39 bus dataset is used. The optimization technique power transfer distribution factor (PTDF) is applied. This technique also searches for available power transmission capacity and optimal flow on the network. The simulation results showed congestion and the number of clusters is reduced. It also prevents black-out events. The computation time of the proposed technique is noted as 6.7 s.

Previous studies showed how to effectively model the stochastic process of appliances load. However, there is no such work to schedule online consumer demand as they may not be aware of energy cost before time. In this regard, in [75], Bahrami *et al.* have worked on long-term scheduling problem of appliances along with the varying behavior of energy. The energy price is modeled as Markov decision problem. This enables the model to observe the behavior of each interactive entity. The Markov perfect equilibrium technique is applied for optimal load scheduling. They also developed an online load scheduling learning (LSL) technique to find out user's equilibrium policy. The LSL showed cost and PAR reduction 28% and 13%, respectively. However, this work is only applied to a single home. This work can be extended to multiple homes along with multiple electricity markets. The literature summary of energy market has been shown in Table 4.6.

4.4.3.6   Mircorgrid

From past few decades, it confirms that network microgrid plays an important role in making an energy-efficient and reliable system. However, due to the unpredictable

Table 4.6   Literature summary of energy

Ref	Technique	Strength	Limitations
Samimi [71]	Stochastic model for active and reactive energy market MILP, demand buy-back DR	Reduced energy cost	Load reduction at specific time
Srinivasan [72]	Optimization for Singapore energy market game-theoretic approach, dynamic pricing scheme	Peak load reduced 10%	Small dataset
Hajebrahimi et al. 2017 [73]	Energy transmission at large scale nonlinear economic model, price elasticity DR	Reduced energy cost	Not implemented on other DR schemes
Dehnavi [74]	Selection of optimal buses, PTDF	Computation time is 6.7 s	
Bahrami [75]	Online long-term appliances scheduling, online LSL	PAR reduced 28%, cost reduced 13%	Single home was considered

nature of renewable energy resources, they impose new challenges on the smart distributed energy system. To address this issue, in some papers, stochastic technique is used. In [76], Nikmehr *et al.* have proposed another scheme for network microgrid to schedule consumers load. In this work, intermittent nature of load and generation unit is considered. They used time-of-use and real-time pricing of DRM. The optimization technique PSO is used for scheduling consumer load under uncertainty scenario. The simulation results showed the execution time of PSO is 241 s, while other stochastic technique showed 2,763 s. The operational cost is reduced to 17.3%, 30.6% with TOU and RTP, respectively.

In [77], a peer-to-peer network consists of consumers generation unit, i.e., PV is considered. They used priced-based DR strategy. The energy-sharing problem is modeled as a dynamic internal pricing scheme which provides supply and demand ratio. In this work, the flexibility of consumer's consumption is considered. The objective function economic cost and user's willingness is used. The performance of the system is evaluated in terms of prosumers cost and sharing of energy. The simulation results showed that total power loss is reduced from 3,321 to 3,187/kW h. The convergence rate is noted as 60 iterations. However, this work was not tested on a distributed network.

In [78], work is done on the smart microgrid and proposed stochastic optimization problem model with an objective function to minimize operational cost and CO_2 emission along with renewable energy resources. In this work, probability density function is used to predict wind speed and solar irradiance. Three types of consumers were considered, i.e., residential, industrial, and commercial. The incentive-based DR strategy is applied with three different case studies, i.e., (1) operational cost and emission; (2) operational cost, emission, and DR; (3) multi-objective function, operational cost, and emission. The simulation results showed that by using DR, the operational cost is reduced by 21% and emission by 14%. The literature summary of microgrid has been shown in Table 4.7.

Table 4.7   Literature summary of microgrid

Ref	Technique	Strength	Limitations
Nikmehr [76]	Load scheduling of network-based microgrid, PSO, real-time, and TOU DR	Cost reduced 17.3%	PAR was ignored
Liu [77]	Peer-to-peer network based microgrid, dynamic pricing scheme	Power loss reduced 4%	Not tested on distributed network
Aghajani [78]	Stochastic optimization problem for smart microgrid, probability density function, incentive-based DR	Operational cost reduced 21%	PAR was ignored

4.5   Open-research problems and discussion

In this section, we are going to discuss different methodologies and techniques on DRM to SG application and their open-research problems. The SG brings many facilities to users, e.g., energy efficiency, customer satisfaction, reduction in energy cost, and load balancing as a few to mention. However, there exist a number of challenges to be researched.

How to handle demand response in SG environment? To address this question, different approaches have been applied to SG applications. We surveyed three types of literature (learning-based approach, complex system, and other techniques) that are applied to handle demand response in SG applications.

4.5.1   Open-research problems in learning system

Learning-based approach involves ANN and RL. ANN is applied to the virtual power plant by Lloret and Valencia in article [35]. This showed 1.5% error rate. However, this needs enough information for prediction of future demand. RL with the combination of a deep belief network is applied by Mocanu *et al.* in article [36] for energy prediction and showed 91.42% accuracy rate. However, this model showed poor performance in different scenarios. Lakić *et al.* [10] applied SA-Q to learn system reservation to offer power at different times. This has increased cost-benefit rate. In article [37], demand response was addressed by Dusparic *et al.* to predict load for next day. This reduced 33% peak usage; however, there is no collaboration among agents. Wen *et al.* [38] worked on DRM for single unit or building. The proposed technique by Zhang *et al.* [39] maximizes social welfare and shows that it converged in 40 iterations.

4.5.2   Open-research problems in complex system

Complex system literature is involved in collaborative CAS, demand-side integration, PSO, and game-theory approach. A collaborative approach was adopted in different literature such as Manickavasagam in article [41], proposed ECC mechanism that consists of the layer-based model. The FLC is used for ECC controlling. However, there is no defined link or relation between monitoring and fuzzy logic. Golpayegani *et al.* in article [18] extends the previous work done on P-MCTS by integrating negotiation and collaboration among agents in P-MCTS. This study resolved the agent conflict issue. However, this model is not able to predict available energy. In an article [42], the authors addressed the prediction of energy issue discussed in the previous literature.

CAS approach is used in articles [4,44]. Kremers *et al.* in article [44] presented a layer-based model for energy transmission and communication. This model offers dynamic load management. The simulation results showed 40% energy consumption. Another approach from the CAS perspective was studied by Thimmapuram *et al.* in article [4] that deals with the elastic demand of the consumers. This work reduced peak load. However, energy cost increased for some end users. Mocci *et al.* in article [45] worked on integrating EVs into the SG. This technique achieved 85% response rate. However, this model was unable to predict the state of the battery. In article [79], authors presented priority banking scheme and offers energy shares to the users. In this work, network loss is reduced by 50%. However, this offers small share to the users. Hurtado *et al.* in article [46] proposed bidding energy management system technique that maximizes the user comfort level. However, this creates an unbalanced situation between energy cost and consumption. Wei *et al.* in article [47] proposed the two-stage two-level model. The simulation results showed that performance has been raised up to 5%. However, this model is not robust to error. Song *et al.* in article [49] proposed non-stationary demand scheme and achieved 50% energy cost reduction. However, there exists a trade-off between cost and PAR. Nunna and Doolla [50] deal with multiple microgrids which allow the user to participate in the market through a priority index number. Simulation results showed that with higher priority, users get energy with low cost.

4.5.3 Open-research problems in other techniques

The other technique category comprises security management, HEM, renewable energy resources, EVs, energy market, and microgrid. We have seen that a number of studies have been carried out to address different problems like minimization of energy cost, user discomfort, peak load, etc. Next, we discuss each work lies under the other technique category.

Regarding security management, different approaches have been presented in the domain of SG to protect user's confidential data from unauthorized users. However, there are still some issues that exist in the current security management approaches like some part of the system is ignored; sometimes corrupt data is not addressed, etc. The security in SG is a critical issue. This needs to be addressed to obtain sustainable and secure smart power system. The data and information transmission within the SG is very large, there is a need for advanced and new enhancement in communication infrastructure of the SG.

In the context of HEM, different smart appliances consume energy. However, the demand patterns and available power are not remaining same at all time. Their value changes with time. This fluctuation always tends to create an unbalance situation between energy demand and available energy. To some extent, this issue is addressed by a number of research works. A study such as presented in [57] worked on the ice-thermal storage system. The main purpose of the work was to control voltage with the conjunction of renewable energy resource. The possible extension of this work can be to use the proposed scheme on the large-scale distributed power system. As in real world, power system presents a complex system. So it needs to work on a large scale to observe the behavior of each component in the system.

In [80][81], the authors have presented work on appliances scheduling. These studies present how scheduling the power demand of different appliances by using different heuristic approaches can be effective. However, these current studies only focused on a single home. The possible extension of the current work may be to test the proposed technique on multiple homes along with different DR strategies to investigate the load as well as energy cost pattern.

Nowadays, the concept of EVs has been introduced in SG domain. The EVs have the capability to store and transmit energy. These EVs are used to store energy whenever energy cost is low from grid unit. Then they sell energy at low cost when the load on the main grid is high. So this reduces load burden on the grid as well as high energy cost. In this context, a number of studies such as in papers [62–64] have proposed different models that show how to effectively use EVs in SG scenarios for load and energy-cost reduction. However, there are still open-researches issues like sometime they ignore energy cost, only feasible under limited voltage, some schemes are not tested on different DR.

Renewable energy resources offer alternate energy resource in the form of wind and PV energy. Users can fulfill their energy demands from these resources. However, their energy production is unpredictable. They only depend on weather condition. From past few decades, different heuristic optimization techniques are used for handling the unpredictable nature of these renewable energy resources. A study presented in [65] worked on PV promotion. However, they just focused on RTP scheme and ignored other DR strategies. They need to study the effectiveness of the current work on other DR strategies. A study in the paper [66] proposed a buy-back scheme for renewable energy resource. In this work, they used a centralized communication infrastructure which is not useful in the case of a blackout. Fault in one part can tend to create disturbance in the whole system. Other studies such as presented in papers [67–69], [82] also worked on renewable energy resources and demonstrated the energy cost reduction. However, they just focused on cost reduction; other parameters like user discomfort and PAR reduction is ignored.

The energy market is responsible for buying energy from power sources and then selling to the consumers. This area of research also studied different literature. They demonstrated the peak load and energy-cost reduction. However, the current work was presented on a small level. Studies such as in papers [76–78] worked on the microgrid. By using optimization techniques, energy cost is successfully reduced. However, they ignored the PAR parameters as well as users comfort level.

4.6 Conclusions

The DRM plays an important role in the SG environment. It offers a broad range of advantages on system operation by reducing energy cost as well as effects on load balancing. In this part, we covered the different approaches applied for DRM in SG and proposed a classification of DRM models according to the techniques used for their implementation. The current literature in SG from DRM aspect is categorized into three main research directions. These research directions are learning-based approach, complex system, and some other different techniques. We finally described each technique and its model in detail. We also highlighted open-research problems exist in each solution.

References

[1] Gungor VC, Sahin D, Kocak T, *et al.* Smart grid technologies: Communication technologies and standards. IEEE Transactions on Industrial Informatics. 2011;7(4):529–539.

[2] Bollinger LA, van Blijswijk MJ, Dijkema GP, *et al.* An energy systems modelling tool for the social simulation community. Journal of Artificial Societies and Social Simulation. 2016;19(1):1.

[3] Siano P. Demand response and smart grids: A survey. Renewable and Sustainable Energy Reviews. 2014;30:461–478.

[4] Thimmapuram PR, Kim J. Consumers' price elasticity of demand modeling with economic effects on electricity markets using an agent-based model. IEEE Transactions on Smart Grid. 2013;4(1):390–397.

[5] Kamyab F, Amini M, Sheykhha S, *et al.* Demand response program in smart grid using supply function bidding mechanism. IEEE Transactions on Smart Grid. 2016;7(3):1277–1284.

[6] Rahman M, Mahmud M, Pota H, *et al.* A multi-agent approach for enhancing transient stability of smart grids. International Journal of Electrical Power & Energy Systems. 2015;67:488–500.

[7] Giraldo J, Mojica-Nava E, Quijano N. Synchronization of isolated microgrids with a communication infrastructure using energy storage systems. International Journal of Electrical Power & Energy Systems. 2014;63:71–82.

[8] Lawrence TM, Boudreau MC, Helsen L, *et al.* Ten questions concerning integrating smart buildings into the smart grid. Building and Environment. 2016;108:273–283.

[9] Haider HT, See OH, Elmenreich W. Residential demand response scheme based on adaptive consumption level pricing. Energy. 2016;113:301–308.

[10] Lakić E, Artač G, Gubina AF. Agent-based modeling of the demand-side system reserve provision. Electric Power Systems Research. 2015;124:85–91.

[11] Babalola A, Belkacemi R, Zarrabian S. Real-time cascading failures prevention for multiple contingencies in smart grids through a multi-agent system. IEEE Transactions on Smart Grid. 2016;9(1):373–385.

[12]   Chen C, Wang J, Qiu F, *et al.*   Resilient distribution system by micro-grids formation after natural disasters. IEEE Transactions on Smart Grid. 2016;7(2):958–966.

[13]   Eriksson M, Armendariz M, Vasilenko OO, *et al.* Multiagent-based distribution automation solution for self-healing grids. IEEE Transactions on Industrial Electronics. 2015;62(4):2620–2628.

[14]   Ghorbani MJ, Choudhry MA, Feliachi A.   A multiagent design for power distribution systems automation. IEEE Transactions on Smart Grid. 2016;7(1): 329–339.

[15]   Kahrobaee S, Rajabzadeh RA, Soh LK, *et al.* A multiagent modeling and investigation of smart homes with power generation, storage, and trading features. IEEE Transactions on Smart Grid. 2013;4(2):659–668.

[16]   de Durana JMG, Barambones O, Kremers E, *et al.* Agent based modeling of energy networks. Energy Conversion and Management. 2014;82:308–319.

[17]   Li Q, Chen F, Chen M, *et al.*   Agent-based decentralized control method for islanded microgrids.   IEEE Transactions on Smart Grid. 2016;7(2): 637–649.

[18]   Golpayegani F, Dusparic I, Taylor A, *et al.* Multi-agent collaboration for conflict management in residential demand response. Computer Communications. 2016;96:63–72.

[19]   Teleke S, Baran ME, Bhattacharya S, *et al.* Rule-based control of battery energy storage for dispatching intermittent renewable sources. IEEE Transactions on Sustainable Energy. 2010;1(3):117–124.

[20]   Nardelli PHJ, Rubido N, Wang C, *et al.* Models for the modern power grid. The European Physical Journal Special Topics. 2014;223(12): 2423–2437.

[21]   Greer C, Wollman DA, Prochaska DE, *et al.* NIST framework and roadmap for smart grid interoperability standards, release 3.0. Special Publication (NIST SP)-1108r3. 2014.

[22]   Nunna HK, Saklani AM, Sesetti A, *et al.* Multi-agent based demand response management system for combined operation of smart microgrids. Sustainable Energy, Grids and Networks. 2016;6:25–34.

[23]   Yu M, Hong SH. Supply–demand balancing for power management in smart grid: A Stackelberg game approach. Applied Energy. 2016;164:702–710.

[24]   Valogianni K, Ketter W. Effective demand response for smart grids: Evidence from a real-world pilot. Decision Support Systems. 2016;91:48–66.

[25]   Fera M, Macchiaroli R, Iannone R, *et al.* Economic evaluation model for the energy demand response. Energy. 2016;112:457–468.

[26]   Labeodan T, Aduda K, Boxem G, *et al.* On the application of multi-agent systems in buildings for improved building operations, performance and smart grid interaction—A survey.   Renewable and Sustainable Energy Reviews. 2015;50:1405–1414.

[27]   Niazi MA. Towards a novel unified framework for developing formal, network and validated agent-based simulation models of complex adaptive systems. PhD Dissertation, University of Stirling, Scotland, UK; 2011.

[28]   Niazi MA. Complex adaptive systems modeling: A multidisciplinary roadmap. Complex Adaptive Systems Modeling. 2013;1(1):1.

[29] Santos G, Pinto T, Praça I, *et al.* MASCEM: Optimizing the performance of a multi-agent system. Energy. 2016;111:513–524.

[30] Haghnevis M, Askin RG, Armbruster D. An agent-based modeling optimization approach for understanding behavior of engineered complex adaptive systems. Socio-Economic Planning Sciences. 2016;56:67–87.

[31] Weiss G. Multiagent systems: a modern approach to distributed artificial intelligence. Cambridge, MA: MIT Press; 1999.

[32] Rayati M, Sheikhi A, Ranjbar AM. Applying reinforcement learning method to optimize an energy hub operation in the smart grid. In: Innovative Smart Grid Technologies Conference (ISGT), 2015 IEEE Power & Energy Society. IEEE; 2015. p. 1–5.

[33] Ruelens F, Claessens BJ, Vandael S, *et al.* Residential demand response of thermostatically controlled loads using batch reinforcement learning. IEEE Transactions on Smart Grid. 2017;8(5):2149–2159.

[34] Li D, Jayaweera SK. Reinforcement learning aided smart-home decision-making in an interactive smart grid. In: Green Energy and Systems Conference (IGESC), 2014 IEEE. IEEE; 2014. p. 1–6.

[35] Hernández L, Baladron C, Aguiar JM, *et al.* A multi-agent system architecture for smart grid management and forecasting of energy demand in virtual power plants. IEEE Communications Magazine. 2013;51(1):106–113.

[36] Mocanu E, Nguyen PH, Kling WL, *et al.* Unsupervised energy prediction in a smart grid context using reinforcement cross-building transfer learning. Energy and Buildings. 2016;116:646–655.

[37] Dusparic I, Harris C, Marinescu A, *et al.* Multi-agent residential demand response based on load forecasting. In: Technologies for Sustainability (SusTech), 2013 1st IEEE Conference on. IEEE; 2013. p. 90–96.

[38] Wen Z, O'Neill D, Maei H. Optimal demand response using device-based reinforcement learning. IEEE Transactions on Smart Grid. 2015;6(5):2312–2324.

[39] Zhang W, Xu Y, Liu W, *et al.* Distributed online optimal energy management for smart grids. IEEE Transactions on Industrial Informatics. 2015;11(3):717–727.

[40] O'Neill D, Levorato M, Goldsmith A, *et al.* Residential demand response using reinforcement learning. In: Smart Grid Communications (SmartGridComm), 2010 First IEEE International Conference on. IEEE; 2010. p. 409–414.

[41] Manickavasagam K. Intelligent energy control center for distributed generators using multi-agent system. IEEE Transactions on Power Systems. 2015;30(5):2442–2449.

[42] Le Cadre H, Bedo JS. Dealing with uncertainty in the smart grid: A learning game approach. Computer Networks. 2016;103:15–32.

[43] Huang H, Li F, Mishra Y. Modeling dynamic demand response using Monte Carlo simulation and interval mathematics for boundary estimation. IEEE Transactions on Smart Grid. 2015;6(6):2704–2713.

[44] Kremers E, de Durana JG, Barambones O. Multi-agent modeling for the simulation of a simple smart microgrid. Energy Conversion and Management. 2013;75:643–650.

[45]   Mocci S, Natale N, Pilo F, *et al.* Demand side integration in LV smart grids with multi-agent control system. Electric Power Systems Research. 2015;123: 23–33.

[46]   Hurtado L, Nguyen P, Kling W. Smart grid and smart building inter-operation using agent-based particle swarm optimization. Sustainable Energy, Grids and Networks. 2015;2:32–40.

[47]   Wei W, Liu F, Mei S. Energy pricing and dispatch for smart grid retailers under demand response and market price uncertainty. IEEE Transactions on Smart Grid. 2015;6(3):1364–1374.

[48]   Chai B, Chen J, Yang Z, *et al.* Demand response management with multiple utility companies: A two-level game approach. IEEE Transactions on Smart Grid. 2014;5(2):722–731.

[49]   Song L, Xiao Y, Van Der Schaar M. Demand side management in smart grids using a repeated game framework. IEEE Journal on Selected Areas in Communications. 2014;32(7):1412–1424.

[50]   Nunna HK, Doolla S. Demand response in smart distribution system with multiple microgrids. IEEE Transactions on Smart Grid. 2012;3(4):1641–1649.

[51]   O'Brien G, El Gamal A, Rajagopal R. Shapley value estimation for compensation of participants in demand response programs. IEEE Transactions on Smart Grid. 2015;6(6):2837–2844.

[52]   Rahman MS, Basu A, Kiyomoto S, *et al.* Privacy-friendly secure bidding for smart grid demand-response. Information Sciences. 2017;379:229–240.

[53]   Moghaddam MHY, Leon-Garcia A, Moghaddassian M. On the performance of distributed and cloud-based demand response in smart grid. IEEE Transactions on Smart Grid. 2017;9:5403–5417.

[54]   Tsai SC, Tseng YH, Chang TH. Communication-efficient distributed demand response: A randomized ADMM approach. IEEE Transactions on Smart Grid. 2017;8(3):1085–1095.

[55]   Wada K, Sakurama K. Privacy masking for distributed optimization and its application to demand response in power grids. IEEE Transactions on Industrial Electronics. 2017;64(6):5118–5128.

[56]   Ghazvini MAF, Soares J, Abrishambaf O, *et al.* Demand response implementation in smart households. Energy and Buildings. 2017;143:129–148.

[57]   Luo X, Lee CK, Ng WM, *et al.* Use of adaptive thermal storage system as smart load for voltage control and demand response. IEEE Transactions on Smart Grid. 2017;8(3):1231–1241.

[58]   Yu CN, Mirowski P, Ho TK. A sparse coding approach to household electricity demand forecasting in smart grids. IEEE Transactions on Smart Grid. 2017;8(2):738–748.

[59]   Li C, Yu X, Yu W, *et al.* Efficient computation for sparse load shifting in demand side management. IEEE Transactions on Smart Grid. 2017;8(1):250–261.

[60]   Wang Z, Paranjape R. Optimal residential demand response for multiple heterogeneous homes with real-time price prediction in a multiagent framework. IEEE Transactions on Smart Grid. 2017;8(3):1173–1184.

[61] Huang H, Cai Y, Xu H, *et al.* A multiagent minority-game-based demand-response management of smart buildings toward peak load reduction. IEEE Transactions on Computer-Aided Design of Integrated Circuits and Systems. 2017;36(4):573–585.

[62] Yao L, Lim WH, Tsai TS. A real-time charging scheme for demand response in electric vehicle parking station. IEEE Transactions on Smart Grid. 2017;8(1):52–62.

[63] Le Floch C, Bansal S, Tomlin CJ, *et al.* Plug-and-play model predictive control for load shaping and voltage control in smart grids. IEEE Transactions on Smart Grid. 2017;1(1):1–10.

[64] Jannati J, Nazarpour D. Optimal energy management of the smart parking lot under demand response program in the presence of the electrolyser and fuel cell as hydrogen storage system. Energy Conversion and Management. 2017;138:659–669.

[65] Wang G, Zhang Q, Li H, *et al.* Study on the promotion impact of demand response on distributed PV penetration by using non-cooperative game theoretical analysis. Applied Energy. 2017;185:1869–1878.

[66] Chiu TC, Shih YY, Pang AC, *et al.* Optimized day-ahead pricing with renewable energy demand-side management for smart grids. IEEE Internet of Things Journal. 2017;4(2):374–383.

[67] Jiang Y, Xu J, Sun Y, *et al.* Day-ahead stochastic economic dispatch of wind integrated power system considering demand response of residential hybrid energy system. Applied Energy. 2017;190:1126–1137.

[68] Hu M, Xiao F. Investigation of the demand response potentials of residential air conditioners using grey-box room thermal model. Energy Procedia. 2017;105:2759–2765.

[69] Amrollahi MH, Bathaee SMT. Techno-economic optimization of hybrid photovoltaic/wind generation together with energy storage system in a stand-alone micro-grid subjected to demand response. Applied Energy. 2017;202:66–77.

[70] Shakeri M, Shayestegan M, Abunima H, *et al.* An intelligent system architecture in home energy management systems (HEMS) for efficient demand response in smart grid. Energy and Buildings. 2017;138:154–164.

[71] Samimi A, Nikzad M, Siano P. Scenario-based stochastic framework for coupled active and reactive power market in smart distribution systems with demand response programs. Renewable Energy. 2017;109:22–40.

[72] Srinivasan D, Rajgarhia S, Radhakrishnan BM, *et al.* Game-theory based dynamic pricing strategies for demand side management in smart grids. Energy. 2017;126:132–143.

[73] Hajebrahimi A, Abdollahi A, Rashidinejad M. Probabilistic multiobjective transmission expansion planning incorporating demand response resources and large-scale distant wind farms. IEEE Systems Journal. 2017;11(2):1170–1181.

[74] Dehnavi E, Abdi H. Determining optimal buses for implementing demand response as an effective congestion management method. IEEE Transactions on Power Systems. 2017;32(2):1537–1544.

[75]   Bahrami S, Wong VW, Huang J. An online learning algorithm for demand response in smart grid. IEEE Transactions on Smart Grid. 2017;9(5): 4712–4725.

[76]   Nikmehr N, Najafi-Ravadanegh S, Khodaei A. Probabilistic optimal scheduling of networked microgrids considering time-based demand response programs under uncertainty. Applied Energy. 2017;198:267–279.

[77]   Liu N, Yu X, Wang C, *et al.* An energy sharing model with price-based demand response for microgrids of peer-to-peer prosumers. IEEE Transactions on Power Systems. 2017;32(5)3569–3583.

[78]   Aghajani G, Shayanfar H, Shayeghi H. Demand side management in a smart micro-grid in the presence of renewable generation and demand response. Energy. 2017;126:622–637.

[79]   Ellabban O, Abu-Rub H. Smart grid customers' acceptance and engagement: An overview. Renewable and Sustainable Energy Reviews. 2016;65:1285–1298.

[80]   Manzoor A, Javaid N, Ullah I, *et al.* An intelligent hybrid heuristic scheme for smart metering based demand side management in smart homes. Energies. 2017;10(9):1258.

[81]   Ahmad A, Khan A, Javaid N, *et al.* An optimized home energy management system with integrated renewable energy and storage resources. Energies. 2017;10(4):549.

[82]   Behboodi S, Chassin DP, Djilali N, *et al.* Transactive control of fast-acting demand response based on thermostatic loads in real-time retail electricity markets. Applied Energy. 2018;210:1310–1320.

Applications of multi-agent systems in smart grid: a survey and taxonomy

Waseem Akram[1] and Muaz A. Niazi[1]

Multi-agent systems (MASs) in the smart-grid area have received a great deal of attention from the research community in recent years. Studies on MAS to the smart grid have brought a number of interesting technical discussions on simulation and modeling of the smart grid and research contributions. Researchers are trying to bring energy efficiency and load balancing in the smart grid. Many of these research works have achieved efficiency in power-system domain, while the social system and consumer satisfaction still need improvement. By focusing on the MAS in smart grid, in this part, we survey the body of knowledge and discuss the challenges of simulation and modeling of MAS in the smart grid. We investigate and group the existing solutions and highlight open-research problems.

5.1 Overview

We first start by giving an overview of the smart-grid concept. Next, we present a detailed review of different literature in the smart-grid domain. This is followed by open research problems and discussions. The chapter ends with a conclusion.

5.2 Introduction

The traditional power system provides one-way power flow, which is responsible for generation and transmission of energy to end users. However, the user demand changes with time (variable demand). The one-way power flow could not deal with variable demand. This problem gained the attention of researchers and introduced smart-grid technology by integrating information and communication technology with the traditional system. The smart grid is a power system consisting of various technologies like a smart meter, ICT, smart homes, generators, storage devices appliances, load, etc.

[1]Computer Science Department, COMSATS Institute of Information Technology, Pakistan

The smart grid is a network composed of distributed nodes, all operations of the system are controlled intelligently and autonomously, in order to achieve efficient energy system [1].

Fuel consumption changes the climate. This change attracted the researchers to introduce renewable energy resources like solar and wind. However, the outcome of these resources is unpredictable due to its fluctuation behavior. To achieve future sustainability, reliability, and resilience features of the smart grid, the research community is attracted to deploy renewable energy resources in power system. The structure of the power system is now shifted to more bottom-up approach. This means that all decisions related to power generation and transmission are taken by various actors (agents) in generation unit in a distributed manner. Various actors interact with the technical system (power system) and they are dependent on each other.

The smart-grid system is made up of two main components, e.g., technical system and social system. The technical system consists of power plants, power lines, load, transformers, and busses. The social system consists of consumers, operators, and electricity retailers. Each component of the social system interacts with each component of the technical system.

The deployment of renewable energy resources needs more coordination, management, and controlling techniques to achieve reliable and efficient system. A MAS is a useful tool for coordination and management of all operations within the smart grid, due to its distributed and autonomous property. MAS is widely used for smart-grid application. They are responsible for the management and control of all smart-grid activities. They can perform various tasks like communication among different agents, fault detection and prevention, power scheduling, voltage controlling, and storing energy.

In previous literature, a number research works have been carried out in the smart-grid domain. However, currently, there is no such work that investigates and analyzes these works. There is a need to find out which technique is feasible in what scenario. In this chapter, we provide a detailed survey and comparison of different techniques available for smart-grid system over the period 2010–16. The aim of this study is to present a comprehensive understanding of the smart-grid domain, its application, as well as the open-research problems that need to be addressed to gain sustainable and reliable system. We have cited a large number of scientific publications round about 100 papers. To the best of our knowledge, this is the first comprehensive survey on MAS in the smart grid. While during our literature review, we found one paper [2] that presents a survey on a specific aspect of MAS in the smart grid. In [2], the author focuses on demand-side management, generation and transmission management. Although the author discussed important issues in the domain. However, there is no discussion about other relevant aspects such as communication, self-healing, power scheduling, and storage management.

In this part, we aim to present more comprehensive and concise overview up to date by targeting five aspects such as communication, demand-side management, fault detection and prevention, power scheduling, and storage and voltage management.

5.3 A review of multi-agent system to smart-grid application

In this section, we are going to discuss briefly different approaches and solutions in the smart grid from MAS perspective. We grouped existing literature into five categories that are communication management, demand–response management, faults controlling, power scheduling, and storage management. The proposed taxonomy is shown in Figure 5.1.

5.3.1 Communication management

In a smart-grid environment, different agents communicate with each other to share information about power demand and capacity. This process is categorized into group communication and learning-based communication.

5.3.1.1 Group communication

In group communication, different entities are connected through a common purpose and interact with each other. In this section, we present some of the previous studies carried out in the smart grid that make use of group communication paradigm. This paradigm is categorized into subcategories according to the techniques used in the study. Next, we discuss each of these techniques.

Hierarchal framework

Li *et al.* [3] presented an agent-based decentralized control scheme for distributed smart-grid network. It consists of two layers. One is the bottom layer that represents a communication network composed of agents that act as controllers and collect information about grid status. Second is the top layer representing a distribution process of the power grid network. The agents at bottom layer control the power produced by distributor grids. This study achieved balance state between power and demand. It also reduced communication complexity and voltage variation.

One-way power communication in smart grid is considered to be slow in response. In [4], Al-Agtash presented a novel agent-based model for two-way power communication in the smart grid. This model provides two-way power flow between user's demand and power generators. This architecture consists of three layers: power generators, middle-ware, and electricity agents. Agents operate in an integrated manner within smart grid. They control and monitor demand variations and selling of power at customer side. These agents provide reliability, security, and stability of the system. Simulation results showed that market price decreased from 80 to 50/mW h. However, there are still some design issues, i.e., API, integrity, and consistency of agents operation.

The decentralized management system in a smart grid makes each part of the system intelligent and autonomous. Palicot *et al.* [5] have presented hierarchal cognitive radio network architecture for the smart grid. The framework focuses on the hierarchal position of each element of the system. The results showed that peak power 55,000 W reduced to 900 W. This method reduced pressure on the system and also reduced the risk of failure.

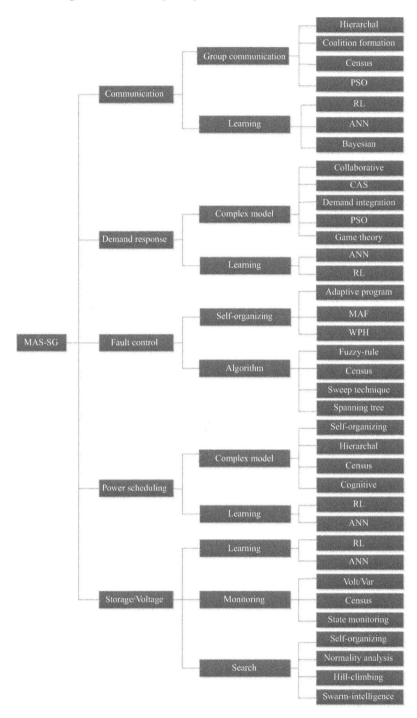

Figure 5.1   Proposed taxonomy of literature review in smart grid

In a smart-grid environment, power demand and supply balance can be achieved by enabling agents to share information. In [6], Larsen *et al.* proposed an information sharing model of imbalance power in the smart grid. In this model, agents communicate with their neighbors and exchange information about the imbalance of power. By comparing this model with the centralized algorithm, it showed that centralized algorithm requires 29 times more time for 150 households than the proposed method. However, in this method, agents keep information about their own imbalance power. They also only communicate directly with their neighbors.

For communication management, in [7], another framework is proposed by Yan *et al.* based on zero correlation zones. In this technique, communication between machine-to-machine was initiated through mutual authentication. This technique maintains physical layer security and reduces traffic overhead. The performance of the system is described in terms of time–efficiency ratio. The results showed that the proposed technique is better in larger nodes and low load impedance scenario. However, with small nodes scenario, it gives worse results.

Coalition formation

Power scheduling and transmission are challenging tasks as the structure of smart grid become complex. In [8], Ye *et al.* presented a multi-agent coalition formation–based dispatch technique. This technique has decentralized behavior, and it does not require any global information. Each node acts as an autonomous agent, negotiates with each other (agents), and takes part in the decision-making process. Agents work in the group, generating outstanding output. This model requires a large time for negotiation between agents. Results demonstrate good average utility.

In [9], Dagdougui and Sacile addressed the issue regarding the optimization of cost and power exchange among the smart microgrid network. They proposed a decentralized control and monitoring strategy of the smart microgrid network. In the study system, each microgrid is able to generate, store, and transmit energy to other microgrids and main grid. This method allows storage devices to operate around a reference value through cooperatively sharing power information among microgrids. For a fixed number of network, this method makes low iteration, but the iteration no. is influenced by the size of the network. It also does not handle the stochastic scenario.

In [10], Nguyen and Flueck addressed the communication latency problem in smart-grid application. Communication latency has a great impact on system performance. In this technique, the system is modeled as a random parameter with probability density function, and sending/receiving messages are generated randomly. The results showed restoration time is 3.983 s. However, this method does not focus on individual communication and communication bandwidth.

Census

As the environment of smart grid becomes complex, it is difficult to manage it using census-based approach due to its lack of information. In [11], Zhang presented a robust incremental cost estimation scheme for power dispatch in the smart grid. It consists of two layers that are executed in parallel. One layer is a gossip update rule that calculates average power mismatch. The second layer is consensus incremental cost estimation that calculates the system incremental cost. This method enhanced the system vulnerability and it requires less information. The results showed that information loss and iteration have a direct relationship. This method gives better results when information loss is 5%.

Particle swarm optimization

Integrating smart grid and smart green have attained great research interest recently. In such area, communication is bidirectional. In [12] by Wang, work is carried out on the communication protocol and proposed a negotiation-based agent adaptive attitude bidding strategy. The agents adjust their behavior in response to the variation in the communication process. This work presents particle swarm optimization agent adaptive bidding strategy (PSO-AABS) used by trader agents for decision in response to opponent behavior. It handles bidirectional communication between smart grid and green building. This method maximizes trader's payoff and negation time. The results showed that the cost of buying reduced 7% and their negation time 27%, while the cost of selling reduced 17% and their negation time 9%. However, customer comfort level is also reduced.

5.3.1.2   Learning-based approach

In this section, we discuss some of the learning techniques that were presented in smart grid in the past for addressing communication aspects of the smart-grid system. These techniques are reinforcement, neural, and Bayesian-learning approach.

Reinforcement learning approach

In [13], Yu *et al.* studied the smart generation control of multi-agent multi-area distributed smart-grid system. This paper presents a novel scheme named correlated equilibrium $Q(\lambda)$ to produce an optimal equilibrium solution for the load control in a distributed network of the power system. In this work, a formulated equilibrium selection function is presented. This technique enhances the overall long-run performance of the system. The results demonstrate fast convergence rate.

In smart-grid infrastructure, each node needs to maintain same frequency and voltage to prevent failures. In [14], Giraldo *et al.* focused on frequency synchronization. It enables the microgrid to remain synchronized even in unknown changes. It showed that the stability of the system is improved with an energy storage system (ESS). This model works with linear time invariant; it does not address nonlinear time invariant.

Artificial neural network

In [15], Saraiva *et al.* focused on the classification of load power in the smart grid. In this work, a classification of the nonlinear load in the smart grid is presented using the artificial neural network (ANN) and MAS. They used smart meter agents in the classification process. The smart meter measures voltage and sends the result to other agents at the substation. Experimental results showed that system accuracy achieved 98.7% and also the overall cost is less. However, this approach is not robust to error.

Table 5.1   Literature summary of communication

Ref	Technique	Strength	Limitations
Li [3]	Decentralized controlled scheme, system dynamic modeling	Reduced communication complexity	Agents only communicate with its neighbors
Palicot [5]	Cognitive radio network architecture, hierarchal framework	Peak power reduced	The retailers are ignored
Larsen [6]	Sharing information about imbalance power, network topology	Balance of power and demand achieved	Only communication between neighbors
Yan [7]	Machine-to-machine communication, zero correlation zone communication protocol	Maintain security of physical system	Worse results in case of small system
Dagdougui and Sacile [9]	Optimization of cost and power in microgrid, network theory	Low convergence rate	Not handle stochastic scenario
Nguyen and Flueck [10]	Communication latency in smart grid, coalition formation	Restoration time is 3.9 s	Not handle individual communication and communication bandwidth
Zhang [11]	Cost estimation scheme, census		Worse results in the case of large information loss
Wang [12]	Adaptive biding communication strategy, PSO	Reduced power cost 7%	User comfort also reduced
Giraldo [14]	Frequency synchronization in smart grid, census	Stability of the system increased	Not handle nonlinear time invariant
Saraiva [15]	Classification of load, ANN	98.7% accuracy	Not robust to error
Misra [16]	Energy trading with incomplete information, Bayesian learning	Utility increased 40%	Not control packet loss

Bayesian learning

Information about prices and demands may be lost during the communication process. The incomplete information affects the performance of smart grid. In [16], Misra *et al.* addressed this issue regarding smart grid. In this work, the agent-based model is proposed using Bayesian learning approach. It consists two types of agents: customer and grid agents. Customer's agents calculate the price given by the grid. Grid's agents calculate demand given by customers based on the probability of their belief. Simulation results showed that utility is increased by 40%. However, this method ignored control packet loss rate. The literature summary of communication management has been shown in Table 5.1.

5.3.2   Demand–response management

Demand–response management is the process of shifting consumer demand from high demand to low demand by giving incentives to the users. This approach is categorized into two approaches which are learning based and complex system approach.

5.3.2.1   Learning-based approach

In smart grid, we found two types of learning-based approaches: neural network and reinforcement learning (RL). These techniques have been applied to address demand–response problem. Next, we discuss these learning-based studies.

Artificial neural network

Hernández *et al.* in [17] proposed a MAS for the virtual power plant. The virtual power plant consists of small elements or a single unit. ANN is applied for efficient control and management of operations in a virtual power plant. The experimental results showed 1.5% error rate. This model works on a small level. It also needs more information for predicting the future state of the system.

Reinforcement learning approach

There needs an intelligent and accurate model for prediction of energy consumption in a smart-grid environment. In [18], Mocanu *et al.* presented a model for energy prediction based on RL. The RL technique works without using any historical data. This model integrates the RL with the deep belief network. It estimates the state space and then finds optimal policy by using the RL. Experimental results showed 91.42% accuracy of the system. However, this model is not implemented for the different level in the smart grid.

In [19], Lakić *et al.* presented an agent-based model using SA-Q learning technique. It learns how much system reserve to offer power at the different time. It also increases the ratio between economic cost and its benefits. The results showed an improvement in performance and economic outcome to users. However, this method is not applied to the multi-agent framework.

In [20], Dusparic *et al.* proposed a multi-agent scheme based on RL for demand–response problem in the smart grid. This method uses current load information and predicts load for the next one day by using load forecasting approach. The agents learn how to fulfill user demands from available energy. The results showed peak usage reduction is 33% and off-peak increased by 50%. However, in this method, there is no collaboration and communication among agents.

Wen *et al.* [21] addressed demand–response problem in smart grid and proposed framework based on RL technique. In this method, demand–response problem is decomposed over each device. This technique performs the self-initiation job and handles many flexible requests. The complexity of the proposed algorithm is linear. The results showed that for a broad range of trade-off parameters, it outperforms. However, this technique focuses on the demand–response problem of a single unit or building.

In [22] by Ruelens *et al.*, research work is carried out on the demand–response problem using a batch RL technique. The batch RL technique covers the inefficient information problem of RL method. This method uses a batch of experiences to find out optimal policy. In this work, two agents, water and residential building agents, are used. For dynamic pricing, to minimize cost a closed loop policy and for the day ahead scheduling open-loop policy is followed. The result showed a reduction in energy cost by 19%, and consumption rate is increased by 4%.

When an economic dispatch and demand response are treated as separate and sequential operation, energy efficiency decreases. Zhang *et al.* [23] presented optimal energy-management strategy in order to maximize social welfare. This method operates through coordination of demand response and economic dispatch. Economic dispatch is provided by generator and demand response by the customers. This method is also used for discovery of the power demand–supply mismatch. The simulation results showed convergence rate of 40 iterations.

Another approach for demand response is studied by O'Neill *et al.* [24] and proposed consumer automated energy system. This technique reduces residential energy cost and usage. This method uses online energy cost estimation and user decision policy. This is the independent approach to energy price and system behavior. In this method, users decide which device will use energy and how much. The results showed 40% cost reduction by using price unaware energy scheduling.

5.3.2.2   Complex system

In this section, we present related work carried out for demand–response management from a complex system perspective. We grouped the existing techniques into collaborative, complex adaptive system, demand-side integration, particle swarm optimization, and game theory approach.

Collaborative approach

MASs are widely used for controlling and managing a smart grid. In [1], Manickavasagam proposed and developed intelligent energy control center (ECC) mechanism for the smart grid. This technique consists of two layers. The one is DER serve as a client and the other is ECC as a server. ECC is controlled and monitored by a fuzzy logic controller (FLC). Communication and negotiation between client servers take place through internet protocol. The simulation results are stored in an excel database acting as a monitoring agent. ECC uses these results for decision-making in DERs. However, communication between results and FLC is not taken into account.

The mismatch between supply and demand reduces system performance. Parallel Monte Carlo tree search (P-MCTS) can produce an optimal solution for power balancing, but it has no coordination support. In [25], Golpayegani *et al.* extended the P-MCTS work by introducing collaborative and coordination concept. Agents negotiate with each other and present their proposal. This method resolves problems of agent's conflict, load-shifting, and charging capacity. The results showed that charge capacity increased from 33% to 50%. However, this model does not deal with prediction of data.

In [26], Le Cadre and Bedo worked on uncertainty in a smart-grid environment and present decentralized hierarchal based on the learning game approach. It is composed of supplier, generator, and consumer agents. Agents forecast demand and production of the grid in a collaborative manner. It determines the price that balance power and demand. The results showed that in a shared information network, faster convergence rate is achieved using cooperative learning as compared with an individual learning.

In [27], Huang *et al.* addressed demand–response issue in a smart-grid environment. This study proposed another approach in which elastic economic dispatch process is modeled. The flexible load cost is used as a control signal. The control signal balances demand and response. In this method, Monte Carlo and interval mathematic technique is used for boundary calculation. This estimates the uncertainty which is the difference between the present and the target value. Simulation results showed that the interval mathematic technique is efficient as compared to Monte Carlo. It also showed less convergence time, e.g., 1% as compared with Monte Carlo technique. However, this technique does not handle probability distribution.

Complex adaptive system

In [28], Kremers *et al.* presented a bottom-up approach for the smart grid. It consists of two layers: physical layer for electrical power transmission and logical layer for communication. This model has the ability to integrate new devices in a smart-grid environment. It provides dynamic load management, power, and communication controlling and monitoring. Experiment results showed 40% reduction factor in energy consumption. However, this model is not capable of handling high-load management.

In [29], Thimmapuram and Kim proposed an agent-based model using elasticity market complex adoption system to a smart-grid domain. This technique handles user elastic demand and lower cost. This method reduced peak load in the range of 8%–5%. However, the cost of energy for some users has increased.

Demand-side integration

Demand-side integration in smart grid results in security, quality, efficiency, and reduction in cost. In [30], Mocci *et al.* proposed a MAS for integration of demand and electric vehicles (EVs). The load agents calculate power demand and act as master agents. The master agents with cooperative agents send power load and global data to the demand side. It achieved demand–response rate of 85%. It also reduced the flow of data. However, this technique is not able to calculate the state of batteries of different storage at the different time.

In [31], Nunna *et al.* proposed a priority banking scheme. It concerns with users' demands. This method gives some share to the users from available resources. It monitors user demand and updates their priority. This method reduced network loss by 50% and also reduced dependency on overall grid. However, this technique provides fewer shares to users.

Particle swarm optimization

Advance power grid shifting from vertical to a horizontal structure which requires efficient management system. In [32], Hurtado *et al.* proposed building energy management system (BEMS). This technique provides interaction between different environments. It uses PSO approach to maximize comfort level and energy efficiency. In this hierarchical infrastructure, lower level agents abstract information and provide to higher level agents. Performance is described by weight factor; fair scenario showed 0.5 weight while bias scenario results in 0.3 weight. However, this technique generates some unbalanced situation.

Game theory approach

The retailer and market price management in smart grid gained great attention in research work. In [33], Wei *et al.* focused on energy price and dispatch problem in a smart-grid environment. This study proposed a two-stage two-level model. Customer demand and price is considered as the first stage using the Stackelberg game approach, while the operation of storage devices is considered as the second stage using linear max–min problem. Then the model is translated into mixed integer linear problem (MILP). The results showed 5% improvement in system performance, and it also increased retailer profit. However, this method is very sensitive and knowledge gathering process is very difficult.

In [34], Chai *et al.* addressed demand–response issue and presented a two-level game approach. This technique handles multiple utility company and users. The utility company is modeled as noncooperative and communication between users as evolutionary. The results showed that proposed technique reduced cost payment from 3,197.7 to 2,425.6 and energy demand is increased from 1,224.9 to 1,478.4. However, this work does not employ constraint on power consumption.

In [35] by Song *et al.*, another framework for optimal nonstationary demand-side management in a smart-grid environment is proposed. In this method, the user selects their energy usage pattern according to their priority and needs. They used a repeated game approach which provides interaction among foresighted price anticipating users. This method showed 50% reduction in energy cost and robustness in error. However, higher threshold value results in a trade-off between cost and peak average ratio.

In [36], Nunna and Doolla carried out research work on management of demand response in multiple microgrid networks. In this work, customers participate in demand–response strategy. This study proposed a priority index approach through which customers participate in the market. This method reduced peak demand. It is found that customers with high priority index get power at low cost.

In [37], O'Brien *et al.* focused on demand–response management in smart-grid application. In this work, demand response is modeled as the game-theoretic environment, and Shapley-value (SV) is used for payment distribution process. RL technique is used to estimate SV. Simulation results showed that for random sampling, 1,000,000 samples take 58.2 s execution time, while for sigmoid sample, 51,129 samples take 6.5 s, this showed that uniform sample balances demand and response. However, this method is not suitable for distribution scheme and its direct estimation is difficult. The literature summary of demand-response management has been shown in Table 5.2.

5.3.3 Fault monitoring

Fault-monitoring process involves the detection and prevention of any fault that occurs in a smart-grid environment. This approach is categorized into self-organizing and algorithm approach.

5.3.3.1 Self-organizing

Self-organization is an activity of the system in which each or some parts of the system arrange themselves based on the local interaction among each component of

Table 5.2   Literature summary of demand–response management

Ref	Technique	Strength	Limitations
Hernández et al. [17]	Energy forecasting in VPP, ANN	1.5% error rate	Works on single unit
Mocanu [18]	Energy prediction for building, RL + belief neural network	91.42% accuracy	Not implemented on different level
E. Lakić [19]	Learning how much system reserve to offer at different time, SA-Q learning	Ration between cost and benefit increased	Not implemented on MAS framework
I. Dusparic [20]	Energy prediction of next day-ahead for EVs, RL+ load forecasting	33% usage reduced	No communication
Wen [21]	Segmentation of each device, RL	Linear complexity	Single unit
Ruelens [22]	Thermostatic load controlling, batch RL	19% cost reduction	Single unit
Zhang [23]	Finding mismatch between demand and supply, economic dispatch	Convergence rate 40 iterations	Only local communication
Golpayegani [25]	Conflict management between Evs, CP-MCTS	Battery capacity increased 17%	No prediction
Kremers [28]	Agent-based model of simple smart grid, CAS	40% consumption reduced, and peak load lies in the range of 5%–8%	Not handling high load
Mocci [30]	Controlling of integrated demand and EVs, DSI	Response rate achieved 85%, network loss rate reduced 50%	No calculation of battery state
Hurtado [32]	Controlling of interoperation of smart building, PSO	Performance in the form of weight factor, which is achieved 0.5	Unbalance situation
Wei [33]	Management of energy price and dispatch problem, MILP	5% performance improved	Difficult to gather information
Chai [34]	Controlling multiple utility centers and end users, two-level game	Cost reduced from 3,197.0 to 2,425.6 RS	No constraints on power consumption
O'Brien [37]	Payment distribution process, Shapley value distribution	Converge in 58.2 s	Not suitable for energy distribution

the system. In this section, we discuss some of the self-organization approaches that have been carried out for addressing fault-monitoring problem.

Adaptive programming scheme

Some naturally accruing events lead to the cascading failure and loss in a smart-grid system. In [38], Babalola *et al.* proposed an adaptive MAS for prevention of cascading failure and loss in the smart grid. The proposed model searches for overloaded transmission lines and then redistribute power to that line. The system decreases the transmitted power in the overloaded lines and brings the lines to in working state. This process successfully halts the cascading failure without load shedding. However, this approach needs major hardware requirement and efficient dispatch power history. Additionally, the algorithm also consists of a large number of constraints.

In [39], Nassar and Salama introduced the dynamic microgrid concept having flexible boundaries. With this feature, the size of the grid can be reduced or extended according to the need. It uses forward–backward sweep technique for power flow. In an emergency situation, self-healing feature is achieved. The result showed good performance when compared with fixed boundary system. However, the computation time of this technique is very large which is 15.106/h.

In [40] by Chen *et al.*, work is done on restoration of the power flow after a natural disaster. In this work, multi-agent coordination control scheme based on a mixed-integer linear program is presented. The proposed system controls on and off status of switched devices. A local communication technique is used for discovery of global information. The global information is used for the optimal decision. The results showed the computation time of this technique 0.265 s. However, this work does not focus on communication range, battery capacity, and the requirement for global information discovery.

Multi-agent framework
Fault detection and its diagnosis avoid loss of synchronous operation in power system. In [41], Rahman *et al.* presented an intelligent agent-based model for system protection in critical time. This model has the ability of autonomous decision-making for circuit breakers and detects a fault in critical time. Simulation results showed the flexibility and stability of the system. However, this model cannot be implemented in the large and complex power system.

Wolf-pack hunting
In [42], Xi *et al.* presented multi-agent wolf-pack hunting approach for the smart-grid system. The wolf-pack idea is derived from a hunting group of a wild wolf pack. The basic idea is to ensure survival in the harsh environment. This model can handle optimal management of power distribution and can operate in load disturbance condition. Experimental results showed that the convergence rate is 51.37%–57.4% and the error rate is 0.5%. The agents exchange information so rapidly and calculate the optimal policy. It increased utilization cost with reduction of generation cost.

5.3.3.2 Algorithmic approach
In the past, studies based on algorithmic approaches such as a fuzzy-rule, census, sweep technique, and spanning tree approach are also presented in the smart-grid domain. Next, we discuss these studies.

Fuzzy-rule
In [43], Elmitwally *et al.* proposed distributed system based on fuzzy rule-based multi-agent approach. Its work mainly focuses on eliminating congestion of smart-grid

components, voltage violation, and cooperative operations. During the experiment, congestion is eliminated in 2.17 s. Voltage is controlled by keeping the operation in the limit. This approach performs voltage adjustment task in 28 s. However, its performance decreased in the case of communication failure.

Census scheme
In [44], Teng *et al.* proposed a restoration framework for an emergency situation in a smart-grid environment. In this method, a dynamic leader agent is used for operation in emergency and disaster situation, and bus agents operate in a normal situation. This method reduced communication time, and communication bandwidth is kept saved during a disaster.

Sweep technique
In [45], Nguyen and Flueck proposed another decentralized distributed agent-based model for power flow problem. It consists of multi-agents having autonomous, local view, and decentralized behavior. Agents use back and forward sweep iteration technique for power flow solving. The results showed computation time 81.96 s.

Spanning tree approach
In [46], Eriksson *et al.* presented a multi-agent distributed algorithm for integrated volt/var control in the smart grid. Agents are collaboratively controlling and managing voltage and capacitor. This method deals with the optimization of voltage profile, reducing system loss, and switching of the capacitor. Two types of agents are used: a switching agent who detects and solves system fault and volt/var control agent who controls power flow. This technique controls voltage above the lower limit but does not handle the voltage below the high limit. The results showed that the average time for solving power flow is 9.4405 s, which demonstrates an efficient technique. However, the solution does not lead to optimum. The literature summary of fault control has been shown in Table 5.3.

Table 5.3   Literature summary of fault control

Ref	Technique	Strength	Limitations
Babalola [38]	Prevention of cascading failure, adaptive programming		Needs large information about system states
Nassar [39]	Dynamic boundaries, forward and backward sweep techniques	Self-healing is achieved	Computation time increased
Chen [40]	Coordination control scheme, MILP	Convergence rate 0.265 s	Not focused on communication range
Rahman [41]	Fault location in critical time, MAF	System stability is achieved	Complex system not handled
Elmitwally [43]	Elimination of congestion, fuzzy-rule	Restoration time 2.17 s	Performance decreased in the case of communication failure

5.3.4   Power scheduling

Power-scheduling process involves in setting power consumption and production at the specific time period. This process is categorized into complex and learning-based approach.

5.3.4.1   Complex system

As we know that the complex system comprises many interconnected objects in the system, these objects interact with each other in a nonlinear manner. Next, we discuss different techniques that have been carried from a complex system perspective to address power-scheduling problem.

Self-organizing

Smart grid requires real-time monitoring to provide reliable services for end users. In [47], Colson and Nehrir proposed a decentralized MAS for real-time power management in smart grid. MAS controls the grid assists based on price, resources, and users' demand. The experimental results show that decentralized MAS are reliable for real-time monitoring in the smart grid. It is also shown that as time continues, the performance of storage degrades due to discharging.

Hierarchal approach

In [48], Hu *et al.* proposed a hierarchal approach based on a MAS for smart-grid operation. This approach integrates the EVs and addresses grid congestion and voltage violation problems. The results showed good performance for power scheduling and control. However, the communication between agents is too complex.

There have been several designs proposed for smart-grid architecture but still facing feasibility and economy problems. In [49], Chao and Hsiung proposed fair energy resource allocation algorithm for electricity trading among smart grid. This technique prevents starvation situation and fatal problem. It also reduces power cost. It achieved 96.25% fairness index even in the high worst case. However, this technique does not take power transmission into account.

Rahman *et al.* [50] have proposed an agent-based model to address voltage stability problem. In this model, agents manage their activities through online information and power flow. They estimate voltage variation by using distributed synchronous compensator. Simulation experiments showed robustness performance of the system. However, communication time delay is observed 15 ms, while voltage stability has improved.

In a smart-grid environment, there need to achieve stability and reduction in operation cost. In [51], Radhakrishnan proposed smart-grid framework based on the multi-agent distributed energy management system. It performs optimal energy allocation and management in smart grid. This model consists of renewable energy sources, storage devices, and generators. It controls power balance by the state of charge of the batteries. Simulation results showed a reduction of total cost from 662.2 to 658.4. However, the performance of the proposed algorithm degrades under some uncertain condition.

Census-based approach

A centralized system is not able to handle flexible power loads to maintain the power balance in a smart-grid environment. In [52], Li *et al.* proposed a look-ahead scheduling model for flexible loads in a smart-grid environment. This model consists of three layers: centralized, distributed, and cooperative control. Load agents perform coordination among agents, and cooperative control strategy is used for communication protocol. This model provides flexible strategies to handle the large flexible load. However, this model is not able to handle uncertainty.

In [53], Guo *et al.* proposed an economic dispatch scheme based on projected gradient concerns with economic dispatch problem. It decomposes centralize optimization into local optimal agents. It deals with the stochastic environment. This scheme presents a finite time average census algorithm. In this method, agents iteratively calculate the solution of the optimal problem. Its communication with agents is limited. This method achieved plug-in-play, and it does not require any private information. It can handle quadratic and non-quadratic cost function. The results showed that overall cost of the system reduced.

Kahrobaee *et al.* in [54] presented the concept of smart home within a smart-grid environment. In this work, home is considered an agent who can buy, sell, and store energy and interact with the grid. This framework consists of home agent based on distributed multi-agent network. The home agent makes autonomous decisions to buy, sell and store energy, it takes a decision based on maximum utility. The home agent decision affects the market price. The results showed home agent decision reduced their energy cost as it buys, sells and generate energy at the same time. However, this method is simple and does not address all issue related to demand and supply.

In [55], Samadi *et al.* addressed uncertainty issues in smart grid and present an optimized algorithm based on the central unit. This technique only needs future demand estimation and minimizing energy cost for each user. The results showed that the peak to average load is 25.5% achieved. It also reduced energy expenses. However, the complexity of the system is increased.

In [56], Gregoratti and Matamoros presented another approach for power flow in a smart-grid environment. The proposed technique controls and manages power flow among multiple microgrids. This technique focuses on protecting private local information, and it is based on sub-gradient cost minimization approach. The results showed limited iteration and faster convergence rate. However, in this work, the communication with the main grid was not considered.

Cognitive-based approach

In [57], Bu and Yu studied green cognitive network in smart-grid application. Cognitive network monitors smart-grid operation and provides information to the control unit. The power allocation is performed based on collected information. Power allocation, price, and efficiency are modeled as three-stage Stackelberg game. Results demonstrated 31.09% cost reduction. However, this technique does not handle the incomplete scenario.

5.3.4.2   Learning-based approach

In the past, power scheduling problem has been addressed by using two types of learning techniques that are reinforcement and neural network.

Reinforcement learning

RL technique is an essential tool for computation and estimation of payoff to achieve game equilibrium in a smart-grid environment. Wang *et al.* [58] presented a scheme based on RL technique for energy trading in the smart grid. This method chooses a random strategy and maximizes the average utility and revenue. The proposed scheme is able to achieve Nash equilibrium. This technique handles incomplete information available and stochastic environment. Information is exchanged through the central unit and protects private information. However, implementation of the finite action learning algorithm is a challenging task in real value action environment.

In [59], Samadi *et al.* worked on load scheduling and power trading in a smart-grid environment. The study considered high penetration renewable resources. They adopt the game theory approach. In this method, users can sell their extra power to their neighbors locally. This method handles the reverse power flow problem. This increases the revenue and decreases energy expenses of the users. The results showed that average energy imported is reduced to 820.2 kW from 1,360.9 kW, and energy cost is reduced to 40.37$ from 60.91$.

Energy hub provides interaction between energy carriers in supply requiring loads. In [60], Sheikhi *et al.* extended the energy hub system. This study proposed cloud-computing concept which consists of a utility provider and customer interaction through the cloud. The cloud takes the input of utility power and produces output to the users. This model provides two-way communications between utility companies and energy hub. The results showed that energy cost is reduced to 33%. However, the proposed system is unable to predict consumer's future demands.

In [61], by Ghorbani *et al.*, fault-detection technique based on the MAS is presented in a smart-grid environment. This technique combines centralize and decentralize features that demonstrate the hierarchal coordination scheme. It consists of zone agents, feeder agents, and substation agents. Zone agents provide services to detect and locate the fault and help feeder agents to restore services using the q-learning technique. This method needs fewer messages for communication and reduced computation time. The results showed that 16 messages are required for communication for 21 agents, while centralized and decentralized scheme required 20 and 38 messages, respectively. However, the number of zone agents and feeder agents remain fixed with the system size which results in more burden and computational time in the complex system.

Venayagamoorthy *et al.* [62] proposed intelligent dynamic energy management system (I-DEMS) based on neural network and RL. They used Bellman equation for the optimal control signal and calculate min and max cost-to-go function. They compared this technique with DEMS based on Decision Tree method, DT is inefficient because it supplies energy based on available power. The result shows that I-DEMS is reliable and it extends battery life, but this technique does not predict battery sate.

In [63], Li and Jayaweera presented a hierarchical architecture for communication between utility company and customers. The proposed technique consists of two stages: initial and real-time interaction. At initial interaction, demand response is controlled which proved load remains flat. However, it also showed that by increasing training period, performance of the studied system decreased.

In [64], Rayati *et al.* proposed smart energy hub concept in smart-grid application. Smart energy hub is used for multipurpose transmission of generator energy, information, and user-demand scheduling. In this method, RL is used for optimal solution and the result demonstrates 26% cost reduction.

There are some implementation challenges in the smart grid due to uncertainty and dynamic price. Kim *et al.* in [65] focused on dynamic price management and proposed agent-based RL technique. In this framework, each customer is considered as an agent and learns policy without any advance knowledge. Utility company monitors customer behavior and schedule power on demand. This method increased system and customer cost. However, error in estimation of cost effects system performance. It also does not handle multiple energy resources and bidirectional communication.

In [66], Wang *et al.* presented broker concept in a smart-grid environment. In this framework, the broker is responsible for predicting user demand and then buys energy from utility company using auction strategy. Each customer is distributed as cluster network. The broker uses MDP and RL technique to predict customer demand. This technique balances energy supply–demand and achieved 24.6% imbalance rate. However, due to broker's involvement, computation time is increased.

Load shedding is used to balance power supply and demand. Central controller allocates power to users using bidding process. Lim and Kim [67] presented a bidding scheme based on q-learning technique. It makes policy when power is less than demand. It starts bidding process and identifies those who want to buy power and submit their bid price and quantity. The results showed that power balance is achieved by the repetitive interaction between agents. It also showed that period of exploration increased as trivial interaction decreased.

It is impractical to have complete information in advance about cost and demand for energy in a smart-grid environment. Zhang *et al.* [68] focused on this issue and proposed price-dependent load-scheduling technique. This technique uses post-decision state (PDS) and Markov decision process (MDP). This method can provide an optimal solution in an unknown environment. In this method, consumers can buy and store energy during peak hours. Energy cost and demand are taken as variable entities. Load-scheduling process is considered as MDP using RL PDS technique. This method needs less information to converge into an optimal solution. The results showed that algorithm converges into optimal solution in 1,122 time slot giving 90% average utility. However, this method does not handle load scheduling in a collaborative environment.

Wu and Liao [69] focused on power-dispatch problem and presented function optimal RL scheme for power-dispatch problem in complex and multidimensional space of smart-grid application. This technique searches in sequence result showed that 32.31% reduction of computation time and voltage stability also increased. However, this technique showed the conflict between energy cost and voltage stability.

Table 5.4 Literature summary of power scheduling

Ref	Technique	Strength	Limitations
Colson and Nehrir [47]	Real-time power management, self-organizing	Reliable system	Performance of storage devices decreased
Chao and Hsiung [49]	Electricity trading, fair energy resource allocation	Reduced power cost	Ignored power transmission
Rahman [50]	Voltage stability in SG, hierarchal approach	Robust performance	Communication time increased
Li [52]	Look ahead scheduling, census	Handle large flexible load	Not handle uncertainty
Kahrobaee [54]	Smart home concept, census	Cost reduced	Not address all issues related to demand and supply
Gregoratti [56]	Power flow in multiple microgrid, census	Fast convergence rate	No communication with main grid
Sheikhi [60]	Cloud computing concept in E-hub, RL	Energy cost reduced by 33%	No prediction of energy demands
Kim [65]	Dynamic price management, RL	Monitor customer behaviors	Not handle multiple energy resources
Wu and Liao [69]	Power dispatch in complex scenario, RL	Computation time reduced by 32.31%	Conflict between energy cost and voltage stability

In [70], Shirzeh *et al.* worked on management of renewable energy resources and storage devices in a smart-grid environment. They proposed a MAS based on a plug-and-play technique for managing and controlling resources in the smart grid. Plug-in-play technique used RL method based on distributed value function to adjust power balance of demand and supply. Results showed 81% reduction in fluctuation by using plug-and-play. The number of iterations is also reduced. However, this method does not deal with the stochastic environment.

Artificial neural network
Integrating wind energy resources with other distributed energy resources is a challenging task. In [71], Motevasel and Seifi addressed this issue and proposed an expert energy-management system. This technique finds optimal set points of energy resources and storage devices. This technique controls forecasting, optimizing, and storage module. ANN is used for forecasting process. Results showed that convergence is achieved in 445 iterations. The literature summary of power scheduling has been shown in Table 5.4.

5.3.5 Storage and voltage management

Storage and voltage management scheme handles storage devices and voltage variation. This scheme is categorized into learning, monitoring, and search-based approach.

5.3.5.1   Learning

Storage and voltage-management problem are addressed by using RL and neural network approach. Next, we discuss these learning techniques and try to explain how different studies addressed the storage and voltage problem in smart-grid domain.

Reinforcement learning

In [72] by Li *et al.*, research work is concerned with the implementation of RL technique for load-balancing problem in the smart grid. The proposed scheme is based on dynamic hierarchal approach. It finds an optimal policy to balance power demand and supply. It handles curse dimensionality problem. It is a fast-learning technique in an unknown environment.

In [73], Salehizadeh and Soltaniyan proposed a fuzzy q-learning technique. It handles multidimensional renewable power in less iteration. With this method, 40% iterations decreased as compared to other techniques. It models electricity in continues range.

Wind energy is uncertain and is a variable energy resource; this effects smart-grid performance. In [74], de Montigny *et al.* addressed this issue and proposed multi-agent architecture. This method calculates import and export losses. It also calculates global-demand forecasting using minute-to-minute strategy. Additionally, it also estimates system performance from historical data. Results obtained through minute-to-minute strategy and showed that number of generating unit start and stop increased by 5%. However, computational time of this method is very large.

Load frequency managing and controlling is a hot topic for research in a smart-grid environment. The linear model is not capable of handling dynamic behavior of the system. In [75], Daneshfar *et al.* addressed this issue and proposed multi-agent RL technique which consists of two agents: estimator and controller. Estimator agent finds frequency error, and controller agent uses genetic optimization for frequency control. This technique showed frequency variation fall to zero through the optimal solution. However, load disturbance is generated by reaching to maximum frequency.

In [76], Wei *et al.* addressed battery-management issues in a smart-grid environment. This study proposed a dual iterative q-learning technique based on adaptive dynamic programming for managing and controlling storage devices. In this method, dual iteration, internal iteration for minimizing power cost, and external iteration for finding Q function to converge into optimum is used. This algorithm converged into optimal solution in 20 iterations. However, the proposed algorithm finds optimal solution indirectly. Initial interaction handles demand response at customer side, and the load is considered as a flat point. Real-time interaction is used for decision-making. This technique used hidden mode MDP. This technique outperforms as training period is increasing. However, in the studied system, a smart home was not considered.

Integrating different types of energy storage devices in smart grid produces implementation challenges. In [77], Qiu *et al.* focused on controlling and managing different types of energy-storage devices. This study proposed RL-based scheme to optimize coordination of energy-storage devices. The results showed that system gradually learns with time and results in an optimal solution. This study also showed that system losses decreased. However, it required large computational time, and it does not support power-sharing feature.

Integrating photoelectric energy with smart grid decreases fossil fuel consumption as well as electricity bill. In [78], Wang *et al.* proposed near-optimal control algorithm for the residential storage system which controls power generation, predicts power consumption, and accounts for various loss components during operation. They applied RL technique for prediction amount of energy in ESS. This technique performs optimization on energy price and energy demand price. Experimental results show that the proposed algorithm outperforms and achieves up to 72% enhancement in electricity-cost reduction compared with baseline storage control algorithm. Limitation of this system is that PV generation system only works in sunlight.

Battery management plays a key role in a smart-grid environment. In [79], Kuznetsova *et al.* presented a two step-ahead RL algorithm for battery scheduling within microgrid architecture. It is composed of local consumers, generator, and storage devices connected to the external grid. This technique predicts and forecasts power demand. It finds optimal actions for battery scheduling. Simulation results showed 3.94% improvement in battery. However, the simulation running is very large.

In [80], Vandael *et al.* addressed day-ahead power scheduling problem for EVs in a smart-grid environment. In this method, charging process is performed by the heuristic scheme. The heuristic scheme is controlling and managing each EV. The system collectively learns cost-effective scheduling strategy for EV charging through RL technique. The results showed that average cost increased by 10%. However, this method has some overloading and over constraint issues.

In [81], Guan *et al.* focused on minimizing energy cost in a smart-grid environment. In this work, RL technique is applied to find an optimal policy to storage devices. This method does not require any future prediction about energy generation and consumption but the partial observable environment. The TD-lambda algorithm is used for convergence to the optimal solution in the non-Markovian environment. Simulation results showed 59.8% reduction in energy cost.

Artificial neural network
Battery management plays a key role in smart grid; it is important to measure the health of batteries during operation. In [82], Landi and Gross proposed two different techniques for estimating battery health in smart-grid application. First one is based on fuzzy logic and the second one is a neural network. These techniques use temperature, charging/discharging, and a number of the cycle as parameters. Results showed 5% error rate.

5.3.5.2 Monitoring

In this section, we discuss different approaches presented for storage and voltage monitoring.

Volt/Var control
In [83], Zhang *et al.* presented a multi-agent distributed algorithm for integrated volt/var control in the smart grid. Agents are collaboratively controlling voltage and capacitor. This method deals with the optimization of voltage profile, reducing system

loss, and switching of the capacitor. Two types of agents are used: switching agents who detect and solve system fault and volt/var control agents who control power flow. This technique controls voltage above the lower limit but does not handle voltage below high limit. The results showed that the average time for solving power flow is 9.4405 s, which demonstrates an efficient technique. However, the solution does not lead to optimum.

Census approach
Researchers are also interested in reducing high-power consumption and demand to reduce cost. In [84], Sharma *et al.* proposed agent-based distributed control model to address this issues. In this model, power-storage devices are used as agents. It achieves convergence in agreement of power consumption. It prevents overcharging and discharging of batteries. Results showed 95% and 85% charging and discharging efficiency, respectively. However, the communication between agents is limited, and it does not predict the state of batteries only its maximum/minimum state.

State monitoring
For dynamic state estimator, in [85], Srivastava *et al.* proposed a MAS for the multi-area power system. This method divides the whole network into subsystem and algorithm executes in parallel. This use two unit's: field and phasor unit run separately. At last, center controller integrates their results. The algorithm follows cubature Kalman filter. Results showed $2.4(10^{-2})$ voltage error. It has been showed that extended Kalman filter is not feasible.

In [86], Teleke *et al.* focused on battery management and proposed rule-based control strategy. This technique monitors and controls charge/discharge limit and battery lifetime. It also utilizes 70% battery capacity. The results showed voltage deviation reduction from 24% to 4%. However, this required high-capacity batteries.

Integrating solar energy in a smart grid make it an active system which required cyber-physical management system. In [87], the author presents a goal-based Holonic MAS. This technique uses nested agent concept and controls power strategy and state estimation. The results showed execution time 93 s and absolute error 0.038%. However, the complexity of the system increased by nested agents.

In [88], Klaimi and Merghem-Boulahia focused on energy-management system and proposed a multi-agent intelligent model for smart-grid application. In this technique, intelligent storage devices are used for storing surplus power. This technique reduced energy cost and access to the grid. Results showed 60% cost reduction.

5.3.5.3   Searching
The searching techniques used for addressing storage and voltage problems include self-organizing, normality analysis, hill-climbing, and swarm intelligence. Next, these search-based techniques are discussed.

Self-organizing
The integrating and monitoring of smart microgrid is at the initial stage, and it needs more research studies. In [89], Vaccaro *et al.* proposed and developed a self-organized standalone smart microgrid framework for solving and controlling smart microgrid

operations. They focused on synchronizing and controlling of smart applications involved in the system. Results and experiments show that dynamic agents are useful in power flow problems. It also helps estimation of state in the smart microgrid. However, this approach does not address the computation and estimation of semantic representation of data.

Integrating EVs with the smart grid is a challenging task for the researcher. In [90], Hu *et al.* presented hierarchal control method that coordinates self-interconnected nodes. In operation constraint, marked-based control is used. In this framework, two level are used: upper bound and lower bound. Upper bound controls power scheduling, and lower bound provides power to EVs. Implementation showed that this method is feasible and there is no power loss.

Normality analysis

In [91], Vallejo *et al.* extended the previous work of intelligent monitoring of substations in the smart grid. They used knowledge-based software agents which are used for data collection and decision-making. This model integrates new agents that can be used in different environment condition. They used web services for results monitoring. The experimental results showed 75% absolute normal voltage. This model improves robustness and provides reconfiguration as well as replication of services. However, this model requires more data and information for intensity control.

Hill climbing

In [92], Xi *et al.* presented win/loss fast policy hill-climbing approach for optimal averaging policy learning for the MAS in the smart grid. This model is applicable in stochastic non-Markov environments. This technique is an independent self-play game and can achieve fast convergence learning rate. Simulation results showed 68% faster learning rate than previous techniques used in literature that are q-learning, q-lambda, etc. However, this model has some challenges to real implementation; it faces security and stability problems.

Modeling of the smart grid is a complex task due to its complex nature, multi-agent behavior, and distributed resources. In [93], de Durana *et al.* presented a model for local multi-carrier energy network in the smart grid. This model allows transmitting of different type energy in the smart-grid network. The simulation results showed rebalancing between energy networks. This model only focuses on smart-grid operation, not on energy generation and load management.

Swarm intelligence

In smart grid, it needs to manage and control frequency around reference value in order to get secure and quality power flow. Unbalance frequency produces unbalance power distribution to the consumers. In [94], Evora *et al.* presented a swarm intelligence agent-based system for frequency management in the smart grid. It used decentralize independent agents and can exchange information in the shared environment. It evaluates three policies: detection, communication, and stability. However, increasing microlevel results non-desire effect.

Direct load control (DLC) is used for control and management of demand at consumer side. In [95], Hernandez *et al.* presented DLC method based on multi-objective

Table 5.5   Literature summary of storage and voltage control

Ref	Technique	Strength	Limitations
de Montigny [74]	Minute-to-minute forecasting, RL	System efficiency improved by 5%	Complexity increased
Bevrani [75]	Dynamic behavior of the system, RL	Zero frequency at optimal solution	Load disturbance at high frequency
Qiu [77]	Management of different type storage, RL	Energy loss decreased	Complexity increased
Landi and Gross [82]	Battery health estimation, ANN	Error rate 5%	High-error rate
Sharma [84]	Disturbance control model, census	Charging efficiency 95%	Only estimates maximum and minimum state of the storage
Pahwa [87]	Nested agent concept for the state estimation, goal-based holonic MAS	93 s execution time	Complexity increased
Vaccaro [89]	Controlling and monitoring microgrid, self-organize	Estimates state of microgrid	Not handles semantic representation of data
Vallejo [91]	Substation monitoring, normality-analysis	75% absolute normal voltage	Required large data
Evora [94]	Frequency management, swarm intelligence	System stability is achieved	Increasing microlevel result non-desire effects on the system

particle swarm optimization for smart-grid environment. Appliances operate when constraints are satisfying at system side. The operation of appliances was obtained by distributing constraints among neighbor nodes. This method reduced user energy demand by 20%. However, this method was applied only to three appliances: refrigerator, light, and freezer. Another drawback of this method is that it must generate a result in a fraction of second, otherwise system stability will effect. The literature summary of storage control has been shown in Table 5.5.

5.4   Open research problems and discussion

In this section, we are going to discuss different methodologies and techniques on MAS in smart grid and their open-research problems. The smart grid brings many facilities to users and energy efficiency, customer satisfaction, reduction in energy cost, and load balancing. However, there exists a number of challenges to be researched and resolved. These challenges involve communication protocols, fault detection, prevention, power scheduling, load balancing, and storage/voltage control.

How to handle communication among multi-agent in smart grid? To address this question, a number of research efforts have been carried in the domain of smart-grid domain. We reviewed two approaches that are group communication and learning-based approach used to address communication challenges in the smart grid. Group communication approach involves hierarchal, coalition formation, census and PSO methods. In [3–5], the hierarchal scheme is used to handle communication among multi-agent. In [3], the layer-based framework was proposed, and this framework reduced communication complexity and voltage variation. In [4], two-way communication scheme was proposed and energy cost decreased from 80 to 50. However, this scheme faces some design issues. In [5], the cognitive-based scheme is used which reduced peak power from 55,000 to 900 W. In [6], another scheme based on information sharing about imbalance power is proposed. However, in this scheme, agents only keep information about their own status. In [7], zero correlation schemes were proposed for the machine-to-machine communication. This scheme maintains security and reduces traffic overhead. However, this approach gives worst result in the case of small scenario. In [8,9], coalition formation approach was used. In [8], decentralize dispatch scheme was proposed. In this scheme, agents perform the task in a group or teamwork. However, this scheme showed large communication time. In [9], the proposed technique does not handle stochastic scenarios. In [10], the author addressed communication latency problem, and the result showed 3.983 s restoration time. However, this scheme does not handle individual communication. The census-based scheme was proposed for cost estimation in [11], this scheme increased system vulnerability. PSO scheme based on adaptive biding technique was proposed in [12,96], this reduced buying cost. However, this also reduced customer comfort level. Learning-based approaches involve RL, ANN, and Bayesian learning approaches in [13–16]. RL technique showed fast convergence rate. However, this only handles linear time-invariant. In [15], ANN learning is used for power classification, and this showed 98.7% accuracy in system performance. However, this technique is not robust to error. Bayesian learning is used in [16] for price and demand calculation in the case of incomplete information. This showed 40% increase in total utility. However, this technique does not control packet loss rate.

How to handle demand response in smart-grid environment? To address this question, different approaches also have been applied in the domain of smart grid. We reviewed two types of approaches that are learning-based and complex system-based approach adapted to handle demand response in smart-grid application. Learning-based approach involves ANN and RL. ANN is applied to a virtual power plant in [17], this showed 1.5% error rate. However, this needs enough information for prediction of future demand. RL is applied in [18] for energy prediction and showed 91.42% accuracy rate. However, this method was not applied on a different level. In [19], SA-Q learning was applied to learn system reservation to offer power at the different time. This increased cost-benefit rate. In [20], demand response was addressed to predict load for the next day. This reduced 33% peak usage; however, there is no collaboration among agents. In [21], demand–response management for a single unit or building was proposed. The proposed technique in [22] maximize social welfare and converged in 40 iterations. In [24], online energy cost estimation is proposed and the result showed 40% energy cost reduction.

The complex system comprises collaborative, complex adaptive system, demand-side integration, PSO, and game-theory approaches. The collaborative-based scheme was discussed in [25–27]. This scheme has open issues regarding communication, prediction, and probability distribution. Complex adaptive system approach was applied in [28,29]. This reduced 40% energy cost and also peak load to 8%–5% range. However, this approach does not handle high load and also the cost of energy increased for some users. Demand-side integration was discussed in [30,31]. The open research issues existing in this scheme are as follows: it does not estimate the state of batteries and offers fewer energy shares to the users. In [32], PSO technique based on BEMS framework has unbalanced situation issue. The game theory approach was also applied to address demand–response problem. This approach has open issues related to sensitivity, information gathering, and the trade-off between cost and PAR. It also not suitable for distribution scheme.

How to detect and prevent a fault in the system? To address this challenge, a number of research efforts have been done and cited in our review work. We grouped these studies into two categories, i.e., self-organizing and algorithmic approach. The self-organizing approach consists of adaptive programming, MAF, and WPH. These approaches can perform self-healing task in an efficient manner. However, there exist some open research problems that are as follows: these required major hardware for implementation, unable to address complex model, cannot address battery capacity, there is no global information discovery.

Algorithmic approaches consist of fuzzy-rule, census, sweep and spanning tree techniques. These studies successfully reduced congestion and communication time. However, there still exist some open research problems that to be addressed. In this scheme, the system performance degrades in the case of failure and no guarantee of an optimal solution.

How to perform power scheduling? We surveyed research work and grouped these work into two categories, i.e., complex system and learning-based model. The complex system consists of self-organizing, hierarchal, census, and cognitive-based approaches. The self-organizing approach is discussed in [47]. This technique showed good performance in term of monitoring; however, performance degrades in discharging periods. The hierarchal scheme is discussed in [48–50], this approach has the ability to handle the starvation problem and achieved 96.5% fairness index. However, this scheme increased complexity and computational time. Census-based approaches are also reviewed in this part for power-scheduling task. In [52], flexibility concept is introduced and provides flexible strategy to perform flexible power transmission. In [53], the central unit is introduced and achieved 25.5% peak to the average rate. In [54], the subgradient concept was used for cost minimization. This showed fast convergence rate; however, there is no communication with the main grid. Pruning strategy was discussed in [56] that prune those agents which are not participating in the communication. This reduced search space size; however, this method is unable to prune those agents which are close to each other.

Learning-based approaches (RL and ANN) adopted to address power scheduling problem in the smart grid. With the adaptation of RL-based approaches, private information was protected from external users. It provides reverse power flow facility, where the user can send extra power back to the main grid. Cloud interaction concept was introduced in [60], where user and utility can interact with each other through the cloud. This reduced energy cost to 33%. ANN-based approach is presented in [71], which integrates wind energy resource with other resources. However, learning-based approaches are still facing open-research problems that are as follows: there is no collaborative learning, the conflict between cost and voltage, and there is no procedure to predict system state.

How to manage and handle storage devices and voltage? To address this problem, a number of research works are discussed and reviewed in this part. We grouped these work into three categories, i.e., learning-based, monitoring, and search-based approaches. Regarding learning-based approach, in [74], minute-to-minute forecasting strategy was applied. This increased the number of generating units. However, the computational time is also increased. Different types of energy storage devices was integrated with the system in [77], this decreased energy loss. In [81], two-step ahead forecasting strategy was applied which showed 3.94% improvement in battery life. ANN-based learning scheme was used in [83] for state estimation and showed 5% error rate.

Monitoring-based techniques consist of volt/var, census, and state monitoring. These techniques control voltage and monitor system state. Agent-based distributed control (ABDC) based on monitoring approach prevents overcharging and discharging of the battery. This method achieved 95% and 85% efficiency in charging and discharging, respectively.

Search-based techniques consist of self-organizing, normality analysis, hill climbing, and swarm intelligence. Self-organizing technique addressed application synchronization problem in [91]. However, this technique is unable to handle semantic data. Normality analysis is used in [92] which integrate EVs. In [93], the knowledge-based scheme was used which provide integration of new agents, reconfiguration, and replication services. However, this scheme requires large data for intensity control.

5.5   Conclusions

As a simulation and modeling perspective, the MAS in smart grid has recently been attracting an increasing attention from the research community. The growing domains of interest in MAS in the domain of smart grid are communication protocols, demand response, self-healing, power scheduling, load balancing and storage-device management. A number of research works have been carried out and developed multi-agent based models for smart grid in abovementioned domains.

In this part, we covered the different approaches adopted in MAS for smart-grid modeling and proposed a classification of MAS models according to the techniques used for their implementation. We finally described each technique and its model. We also highlighted open research problems exist in each solution.

The basic objective of MAS in smart-grid modeling is load balancing, to bring balance or equilibrium between users demand and generation capacity. In another word, MAS in smart-grid modeling deals with energy-optimization process. As for the authors are concerned, this is the first article which clearly highlights open research problem in MAS in the smart grid that covers a large number of different research studies.

The aim of this survey was to allow a comprehensive understanding of the various emerging development in the field of the smart grid, the different approaches, their advantages, and limitations. We hope it will be a good guideline and a starting point to those researchers coming to this field and desiring to increase their knowledge in smart-grid domain from MAS perspective.

References

[1]   Manickavasagam K.   Intelligent energy control center for distributed generators using multi-agent system.   IEEE Transactions on Power Systems. 2015;30(5):2442–2449.

[2]   Siano P.   Demand response and smart grids: A survey.   Renewable and Sustainable Energy Reviews. 2014;30:461–478.

[3]   Li Q, Chen F, Chen M, *et al.* Agent-based decentralized control method for islanded microgrids. IEEE Transactions on Smart Grid. 2016;7(2):637–649.

[4]   Al-Agtash S. Electricity agents in smart grid markets. Computers in Industry. 2013;64(3):235–241.

[5]   Palicot J, Moy C, Résimont B, *et al.* Application of hierarchical and distributed cognitive architecture management for the smart grid.   Ad Hoc Networks. 2016;41:86–98.

[6]   Larsen GK, van Foreest ND, Scherpen JM. Power supply–demand balance in a smart grid: An information sharing model for a market mechanism. Applied Mathematical Modelling. 2014;38(13):3350–3360.

[7]   Yan Y, Qian Y, Hu RQ.   A secure and efficient scheme for machine-to-machine communications in smart grid.   In: Communications (ICC), 2012 IEEE International Conference on. IEEE; 2012. p. 167–172.

[8]   Ye D, Zhang M, Sutanto D.   Decentralised dispatch of distributed energy resources in smart grids via multi-agent coalition formation. Journal of Parallel and Distributed Computing. 2015;83:30–43.

[9]   Dagdougui H, Sacile R.   Decentralized control of the power flows in a network of smart microgrids modeled as a team of cooperative agents.   IEEE Transactions on Control Systems Technology. 2014;22(2):510–519.

[10] Nguyen CP, Flueck AJ. Modeling of communication latency in smart grid. In: Power and Energy Society General Meeting, 2011 IEEE. IEEE; 2011. p. 1–7.

[11] Zhang Y, Rahbari-Asr N, Chow MY. A robust distributed system incremental cost estimation algorithm for smart grid economic dispatch with communications information losses.   Journal of Network and Computer Applications. 2016;59:315–324.

[12] Wang Z, Wang L. Adaptive negotiation agent for facilitating bi-directional energy trading between smart building and utility grid. IEEE Transactions on Smart Grid. 2013;4(2):702–710.

[13] Yu T, Wang H, Zhou B, *et al.* Multi-agent correlated equilibrium Q (λ) learning for coordinated smart generation control of interconnected power grids. IEEE Transactions on Power Systems. 2015;30(4):1669–1679.

[14] Giraldo J, Mojica-Nava E, Quijano N. Synchronization of isolated micro-grids with a communication infrastructure using energy storage systems. International Journal of Electrical Power & Energy Systems. 2014;63:71–82.

[15] Saraiva FdO, Bernardes WM, Asada EN. A framework for classification of non-linear loads in smart grids using artificial neural networks and multi-agent systems. Neurocomputing. 2015;170:328–338.

[16] Misra S, Bera S, Ojha T, *et al.* ENTICE: Agent-based energy trading with incomplete information in the smart grid. Journal of Network and Computer Applications. 2015;55:202–212.

[17] Hernández L, Baladron C, Aguiar JM, *et al.* A multi-agent system architecture for smart grid management and forecasting of energy demand in virtual power plants. IEEE Communications Magazine. 2013;51(1):106–113.

[18] Mocanu E, Nguyen PH, Kling WL, *et al.* Unsupervised energy prediction in a smart grid context using reinforcement cross-building transfer learning. Energy and Buildings. 2016;116:646–655.

[19] Lakić E, Artač G, Gubina AF. Agent-based modeling of the demand-side system reserve provision. Electric Power Systems Research. 2015;124:85–91.

[20] Dusparic I, Harris C, Marinescu A, *et al.* Multi-agent residential demand response based on load forecasting. In: Technologies for Sustainability (SusTech), 2013 1st IEEE Conference on. IEEE; 2013. p. 90–96.

[21] Wen Z, O'Neill D, Maei H. Optimal demand response using device-based reinforcement learning. IEEE Transactions on Smart Grid. 2015;6(5):2312–2324.

[22] Ruelens F, Claessens BJ, Vandael S, *et al.* Residential demand response of thermostatically controlled loads using batch reinforcement learning. IEEE Transactions on Smart Grid. 2017;8(5):2149–2159.

[23] Zhang W, Xu Y, Liu W, *et al.* Distributed online optimal energy management for smart grids. IEEE Transactions on Industrial Informatics. 2015;11(3): 717–727.

[24] O'Neill D, Levorato M, Goldsmith A, *et al.* Residential demand response using reinforcement learning. In: Smart Grid Communications (SmartGridComm), 2010 First IEEE International Conference on. IEEE; 2010. p. 409–414.

[25] Golpayegani F, Dusparic I, Taylor A, *et al.* Multi-agent collaboration for conflict management in residential demand response. Computer Communications. 2016;96:63–72.

[26] Le Cadre H, Bedo JS. Dealing with uncertainty in the smart grid: A learning game approach. Computer Networks. 2016;103:15–32.

[27] Huang H, Li F, Mishra Y. Modeling dynamic demand response using Monte Carlo simulation and interval mathematics for boundary estimation. IEEE Transactions on Smart Grid. 2015;6(6):2704–2713.

[28] Kremers E, de Durana JG, Barambones O. Multi-agent modeling for the simulation of a simple smart microgrid. Energy Conversion and Management. 2013;75:643–650.

[29] Thimmapuram PR, Kim J. Consumers' price elasticity of demand modeling with economic effects on electricity markets using an agent-based model. IEEE Transactions on Smart Grid. 2013;4(1):390–397.

[30]   Mocci S, Natale N, Pilo F, *et al.* Demand side integration in LV smart grids with multi-agent control system. Electric Power Systems Research. 2015;125: 23–33.

[31]   Nunna HK, Saklani AM, Sesetti A, *et al.* Multi-agent based demand response management system for combined operation of smart microgrids. Sustainable Energy, Grids and Networks. 2016;6:25–34.

[32]   Hurtado L, Nguyen P, Kling W. Smart grid and smart building inter-operation using agent-based particle swarm optimization. Sustainable Energy, Grids and Networks. 2015;2:32–40.

[33]   Wei W, Liu F, Mei S. Energy pricing and dispatch for smart grid retailers under demand response and market price uncertainty. IEEE Transactions on Smart Grid. 2015;6(3):1364–1374.

[34]   Chai B, Chen J, Yang Z, *et al.* Demand response management with multiple utility companies: A two-level game approach. IEEE Transactions on Smart Grid. 2014;5(2):722–731.

[35]   Song L, Xiao Y, Van Der Schaar M. Demand side management in smart grids using a repeated game framework. IEEE Journal on Selected Areas in Communications. 2014;32(7):1412–1424.

[36]   Nunna HK, Doolla S. Demand response in smart distribution system with multiple microgrids. IEEE Transactions on Smart Grid. 2012;3(4):1641–1649.

[37]   O'Brien G, El Gamal A, Rajagopal R. Shapley value estimation for compensation of participants in demand response programs. IEEE Transactions on Smart Grid. 2015;6(6):2837–2844.

[38]   Babalola A, Belkacemi R, Zarrabian S. Real-time cascading failures prevention for multiple contingencies in smart grids through a multi-agent system. IEEE Transactions on Smart Grid. 2016;9(1):373–385.

[39]   Nassar ME, Salama MM. Adaptive self-adequate microgrids using dynamic boundaries. IEEE Transactions on Smart Grid. 2016;7(1):105–113.

[40]   Chen C, Wang J, Qiu F, *et al.* Resilient distribution system by microgrids formation after natural disasters. IEEE Transactions on Smart Grid. 2016;7(2):958–966.

[41]   Rahman M, Mahmud M, Pota H, *et al.* A multi-agent approach for enhancing transient stability of smart grids. International Journal of Electrical Power & Energy Systems. 2015;67:488–500.

[42]   Xi L, Zhang Z, Yang B, *et al.* Wolf pack hunting strategy for automatic generation control of an islanding smart distribution network. Energy Conversion and Management. 2016;122:10–24.

[43]   Elmitwally A, Elsaid M, Elgamal M, *et al.* A fuzzy-multiagent self-healing scheme for a distribution system with distributed generations. IEEE Transactions on Power Systems. 2015;30(5):2612–2622.

[44]   Teng F, Sun Q, Xie X, *et al.* A disaster-triggered life-support load restoration framework based on multi-agent consensus system. Neurocomputing. 2015;170:339–352.

[45]   Nguyen CP, Flueck AJ. A novel agent-based distributed power flow solver for smart grids. IEEE transactions on Smart Grid. 2015;6(3):1261–1270.

[46]   Eriksson M, Armendariz M, Vasilenko OO, *et al.* Multiagent-based distribution automation solution for self-healing grids. IEEE Transactions on Industrial Electronics. 2015;62(4):2620–2628.

[47]   Colson CM, Nehrir MH. Comprehensive real-time microgrid power management and control with distributed agents. IEEE Transactions on Smart Grid. 2013;4(1):617–627.

[48]   Hu J, Morais H, Lind M, *et al.* Multi-agent based modeling for electric vehicle integration in a distribution network operation. Electric Power Systems Research. 2016;136:341–351.

[49]   Chao HL, Hsiung PA. A fair energy resource allocation strategy for micro grid. Microprocessors and Microsystems. 2016;42:235–244.

[50]   Rahman M, Mahmud M, Oo A, *et al.* Agent-based reactive power management of power distribution networks with distributed energy generation. Energy Conversion and Management. 2016;120:120–134.

[51]   Radhakrishnan BM, Srinivasan D. A multi-agent based distributed energy management scheme for smart grid applications. Energy. 2016;103: 192–204.

[52]   Li Y, Yong T, Cao J, *et al.* A consensus control strategy for dynamic power system look-ahead scheduling. Neurocomputing. 2015;168:1085–1093.

[53]   Guo F, Wen C, Mao J, *et al.* Distributed economic dispatch for smart grids with random wind power. IEEE Transactions on Smart Grid. 2016;7(3): 1572–1583.

[54]   Kahrobaee S, Rajabzadeh RA, Soh LK, *et al.* A multiagent modeling and investigation of smart homes with power generation, storage, and trading features. IEEE Transactions on Smart Grid. 2013;4(2):659–668.

[55]   Samadi P, Mohsenian-Rad H, Wong VW, *et al.* Tackling the load uncertainty challenges for energy consumption scheduling in smart grid. IEEE Transactions on Smart Grid. 2013;4(2):1007–1016.

[56]   Gregoratti D, Matamoros J. Distributed energy trading: The multiple-microgrid case. IEEE Transactions on Industrial Electronics. 2015;62(4):2551–2559.

[57]   Bu S, Yu FR. Green cognitive mobile networks with small cells for multimedia communications in the smart grid environment. IEEE Transactions on Vehicular Technology. 2014;63(5):2115–2126.

[58]   Wang H, Huang T, Liao X, *et al.* Reinforcement learning in energy trading game among smart microgrids. IEEE Transactions on Industrial Electronics. 2016;63(8):5109–5119.

[59]   Samadi P, Wong VW, Schober R. Load scheduling and power trading in systems with high penetration of renewable energy resources. IEEE Transactions on Smart Grid. 2016;7(4):1802–1812.

[60]   Sheikhi A, Rayati M, Ranjbar A. Dynamic load management for a residential customer: Reinforcement learning approach. Sustainable Cities and Society. 2016;24:42–51.

[61]   Ghorbani MJ, Choudhry MA, Feliachi A. A multiagent design for power distribution systems automation. IEEE Transactions on Smart Grid. 2016;7(1): 329–339.

[62] Venavagamoorthy GK, Sharma RK, Gautam PK, *et al.* Dynamic energy management system for a smart microgrid. IEEE Transactions on Neural Networks and Learning Systems. 2016;27(8):1643–1656.

[63] Li D, Jayaweera SK. Reinforcement learning aided smart-home decision-making in an interactive smart grid. In: Green Energy and Systems Conference (IGESC), 2014 IEEE. IEEE; 2014. p. 1–6.

[64] Rayati M, Sheikhi A, Ranjbar AM. Applying reinforcement learning method to optimize an Energy Hub operation in the smart grid. In: Innovative Smart Grid Technologies Conference (ISGT), 2015 IEEE Power & Energy Society. IEEE; 2015. p. 1–5.

[65] Kim BG, Zhang Y, van der Schaar M, *et al.* Dynamic pricing and energy consumption scheduling with reinforcement learning. IEEE Transactions on Smart Grid. 2016;7(5):2187–2198.

[66] Wang X, Zhang M, Ren F, *et al.* GongBroker: A broker model for power trading in smart grid markets. In: Web Intelligence and Intelligent Agent Technology (WI-IAT), 2015 IEEE/WIC/ACM International Conference on. vol. 2. IEEE; 2015. p. 21–24.

[67] Lim Y, Kim HM. Strategic bidding using reinforcement learning for load shedding in microgrids. Computers & Electrical Engineering. 2014;40(5): 1439–1446.

[68] Zhang Y, van der Schaar M. Structure-aware stochastic load management in smart grids. In: INFOCOM, 2014 Proceedings IEEE. IEEE; 2014. p. 2643–2651.

[69] Liao H, Wu Q, Jiang L. Multi-objective optimization by reinforcement learning for power system dispatch and voltage stability. In: Innovative Smart Grid Technologies Conference Europe (ISGT Europe), 2010 IEEE PES. IEEE; 2010. p. 1–8.

[70] Shirzeh H, Naghdy F, Ciufo P, *et al.* Balancing energy in the smart grid using distributed value function (DVF). IEEE Transactions on Smart Grid. 2015;6(2):808–818.

[71] Motevasel M, Seifi AR. Expert energy management of a micro-grid considering wind energy uncertainty. Energy Conversion and Management. 2014;83:58–72.

[72] Li FD, Wu M, He Y, *et al.* Optimal control in microgrid using multi-agent reinforcement learning. ISA Transactions. 2012;51(6):743–751.

[73] Salehizadeh MR, Soltaniyan S. Application of fuzzy Q-learning for electricity market modeling by considering renewable power penetration. Renewable and Sustainable Energy Reviews. 2016;56:1172–1181.

[74] de Montigny M, Heniche A, Kamwa I, *et al.* Multiagent stochastic simulation of minute-to-minute grid operations and control to integrate wind generation under AC power flow constraints. IEEE Transactions on Sustainable Energy. 2013;4(3):619–629.

[75] Daneshfar F, Bevrani H. Load-frequency control: A GA-based multi-agent reinforcement learning. IET Generation, Transmission & Distribution. 2010;4(1):13–26.

[76] Wei Q, Liu D, Shi G. A novel dual iterative *Q*-learning method for optimal battery management in smart residential environments. IEEE Transactions on Industrial Electronics. 2015;62(4):2509–2518.

[77] Qiu X, Nguyen TA, Crow ML. Heterogeneous energy storage optimization for microgrids. IEEE Transactions on Smart Grid. 2016;7(3):1453–1461.

[78] Wang Y, Lin X, Pedram M. A near-optimal model-based control algorithm for households equipped with residential photovoltaic power generation and energy storage systems. IEEE Transactions on Sustainable Energy. 2016;7(1):77–86.

[79] Kuznetsova E, Li YF, Ruiz C, *et al.* Reinforcement learning for microgrid energy management. Energy. 2013;59:133–146.

[80] Vandael S, Claessens B, Ernst D, *et al.* Reinforcement learning of heuristic EV fleet charging in a day-ahead electricity market. IEEE Transactions on Smart Grid. 2015;6(4):1795–1805.

[81] Guan C, Wang Y, Lin X, *et al.* Reinforcement learning-based control of residential energy storage systems for electric bill minimization. In: Consumer Communications and Networking Conference (CCNC), 2015 12th Annual IEEE. IEEE; 2015. p. 637–642.

[82] Landi M, Gross G. Measurement techniques for online battery state of health estimation in vehicle-to-grid applications. IEEE Transactions on Instrumentation and Measurement. 2014;63(5):1224–1234.

[83] Zhang X, Flueck AJ, Nguyen CP. Agent-based distributed volt/var control with distributed power flow solver in smart grid. IEEE Transactions on Smart Grid. 2016;7(2):600–607.

[84] Sharma DD, Singh S, Lin J. Multi-agent based distributed control of distributed energy storages using load data. Journal of Energy Storage. 2016;5:134–145.

[85] Sharma A, Srivastava SC, Chakrabarti S. Multi-agent-based dynamic state estimator for multi-area power system. IET Generation, Transmission & Distribution. 2016;10(1):131–141.

[86] Teleke S, Baran ME, Bhattacharya S, *et al.* Rule-based control of battery energy storage for dispatching intermittent renewable sources. IEEE Transactions on Sustainable Energy. 2010;1(3):117–124.

[87] Pahwa A, DeLoach SA, Natarajan B, *et al.* Goal-based holonic multiagent system for operation of power distribution systems. IEEE Transactions on Smart Grid. 2015;6(5):2510–2518.

[88] Klaimi J, Merghem-Boulahia L, Rahim-Amoud R, *et al.* An energy management approach for smart-grids using intelligent storage systems. In: Digital Information and Communication Technology and its Applications (DICTAP), 2015 Fifth International Conference on. IEEE; 2015. p. 26–31.

[89] Vaccaro A, Loia V, Formato G, *et al.* A self-organizing architecture for decentralized smart microgrids synchronization, control, and monitoring. IEEE Transactions on Industrial Informatics. 2015;11(1):289–298.

[90] Hu J, Saleem A, You S, *et al.* A multi-agent system for distribution grid congestion management with electric vehicles. Engineering Applications of Artificial Intelligence. 2015;38:45–58.

[91]   Vallejo D, Albusac J, Glez-Morcillo C, *et al.* A multi-agent approach to intelligent monitoring in smart grids. International Journal of Systems Science. 2014;45(4):756–777.

[92]   Xi L, Yu T, Yang B, *et al.* A novel multi-agent decentralized win or learn fast policy hill-climbing with eligibility trace algorithm for smart generation control of interconnected complex power grids. Energy Conversion and Management. 2015;103:82–93.

[93]   de Durana JMG, Barambones O, Kremers E, *et al.* Agent based modeling of energy networks. Energy Conversion and Management. 2014;82:308–319.

[94]   Evora J, Hernandez JJ, Hernandez M, *et al.* Swarm intelligence for frequency management in smart grids. Informatica. 2015;26(3):419–434.

[95]   Evora J, Hernandez JJ, Hernandez M. A MOPSO method for direct load control in smart grid. Expert Systems with Applications. 2015;42(21):7456–7465.

[96]   Cheng X, Cao R, Yang L.   Relay-aided amplify-and-forward powerline communications. IEEE Transactions on Smart Grid. 2013;4(1):265–272.

Chapter 6

Shortest path models for scale-free network topologies: literature review and cross comparisons

Agnese V. Ventrella[1,2], Giuseppe Piro[1,2], and Luigi Alfredo Grieco[1,2]

The term *Internet* refers to the global network infrastructure, connecting more than 15 billions of devices around the world. At the time of this writing, it supports a massive distribution of information, which reaches around 1.5 ZB per year. These estimates, however, are continuously growing: by 2021, the annual global traffic will grow to 3.3 ZB per year [1,2]. Therefore, the Internet appears as a complex system that is continuously evolving during the time.

The knowledge about the Internet topology has always been considered an important aspect for researchers, industries, and service providers. It, in fact, is extremely useful to evaluate network resilience [3], analyze topological properties and their evolution [4], predict and improve the performance of communication protocols and the effectiveness of routing algorithms [5], solve specific problems involving a particular topological structure (i.e., how to distribute storage across routers in order to obtain an optimal caching allocation) [6], and so on. Thus, analytical models showing Internet characteristics (like average shortest path and shortest path distribution) and simulation tools able to reproduce Internet-like topologies are key instruments for most of the research activities in this context.

Nevertheless, the complexity and the dynamism of the overall Internet architecture make the study of the Internet topology as one of the hottest and hardest research topic to solve [7]. First of all, it is important to have a clear definition of *topology*. According to the Open System Interconnection (OSI) model, a topology represents a simplified way to depict the interconnections among communication entities [8]. But, what communication entities refers to, is not completely clear a priori. The scientific literature, for instance, considers three main levels of granularity, namely *interface level*, *router level*, and *Autonomous System (AS)-level* [9–11]. Thus, a network topology may expose different information according to the level of granularity taken into account. Second, differently from other large networks, like public switched telephone

[1]Department of Electrical and Information Engineering (DEI), Politecnico di Bari, Bari, Italy
[2]CNIT, Consorzio Nazionale Interuniversitario per le Telecomunicazioni, Italy

network, the Internet did not grow according to a topological design developed by some central authority or administration [12]. Hence, huge dimension, rapid change, and lack of publicly available information inevitably make hard to capture a complete snapshot of the overall network infrastructure [13].

To solve this issue, several methodologies were introduced to infer topology information, based on both active and passive approaches. These mechanisms must be properly configured and adapted when applied to interface router and AS levels of granularity. At the same time, however, it is also important to consider the set of limitations they introduce, thus being able to better estimate the level of accuracy of retrieved data [10,14–17].

Starting from inferred data, it is possible to formulate mathematical models able to capture statistical characteristics of the Internet. Graph theory is widely used to reach this goal [18]. In fact, many models were already developed, which refer to regular, random, small world, and the most recent power-law and scale-free graphs [11,19–22]. Among them, however, the scale-free graph is widely accepted as the best model able to represent Internet-like topologies. A number of network simulators already implement these models and are able to reproduce Internet-like topologies that can be used in a variety of research activities.

Another important step forward in the study of the Internet topology is the modeling of the shortest path connecting any peers attached to the communication systems. The scientific literature already provides models for both average shortest path and distribution of the shortest path length [23–26].

Based on these premises, the present book chapter aims at providing an overview of Internet-like topologies, by covering a broad set of aspects, including the level of granularity, methodologies useful to retrieve topology information, simulation tools, and analytical models. Then, the accuracy of reference models for the distribution of the shortest path length (i.e., Gamma, Lognormal, and Weibull distributions) is evaluated through a massive simulation campaign, carried out by using the Boston university Representative Internet Topology gEnerator (BRITE) tool [27]. From one side, obtained results demonstrate that the available models are able to catch the average value and the distribution of the shortest path over a very broad set of conditions. But, from another side, they also highlight an unresolved issue: they require a case-by-case tuning of model parameters.

The rest of this chapter is organized as in the following. Section 6.1 presents the main levels of granularity of the Internet topology and reviews active and passive methodologies useful to collect data. Section 6.2 discusses Internet topology models based on the graph theory and provides an overview of topology generator tools. Section 6.3 investigates, through computer simulations, the accuracy of analytical models developed for scale-free networks and identifies useful applications of the shortest path distribution. Finally, Section 6.4 draws the conclusions.

6.1   Mapping the Internet topology

The scientific literature generally describes the Internet topology through different levels of granularity. In all the cases, however, the graph theory is deeply adopted as a key instrument that captures well the required details of the overall network architecture [10,28]. Such a consideration is also valid for Internet-like topologies, like restricted portion of the Internet handled by a single Internet Service Provider (ISP). In fact, at the time of this writing, it is common to represent the Internet topology as an undirected graph, G. More specifically, this graph is further characterized by the ordered pair $G = (N, E)$, where N refers to a set of vertices (also called nodes or points), connected by a set of E edges (also called arcs or lines) [29–31]. Without loss of generality, it is possible to assume that devices belonging to the global network infrastructure establish a bidirectional relationship. Therefore, the graph is considered undirected because edges do not have any orientation.

The roles covered by both nodes and edges belonging to an Internet-like topology strictly depends on the level of granularity selected to model the network itself. Conventional approaches include *interface level*, *router level*, and *Autonomous System (AS) level* [9–11] (see the preliminary overview depicted in Figure 6.1). It is important to note that details about network topology, routing policies, peering relationships, and resilience are commercially sensitive, could expose potential vulnerability to attackers, and reveal resilience planning. Accordingly, they are not publicly available. At the same time, the network is dynamic and constantly evolving because of failures, maintenance, and upgrades. For these reasons, information regarding both global structure and local properties of the Internet cannot be retrieved in an easy way. Nevertheless, dedicated approaches can be used to partially solve this problem. They can be divided into two kinds of methodologies, namely, *passive* and *active* [32]. The passive method learns the presence of nodes and their interactions by simply collecting the information flowing over a wire and generated by other communication protocols (which work for different purposes). The active method, instead, supposes to send dedicated packets (i.e., probe messages) to target devices into the network and to collect the related responses.

The following paragraphs describe the three levels of granularity introduced above. At the same time, they also present the most important passive and active methodologies used to retrieve and study Internet or Internet-like network topologies. For each single strategy, pros and cons are evaluated too (see the summary reported in Table 6.1). Finally, they also provide an overview regarding geographic network topologies.

6.1.1 Interface level

The current Internet is based on the host-centric communication paradigm, and the data exchange is handled through the well-known Transmission Control Protocol (TCP)/Internet Protocol (IP) stack. In particular, the IP protocol implements networking functionalities. Among its other specifications, it identifies any network interface of hosts, servers, and routers, through an IP address. Today, for instance, two versions of the IP protocol can be used: IPv4 and IPv6. The former uses a 32-bit address scheme. The latter adopts a 128-bit address scheme [33].

The interface level of granularity depicts the Internet topology by paying attention to network interfaces having an IP address, as well as to their peer-to-peer connections (see Figure 6.1(a)). In this way, a node of the graph describing the considered

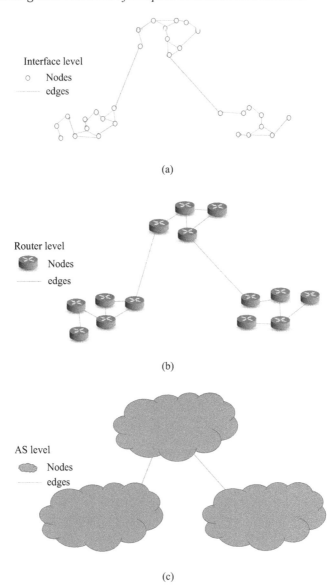

Figure 6.1   Internet topology at three main levels of granularity: (a) Interface level, (b) Router level, and (c) AS level

topology maps a given network interface and edges refer to direct connections between nodes [10]. Routers with multiple configured network interfaces are mapped to multiple logical nodes. Thus, the resulting interface-level topology embraces a number of nodes equal to the number of active network interfaces with an IP address and a number of edges equal to the amount of direct connections established at the network layer.

Table 6.1   Methodologies used to retrieve and study Internet-like topologies and their related issues

Level of granularity	Learning technique	Related issues
Interface level	Active: Traceroute	Absence of ICMP enabled routers; presence of load-balancing strategies
	Active: IP options	Absence of routers supporting IP options
	Active: Subnet discovery	Possibility to have incomplete data
Router level	Active: Alias resolution	Possibility to have incomplete data
	Passive: Internet Routing Registry	Limited access to the database; absence of routers supporting IPv4 multicast
AS level	Passive: BGP	Limited capabilities of monitors
	Passive: Internet Routing Registry	Stale or incomplete data
	Active: Traceroute	Absence of ICMP enabled routers; presence of load balancing strategies

Interface-level topologies are generally learned through active methodologies, based on the *traceroute* tool, the usage of *IP options*, and *subnet discovery*.

6.1.1.1   Active methodology based on *traceroute*

At the time of this writing, *traceroute* is one of the most popular tool adopted to acquire topology details [34]. It was originally written to detect communication problems present within a network, such as routing loops and black holes, as well as to locate where those failures occur. Subsequently, it has been used to pursue other purposes, including the active discovery of Internet-like topologies. This tool is available for most of the operating systems, including Apple macOS, Unix systems, and Microsoft Windows. In the latter case, however, it is generally known by a different name, that is, *tracert*. As default, *traceroute* works with IPv4. But an updated version for IPv6 is also available: *traceroute6* for Apple macOS and Unix systems; *tracert6* for Microsoft Windows [33].

From a technical point of view, *traceroute* relies on the Internet Control Message Protocol (ICMP) [35,36], which represents a messaging protocol working alongside the IP protocol and offering the support for routing operations, network diagnostic, and error notification. With *traceroute*, ICMP is used to calculate the forwarding path and the communication delay between a source node (i.e., who runs the tool) and a target network interface (i.e., an interface belonging to the studied network). To this end, a train of messages is delivered through the User Datagram Protocol (UDP) with a variable value of the time-to-live (TTL) field of the IP header. To ease the comprehension to a broader set of readers, it is important to point out that the TTL field of the IP header is used to limit the lifetime of an IP datagram within the network. For example, if the TTL value is equal to x, it means that the corresponding

IP datagram can pass, at most, through *x* consecutive routers before being discarded. This is because, every intermediate router decrements the TTL value by 1 unit before triggering the forwarding process. Therefore, as soon as the TTL value reaches the value 0, the corresponding IP datagram is no more forwarded toward the destination interface, but an ICMP Time Exceeded message is sent back to the source node for notification purposes.

Starting from these premises, *traceroute* works as follows. At the beginning, the device that runs the tool issues a group of ICMP messages, whose TTL value is set to 1. Note that more than one message is sent at each step because the procedure intends to collect statistical information related to communication delays (such as minimum, maximum, and average value of the round trip time, generally expressed in milliseconds). These initial packets reach only the node directly connected to the sender, before being discarded. The ICMP Time Exceeded messages generated by this node are used by the sender to infer details about the first network interface of the forwarding path. Then, a new set of ICMP messages is sent with a TTL value set to 2. In line with the process described above, the sender can now learn information about the second hop of the forwarding path toward the destination. This process is repeated until the destination node is reached. At the end, the sender collects some details of the network topology, on a hop-by-hop basis [11].

It is important to remark that two main limitations affect *traceroute* [14]. First, if some routers do not implement ICMP, the acquired forwarding path will not consider some of the intermediate network interfaces. Second, in the event that a intermediate router implements a load-balancing strategy, *traceroute* will generate results referring to multiple paths through which packets are sent. Thus, the acquired forwarding path will include additional network interfaces and the learned network topology could not exactly capture the reality.

6.1.1.2   IP options and subnet discovery

Two further measurement techniques exploit IP packet options field and subnet discovery [10].

The options available within the IP header could be useful to support additional functionalities, such as the packet routing toward a path that is different from the usual one or the registration of specific information related to the network topology. For instance, the *source routing* option allows the discovery of new paths. When it is enabled, in fact, the sender can choose at most nine routers that the packet is supposed to go through before reaching the destination. Additionally, the *record route* option can be used to allow routers involved in the forwarding process to store their IP addresses within a dedicated list available in another option field of the IP header. These information are used by the destination node to learn a multi-hop path connecting it to the sender. A drawback is that these options are not supported by all routers.

The subnet discovery technique is based on the subnetting concept. A subnet is a layer 2 subdivision of an IP network, where all the devices are addressed with a common most significant bit-group (e.g., IP prefix). This technique exploits the IP prefix to detect the subnet boundaries and reveals the pingable IP addresses available in the subnet. Achieved results are then used to build the network topology.

This technique can suffer from incomplete data because of relationship policies and routing preferences that make the packet observe only some paths and missing other ones.

6.1.2 Router level

Differently from the interface level of granularity, when the router-level approach is chosen, each router is mapped to only one node of the graph, without considering the possibility that more network interfaces with different IP address can coexist in the same device (see Figure 6.1(b)). In other words, a node is viewed as an aggregation of network interfaces that belong to a single device. Therefore, the router level of granularity describes how routers are connected to each other within an Internet-like topology. In the resulting graph, nodes represent routers and edges indicate networking connectivity among them [10,11].

The details about router-level topologies could be achieved by means of *alias resolution* and *recursive router discovery* techniques.

6.1.2.1 Alias resolution techniques

Alias resolution is an active method, still based on the *traceroute* tool. While *traceroute* is used to infer the forwarding path on a hop-by-hop basis (as previously discussed), additional methods are implemented for properly mapping network interfaces to the right nodes of the topology.

One possibility is based on the fingerprint technique [17]. Here, a device interested to build a router-level representation of a network focuses the attention on a remote network interface, having a known IP address. Such a device issues fake UDP or TCP packets to that IP address, by setting the destination port address to unused values. As expected, the remote network interface replies with an ICMP Port Unreachable error message. In the case that the received error message contains an IP source address that is different from the contacted one, the device performing the alias resolution technique recognizes that these IP addresses refer to network interfaces configured in the same router. Therefore, according to the router level of granularity, these interfaces will be aggregated into the resulting topology representation.

Another approach is referred to as IP-identification fingerprint method. During this learning procedure, a device identifies two potential aliases of IP addresses. Then, it sends to both interfaces a UDP probe packet, by setting the destination port number to a high (and unused) value. In both cases, the two destination interfaces reply with an ICMP Port Unreachable error message. The device collects the IDs of the received messages, that are, for instance, id_1 and id_2. A third fake packet is sent to the destination IP address from which it has received the first ICMP Port Unreachable error message. Again, the remote destination interface answers with a new ICMP Port Unreachable error message with ID equal to id_3. Now, in the case that $id_1 < id_2 < id_3$ and the difference $id_3 - id_1$ is below a given threshold, the IP-identification fingerprint method assumes that the contacted IP addresses are aliased. Thus, their related network interfaces are aggregated in the resulting topology representation [37].

Finally, analytical techniques could also be introduced for further solving alias resolution. Here, the common IP address assignment scheme to infer IP aliases, that belong to two opposite paths, is used. After having identified the subnets, aliases are inferred by analyzing path segments [11].

It is possible to conclude that alias resolution techniques are generally considered accurate. But, sometimes retrieved data could be incomplete. The reason is that *traceroute* can fail when nodes are disconnected, turned off, or configured to not respond to probe packets [17].

6.1.2.2 Recursive router discovery

Recursive router discovery is a passive method that exploits the capability of routers to be queried in order to retrieve information about their neighbors.

In Local Area Networks (LANs), the Simple Network Management Protocol (SNMP) is frequently used for handling network monitoring and for storing collected data in a dedicated Management Information Base (MIB). In particular, SNMP-enabled routers store the list of neighbor interfaces within the ipRoute table of the MIB. Information collected through SNMP could be useful to build a router-level description of an Internet-like topology. From one side, this approach may provide accurate data. But, from another side, its usage is highly restricted because the MIB is accessible only by network administrators [10].

Another solution implementing the recursive router discovery approach is MRINFO, which is based on the Internet Group Management Protocol (IGMP). IGMP was initially standardized to allow hosts and adjacent routers to establish multicast group memberships. With MRINFO, an IGMP Ask Neighbors message is issued in order to receive the list of all the router's interfaces and their related neighbors. The answer is reported in the IGMP Neighbors Reply message [16]. Unfortunately, this technique can only be used with IPv4 multicast enabled routers.

6.1.3 AS level

Before introducing the latest level of granularity useful to describe Internet-like topologies, it is important to remark that the global network infrastructure appears as a connection of several autonomous systems (ASs). Each AS is made up by a group of routers deployed by one or more network operators, on behalf of a single administrative entity [38]. For instance, an AS can refer to the network of a large company, a university, a network service provider, and so on. Typically, individual users, small enterprise networks, and ASs located at the edge of the Internet can join the global network through other ASs, namely, ISP. In turn, ISPs may obtain the same service from one or more upstream ISPs. Each AS is uniquely identified by an AS number (ASN). Originally, it was defined as a 16-bit integer (by admitting a maximum of 65.536 assignments). Then a 32-bit ASN has been introduced in order to uniquely identify a higher number of ASs [39]. In addition, ASs are divided into two categories: transit and stub. A transit AS is part of the core network and usually carries traffic between isolated domains, managed by different administrative entities. A stub AS, instead, provides Internet connectivity to end users. Thus, from one side, it is connected to end users. From another side, it is connected to the rest of the Internet through one or more transit ASs. Sometimes, the administrator of a given AS can change its own traffic relationship with other providers, thus modifying the overall network architecture and making the resulting topology constantly evolving.

The AS level of granularity, also known as inter-domain description, depicts the Internet architecture as a group of interconnected ASs. Accordingly, it brings to an undirected graph where each node identifies one AS and edges represent the logical peering relationship between two adjacent ASs (see Figure 6.1(c)). Despite its coarse level of details, the AS level of granularity is frequently leveraged to study, control, optimize, and implement inter-domain routing, mechanisms for the provisioning of the quality of service, and customer-provider and peering relationships between ISPs.

Also in this case, both passive and active mechanisms can be used to infer information related to the AS level topology. The first mechanism basically collects data generated by the Border Gateway Protocol (BGP) [40] or provided by the Internet Routing Registry [41]. The second one investigates forwarding paths through *traceroute*.

6.1.3.1   Passive methodology based on BGP and Internet Routing Registry

BGP is the current de-facto standard for inter-domain routing. It allows the exchange of routing information between ASs without revealing detailed and internal information about their own networks. In particular, a BGP-compliant router obtains information about existing routes from its BGP neighbors. The obtained routes are processed and shared with to other BGP-compliant router according to specific routing policies. Routes selection generally preserves system scalability. At the end of the process, if the protocol converges, a stable routing solution is found. But, identified routes are generally far from the shortest path. BGP data are gathered by route monitors or collectors, that are specific devices deployed around the globe by some international projects such as University of Oregon's Route-Views [42] or RIPE Routing Information Service [43]. These projects were originally used by ISPs to debug and optimize their networks [44].

Anyway, it is important to note that BGP cannot provide a complete view of ASs because significant information could be missed [8]. This is due to the following motivations: (1) monitors can only see what the connected routers choose to send; (2) monitors are not present in each location; (3) the location of these monitors is not randomly distributed across the Internet; (4) the connections between BGP monitors and routers are not completely reliable because of session resets, collector down time, and missing updates [15].

Another passive method to infer an AS level topology refers to the look up of the Routing Assets Database provided by the Internet Routing Registry, stored in dedicated File Transfer Protocol servers. This database includes information about routing policies, regulation, and peering provided by the ASs themselves. Specifically, the *whois* command can be used to retrieve these information [11,15]. Some limitations that characterize this approach are due to the fact that the stored information can be stale or incomplete.

6.1.3.2   Active methodology based on *traceroute*

Also for the AS level of granularity, *traceroute* can be used to retrieve details about forwarding paths. Data retrieved through *traceroute*, however, must be further processed in order to map network interfaces to the corresponding ASs. In fact, consecutive IP addresses that belong to two different adjacent ASs reveal the connectivity between ASs. The issues associated to *traceroute* and discussed in the previous paragraphs are valid also in this context.

6.1.4   *Geographic network topologies*

All the methodologies discussed above do not provide any reference to the physical location of nodes on the map. Without any doubt, such information can be added to the network topologies, obtained through any level of granularity, in order to increase their usability. For instance, geographical information allows one to (1) simplify the network troubleshooting and the detection of attacks and congestions, (2) guarantee resilience of interconnections in case of disaster scenario, (3) provide location information to Internet services that require them, and (4) provide a visual representation of the Internet [11,45].

For sure, the definition of a geographic network topology depends on the selected level of granularity. Network interfaces and routers can be immediately mapped to a precise location on a map, i.e., to a pair of coordinates. Instead, nodes belonging to the AS level topology do not refer to a single entity and to a unique location, because an AS gathers routers under a common administrator. Therefore, when a geographical information is assigned to an AS, it is just used to coarsely identify the geographic region covered by the AS.

To achieve this further level of detail, active or passive measurement methods can be exploited. Active IP geolocation techniques are typically based on delay measurements that offer good levels of accuracy. But their drawbacks include scalability, high measurement overhead, and very high response time. Passive approaches, such as database-driven geolocation, are faster. They usually consist of a database-engine, e.g., Structured Query Language (SQL)/MySQL, containing records for a range of IP addresses. Nevertheless, this database is difficult to manage and update, and its accuracy is not so high because of the lack of information about the operations used to build it [46].

6.2   Internet models based on the graph theory

Nowadays all the techniques described in the previous section are continuously used to experimentally collect useful data related to Internet or Internet-like networks, as well as to study their characteristics. Some results are publicly available. For instance, reference datasets are provided by Centre for Applied Internet Data Analysis (CAIDA) [47] and Internet Topology Zoo [48]. Starting from these data, it is possible to formulate mathematical models based on graph theory that capture the main facets of Internet-like topologies (from one side) and allow users to reproduce them through computer simulations (from another side).

The following paragraphs introduce fundamental notions inherited from graph theory that are at the basis of the aforementioned models. Then, they present the most important analytical models describing the Internet or restricted portion of it. Finally, they provide an overview of topology generators that are able to reproduce Internet-like topologies based on the aforementioned models.

6.2.1 Fundamental notions from the graph theory

As already anticipated in the Introduction, the Internet topology can be described through an undirected graph, $G = (N, E)$, where N refers to a set of vertices connected by a set of E edges. Nevertheless, such a graph can be further characterized by additional parameters that include [29,49] the following:

- *Node degree, k*: It represents the number of edges incident to a vertex. This parameter allows one to capture the connectivity characteristics of the topology. In particular, networks with higher k register an average better connection that results in a higher robustness to failures.
- *Shortest path, $d(u, v)$*: It identifies the shortest distance between two vertices u and v. More details about shortest path characteristics will be provided in the following sections.
- *Diameter, δ*: It defines the longest shortest path between any node pair u and v, and it is expressed as

$$\delta = \sup_{(u,v)} \ d(u, v) \tag{6.1}$$

- *Clustering coefficient*: It is a measure of the degree to which nodes in a graph tend to cluster together. Two measurements of clustering coefficient can be considered: global and local. Moreover, two definitions of global clustering coefficient are possible. The first one is based on triplets of nodes: three connected nodes form a triplet and three triplets form a triangle. Therefore, the global clustering coefficient is the number of closed triplets over the total number of triplets (both open and closed). The alternative definition evaluates the global clustering coefficient as the mean of the local clustering coefficients related to all the vertices. The local clustering coefficient is defined as the ratio between the number of existing edges within the neighborhood of a vertex and the number of possible edges within the neighborhood of the same vertex. The higher the local clustering of a node, the more interconnected are its neighbors.
- *Betweenness*: It is a measure of the node centrality, calculated as the fraction of shortest paths between node pairs that pass through the node of interest. It is inversely related to the robustness of the graph when a node is removed. In fact, the higher the number of paths that pass through a node, the higher is the damage that will be done when that node is removed. Moreover, it can provide a measure of the traffic load that a node must handle or the influence that an individual node has in the spread of information within the network.
- *Spectrum*: It represents the set of eigenvalues of the adjacency matrix of the graph. It allows a user to measure the overall characteristics of the network and its robustness.

6.2.2 Topology models

The most important Internet topology models proposed in the literature include *regular* and *well-known* topologies, *random* and *small-world* topologies, *power-law* and *scale-free* topologies, and *hierarchical* topologies.

6.2.2.1 Regular and well-known topology models

According to [11], *regular* and *well-known* topologies represent the simplest model used to describe a network. It cannot be applied to the overall Internet architecture. But it only serves to investigate a restricted part of a network (as described below).

The term *regular topology* refers to elementary network architectures, including mesh, rings, trees, stars, and lattice. Therefore, resulting models only support the simulation of a very limited portion of the Internet network (like a LAN) or other basic infrastructures. Thanks to the simplicity of these topologies, they do not require complex generator tools.

The term *well-known topology* refers to specific real networks, such as GÉANT or National Science Foundation Network (NSFnet) backbone. In particular, GÉANT interconnects the European National Research and Educational Network and provides research data communication across the continent [50]. NSFnet backbone interconnected six supercomputer sites, several regional networks, and ARPANET [51].

6.2.2.2 Random and small-world topology model

Looking at the overall Internet, it was originally described through the random graph theory developed by Erdős and Rènyi [19]. Let N, E, and k be the number of nodes, the number of edges, and the average degree of a given network. The random topology model assumes that an average degree k, is equal to:

$$k = \frac{2E}{N},$$

(6.2)

It is kept constant and it is assumed that every pair of nodes is connected with a probability p equal to

$$p = \frac{k}{N}.$$

(6.3)

The resulting model generates network topologies having a small average shortest path length and a small clustering coefficient. Moreover, it does not capture all the characteristics of a real Internet-like topology, such as the presence of hubs.

The limitations of Erdős and Rènyi model were overcome by Watts and Strogatz model [20]. It still allows the generation of network models with a small average shortest path length. But, differently from the previous one, it registers a large clustering coefficient. Among its limitations, the Watts and Strogatz model generates network topologies with an unrealistic degree distribution. It is possible to generate a Watts–Strogatz network by starting from a regular ring lattice with N nodes. Each node is connected to the same number of $2m$ nearest neighbors. Then, each edge has to be removed according to a uniform and independent probability p. This edge has to be rewired in order to connect a pair of nodes uniformly and randomly chosen.

There is also a the Newman–Watts variant of the Watts–Strogatz network that does not include the removal of the edges from the underlying lattice in the building process. In this model, edges are only added between pairs of nodes in the same way as in a Watts–Strogatz network [52].

6.2.2.3   Power-law topology models

The work presented in [21] demonstrated for the first time a new set of properties of the Internet. Specifically, the work considered three different snapshots of the Internet referring to an AS level representation of 1997, an AS level representation of 1998, and a router level representation of 1995. All the topology details were inferred from RouteViews BGP tables. The conducted study identified three specific power laws:

- *Rank exponent*: the out-degree of a node is proportional to its rank to the power of a constant. Let k_v and r_v be the out-degree and the rank of a node v, then the following relation exists: $k_v \propto r_v^R$. The exponent R is obtained by performing a linear regression on k_v, by plotting a log–log graph.
- *Out-degree exponent*: the frequency of an out-degree is proportional to the out-degree to the power of a constant. Let $f(k)$ be the fraction of nodes with degree k, then $f(k) \propto k^O$. The exponent O is obtained by performing a linear regression on $f(k)$ when plotted on a log–log graph.
- *Eigen-exponent*: the eigenvalues λ_i of the adjacency graph, sorted in a decreasing order, are proportional to the order i to the power of a constant, according to the relation $\lambda_i \propto i^E$. The exponent E is obtained by performing a linear regression on λ_i when plotted on a log–log graph.

Moreover, [21] studied the neighborhood size within some distances. Also in this case, the relation follows a power law, but it was considered an approximation because of the small number of samples. In particular, let $P(h)$ be the total number of pairs of nodes within h hops. $P(h)$ is proportional to the number of hops to the power of a constant H, according to the relation $P(h) \propto c^H$ when $h \ll \delta$, where δ is the diameter of the network and $c = N + 2E$.

After [21], several researchers supported these findings and tried to further understand the origin of the power law [53,54]. A very important contribution was provided by the Barabasi–Albert model [22]. At the same time, the literature also proposes opposing theories. For instance, Chen *et al.* [55] argued that an AS level topology does not include all the Internet connectivity. In fact, at least 20%–50% of the physical links are missing. Therefore, the node degree distribution does not follow a strict power-law relationship.

6.2.2.4   Scale-free topology model

Barabasi–Albert formulated the scale-free model, which follows three main properties [22]. First, the network is not static, but it evolves over the time. Second, any new added node will be connected to an existing vertex with a probability depending on the connectivity of the vertex (specifically, the higher the connectivity of the vertex, the higher the chances that the node will be chosen as attachment point by the joining node). This mechanism is known as preferential attachment or *rich get richer* phenomenon [56,57]. Finally, node degree distribution asymptotically settles to a power law.

This means that the node degree has a heavy-tailed distribution. The coexistence in the same network of nodes with widely different degrees is expressed by the term *scale-free* that suggests the lack of an internal scale. This feature distinguishes scale-free networks from lattices, in which all nodes have the same degree, or from random networks, whose degrees vary in a narrow range.

Moreover, another important characteristic of scale-free networks concerns the average shortest path between two vertices of the topology. Its value is small, such as in the small-word models, and it will be discussed in the following section.

The process to generate a scale-free topology entails an evolving network over a discrete time domain: at every timestep, a new vertex is added with $m \leq m_0$ edges, where m_0 is the initial small number of vertices deployed in the system.

Note that power-law random graphs (PLRGs) and scale-free networks are not synonym: while the former is static, the latter evolves during time. Moreover, a PLRG has a pre-given number of nodes and edges which follow a power-law degree sequence. Instead, in the Albert–Barabasi scale-free network, nodes and edges are self-organized in order to asymptotically reach the power-law degree distribution [58].

6.2.2.5   Hierarchical methods

The N-level hierarchical method envisages the generation of the Internet topology by iteratively expanding individual nodes into other graphs [11]. First of all, a connected graph is generated, then each node is substituted by a connected graph. The edges belonging to the original graph are connected to the nodes of the new graphs. This process continues N times. The scale of the final graph is the product of the scales of the individual levels.

6.2.3   Topology generator tools

Many tools implement the aforementioned models. They can be used to reproduce Internet-like topologies for computer simulations. In fact, they are extremely important because several times research activities cannot be carried out on real networks because of dimension, control, and permissions issues. Topology generators should fulfill the following characteristics [11]:

- *Representativeness*: the input arguments should produce accurate topologies.
- *Inclusiveness*: the generator should include different methods and models because of the lack of a universally accepted model.
- *Flexibility*: topologies should not have a limitation on the size (i.e., maximum number of nodes).
- *Extensibility*: users should be able to extend the tool with additional features.
- *Interoperability*: generated topologies should be in a format that is able to be processed by other simulator tools.

- *Efficiency*: the tool should be able to generate large topologies by preserving the required statistical characteristics and by using a reasonable CPU and memory consumption.
- *User friendliness*: the usage of the tool should be easy to learn.

6.2.3.1   Random topology generator tools

Waxman developed one of the first topology generators [59]. It implements an extended version of the Erdős and Rènyi random model, where nodes are randomly located on the Cartesian plane and connectivity among a node pair is generated according to a probability that is a function of the Euclidean distance that separates them in the plane.

6.2.3.2   Power-law topology generator tools

Inet [60] and PLRG [61] are two generators that produce power-law topologies. Both of them start the generation process assigning a degree from a power-law distribution to nodes. Then, they interconnect nodes by using different approaches [54]. Inet, first, creates a spanning tree with the nodes that have a degree greater than one. Then, it connects the remaining nodes with degree one to the spanning tree according to a linear preference, i.e., preferentially to nodes that have a higher degree. Instead, PLRG, first, increases the number of nodes by duplicating each one for a value equal to the degree assigned to it. Then, it interconnects all the clones in a uniform and random way. Note that graphs generated by PLRG can be disconnected and can contain self-loops and duplicate links. Therefore, the actual graph used for research purposes is obtained by extracting the giant connected component (that is always present according to [61]), and eliminating self-loops and merging duplicate links.

6.2.3.3   Scale-free topology generator tools

BRITE is one of the most widespread Internet topology generator [27,62]. It jointly supports Barabasi–Albert, Waxman, and hierarchical topology models. With reference to the Barabasi–Albert topology model, BRITE reproduces the incremental growth and preferential connectivity that characterize the scale-free approach. In particular, it allows the user to choose the parameter m, i.e., the number of neighbors of each new node that is added to the graph during the topology generation process. Higher values of m produce denser topology. The new node v will connect to a potential neighbor node i with a probability $\frac{k_i}{\sum_{j \in C} k_j}$, where k_i is the current out-degree of node i and C is the set of candidate neighbor nodes. This means that a new node added to the network will select with higher probability those nodes that have a higher number of connections.

Moreover, BRITE places nodes on the plane in a random or heavy-tailed way. In the first case, nodes are simply randomly distributed on the plane. When the heavy-tailed distribution is used, the plane is divided into $HS \times HS$ high-level squares. Then, each square is further subdivided into smaller $LS \times LS$ low-level squares. A number of nodes, drawn from a heavy-tailed distribution (bounded Pareto distribution), is attributed to each high-level square; then, these nodes are randomly located in each low-level square, BRITE also provides a bandwidth value to each link according to four distributions:

- *Constant*: All links have the same value.
- *Uniform*: Bandwidth values are assigned according to a uniform distribution between two input values.
- *Exponential*: Bandwidth values are assigned according to an exponential distribution with mean equal to an input value.
- *Heavy-tailed*: Bandwidth values are assigned according to a heavy-tailed distribution (Pareto with shape 1.2) with minimum and maximum values equal to two input values.

Finally, BRITE assigns geographical coordinates to each node.

6.2.3.4   Hierarchical topology generator tools

The Georgia Tech Internetwork Topology Model (GT-ITM) topology generator can be used to produce a two-level method, also known as transit-stub model [63,64]. GT-ITM first creates a connected random graph by using the Waxman method or a variant of it. Each generated node represents a transit domain. Then, these nodes are expanded in order to form another connected random graph that represents the backbone topology of the transit domain. For each node of the transit domain, stub domains are attached by generating a certain number of random graphs.

6.3   Shortest path models

Topology models are very relevant to the Internet performance evaluation. In this context, it is essential to characterize the distribution of shortest paths in order to gain precious insights on the network behavior. In reality, the communication between two peers is not the shortest path offered by the network topology. Actual multi-hop paths are generally longer than the shortest. This phenomenon is known as *path inflation*. It can be due to routing policies and to traffic engineering techniques that spread the load among more links in the topology [65,66]. But, because of the lack of more accurate models that are able to provide multi-hop communication paths, the shortest path model is still considered as a reference approach for estimating network performances [67].

This section proposes a cross-comparison of shortest path models currently available in the literature, while focusing the attention to scale-free networks. The study highlights pros and cons of these models, which emerge when they are applied to different networks.

6.3.1   Parameters definition

The distribution of shortest path lengths indicates the number of shortest paths among all node pairs. It allows a statistical characterization of the network by using average path length and graph diameter. For scale-free networks, the average shortest path length, $\bar{d}$, is approximately equal to

$$\bar{d} \approx \log N, \tag{6.4}$$

where N represents the number of nodes in the topology. In particular, this formula refers to the scale-free network that are built by adding each new vertex to m other nodes with $m = 1$. Otherwise, if $m > 1$, the average shortest path, $\bar{d}$, is asymptotically equal to

$$\bar{d} \sim \frac{\log N}{\log \log N} \tag{6.5}$$

6.3.2 Shortest path models

The definition of analytical models, that are able to describe the shortest path lengths, is still an open issue. Models proposed in the literature include *Gamma distribution*, *Weibull distribution*, and *Lognormal distribution*.

6.3.2.1 Gamma distribution

According to [24–26], the shortest path distribution can be modeled through a Gamma distribution. The probability density function, that describes the shortest path distribution according to the Gamma distribution model, is

$$f(x; \theta, \eta) = \frac{1}{\theta^\eta} \frac{1}{\Gamma(\eta)} x^{\eta-1} e^{x/\theta} \tag{6.6}$$

where $\Gamma(\cdot)$ is the gamma function and $\theta > 0$ and $\eta > 0$ are scale and shape parameters, respectively. This model indicates that the distance distribution of all nodes consists of two regimes. The former is characterized by a rapid growth. The latter refers to an exponential decay.

6.3.2.2 Weibull distribution

The work [23] applies the extreme value theory [68] to find the most appropriate model to describe the shortest path distribution. Starting from the Fisher–Tippett–Gnedenko theorem, three distributions were found: Gumbel, Frechet, and Weibull [69]. Among them, the Weibull distribution emerged as the most suitable one because the sampled distribution has to have a finite lower limit. This is the case of path lengths with a lower bound equal to zero. The probability density function describing the shortest path distribution according to the Weibull model is

$$f(x; \lambda, \kappa) = \frac{\kappa}{\lambda^\kappa} x^{\kappa-1} e^{-(x/\lambda)^\kappa} \tag{6.7}$$

where $\lambda > 0$ and $\kappa > 0$ are scale and shape parameters, respectively.

6.3.2.3 Lognormal distribution

The Lognormal distribution model was presented in [23]. It uses a probability density function defined as in the following:

$$f(x; \mu, \sigma) = \frac{1}{x\sigma\sqrt{2\pi}} e^{(\log x - \mu)^2/2\sigma^2}, \tag{6.8}$$

where $-\inf < \mu < +\inf$ is the logarithm of the mean and $\sigma > 0$ is the logarithm of the standard deviation.

6.3.3 Cross-comparison among shortest path models

The accuracy of the reference models for the distribution of the shortest path length (i.e., Gamma, Lognormal, and Weibull distributions) is evaluated through a massive simulation campaign, carried out by using the BRITE tool [27]. The study focuses on scale-free network topologies with different number of nodes and different values of the node degree. Specifically, the number of nodes N is set to 5,000, 10,000, and 20,000. The node degree m is set to 1, 2, and 3. For each set of parameters, 30 different topology realizations are generated and evaluated to produce final results.

For each distribution, related parameters are estimated through curve fitting. The resulting probability density function (pdf) and cumulative distribution function (cdf) are shown in Figures 6.2–6.4. Obtained results demonstrate that the average shortest path and the network diameter increase with the number of nodes of the topology. This behavior reflects the theoretical formulation reported in both Eqs. (6.4) and (6.5). When m increases, the average shortest path and the diameter decreases. In this case, in fact, each node added to the topology will be connected to a higher number of existing nodes. Therefore, the overall path lengths will be reduced. Moreover, the probability of the average shortest path becomes higher. These behaviors are further confirmed by results reported in Table 6.2, showing the theoretical and simulated average shortest path and the average diameter obtained through computer simulations.

More in general, however, all the curves seem to well describe what is provided by computer simulations. But, to better study and compare the accuracy of the considered models, the Kolmogorov–Smirnov test is used [70]. It evaluates the maximum absolute difference between the cumulative distribution functions generated through simulations and the cumulative distribution functions of the theoretical models. The results reported in Table 6.3 clearly demonstrate that all the models provide a good fitting. In particular, the Gamma distribution shows the lowest error when $m = 1$, and the Weibull distribution reaches the best results when $m = 2$ and $m = 3$. In all cases, the lognormal distribution registers the worst behavior.

It is possible to conclude that the conducted study clearly demonstrate that the available models are able to catch the average value and the distribution of the shortest path distribution over a very broad set of conditions. Unfortunately, the parameters of each distribution must be properly set through curve fitting. Accordingly, the models require a case-by-case tuning of parameters.

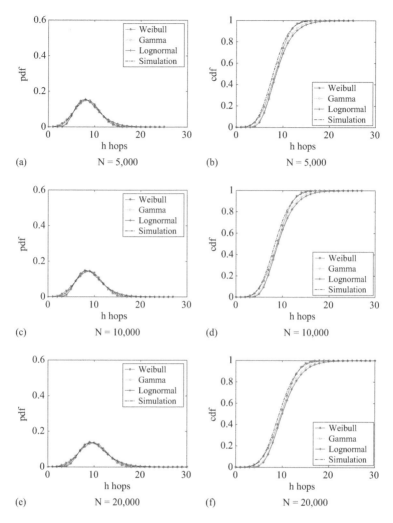

Figure 6.2   Probability density function and cumulative distribution function of the shortest path length, obtained for m = 1 and different values of N: (a) N = 5,000, (b) N = 5,000, (c) N = 10,000, (d) N = 10,000, (e) N = 20,000, and (f) N = 20,000

6.3.4   Shortest path models applications

The knowledge of the shortest path distribution is useful for the evaluation of end-to-end communication performance. In particular, it can be used in the following applications:

- *Hop-count filtering*: Techniques, such as [71,72], exploit the hop-count distribution to detect and discard spoofed IP packets. In detail, the detection is effective

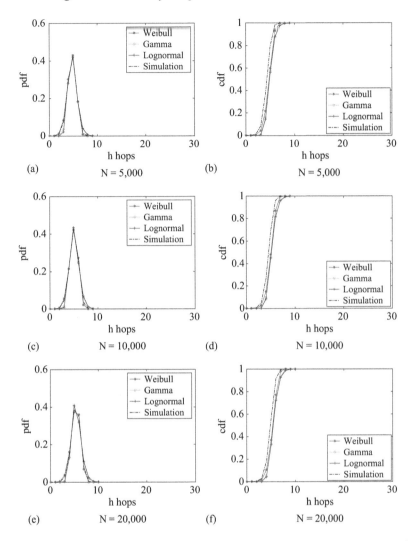

Figure 6.3   *Probability density function and cumulative distribution function of the shortest path length, obtained for m = 2 and different values of N:*
(a) N = 5,000, (b) N = 5,000, (c) N = 10,000, (d) N = 10,000,
(e) N = 20,000, (f) N = 20,000

if the source IP addresses have the same hop-count value as that of an attacker. Therefore, it is important to check if the hop-count distributions are not clustered around a single value at various locations of the network. The hop count will be more effective if the standard deviation is high.

• *Epidemic spreading*: Shortest path distribution can be exploited in the study of epidemic spreading models [26,73]. In particular, path-length statistics are closely

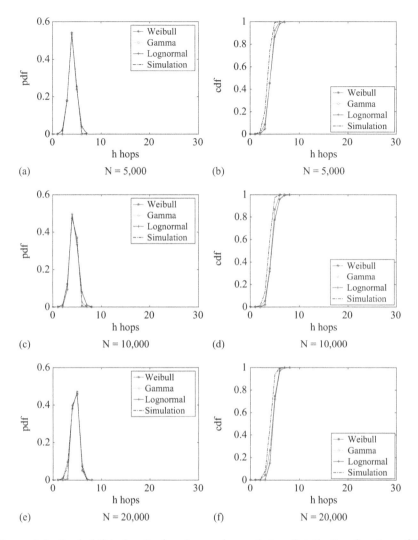

Figure 6.4   *Probability density function and cumulative distribution function of the*
shortest path length, obtained for m = 3 and different values of N:
(a) N = 5,000, (b) N = 5,000, (c) N = 10,000, (d) N = 10,000,
(e) N = 20,000, and (f) N = 20,000

related to velocities or durations of network-spreading processes. Moreover, the
host position in the routing path influences whether the host will be susceptible
to the epidemic effects or actually infected.

- *Triangular routing*: The hop-count distribution is a useful tool to study network performance in scenarios that require the so-called triangular routing. This
  mechanism is used when a direct path between source and destination cannot

Table 6.2   Theoretical and simulated average shortest path and average diameter obtained through computer simulations for different values of m

	$m = 1$		
	$N = 5,000$	$N = 10,000$	$N = 20,000$
Theoretical $\bar{d}$	8.5172	9.2103	9.9035
Simulated $\bar{d}$	8.5062	9.0178	10.0380
Simulated $\bar{\delta}$	22.0000	24.3667	26.6667
$m = 2$			
Theoretical $\bar{d}$	3.9761	4.1482	4.3192
Simulated $\bar{d}$	4.7476	5.0263	5.3117
Simulated $\bar{\delta}$	8.4000	9.0000	9.0667
$m = 3$			
Theoretical $\bar{d}$	3.9761	4.1482	4.3192
Simulated $\bar{d}$	4.0454	4.2780	4.5037
Simulated $\bar{\delta}$	7.0000	7.0333	7.1000

Table 6.3   Goodness of fit obtained through the Kolmogorov–Smirnov test for the three shortest path distribution models, for different values of N and m

	$m = 1$		
	$N = 5,000$	$N = 10,000$	$N = 20,000$
Weibull	0.0209	0.0277	0.0322
Gamma	0.0143	0.0102	0.0105
Lognormal	0.0546	0.0433	0.0416
$m = 2$			
Weibull	0.0049	0.0125	0.0015
Gamma	0.0876	0.0677	0.0562
Lognormal	0.1687	0.1196	0.0928
$m = 3$			
Weibull	0.0103	0.0308	0.0035
Gamma	0.0572	0.0476	0.1309
Lognormal	0.0943	0.0791	0.2563

be established, thus requiring a gateway. Examples of triangular routing are the following:

1. *MQ Telemetry Transport (MQTT)* : In the MQTT protocol [74], the gateway is identified by the broker that acts as intermediary between publishers and subscribers. It receives topics produced by publishers and relays them to the subscribers that are interested to them.

2. *Information centric networks (ICN)*: ICN architectures, such as Publish-Subscribe Internet Technology (PURSUIT) [75], use rendezvous nodes as relaying nodes. Producers that want to advertise a content, or consumers that want to subscribe to a content, do not issue messages directly to the rendezvous nodes responsible of that content but to their local rendezvous nodes that act as a gateway.

3. *Cloud of Things*: In the Cloud of Things [76], the cloud acts as a relay node between Internet of Things sensors and applications. Data collected by sensors are pushed to the remote data center, where they are stored and processed. When a consumer application is interested in some resources, it retrieves the contents from the cloud.

4. *Network Address Translation (NAT)*: To overcome problems due to NAT, peer-to-peer (P2P) networks and Voice over IP (VoIP) applications use a communication-relaying protocol called Traversal Using Relays around NAT (TURN) [33]. According to this protocol, a client asks a server that resides on the public Internet to act as a packet relay by using public address and port.

5. *Mobile IPv4*: In Mobile IPv4 [77], the home agent fills the role of the gateway. It intercepts datagrams addressed to the mobile node and delivers them through the care-of address that is the temporary IP address of the mobile device.

6.4 Conclusion

The chapter focuses on Internet topology models. First, it investigated topology representations provided at different levels of granularity. Second, it reviewed topology models based on the graph theory and related topology generator tools. Third, it studied analytical models showing average shortest path and distribution of the shortest path length for scale-free networks. In fact, starting from these concepts, the accuracy of reference models for the shortest path are studied and compared through computer simulations. Obtained results demonstrate that available models are able to catch the average value and the distribution of the shortest path distribution over a very broad set of conditions. More specifically, when $m = 1$, the Gamma distribution shows the lowest distance for all N values. When $m = 2$ and $m = 3$, instead, the Weibull distribution provides the lowest error. The lognormal distribution always registers the worst behavior. On the contrary, all the evaluated models require a case-by-case tuning of their parameters, which represents an important limit to be solved in future research activities.

Acknowledgment

This work was partially founded by PON projects founded by the Italian MIUR, including Pico&Pro (code: ARS01_01061) AGREED (code: ARS01_00254), FURTHER (code: ARS01_01283), and RAFAEL (code: ARS01_00305), and by the research project E-SHELF (code: OSW3NO1) founded by the Apulia Region – Italy.

References

[1]   Cisco. The Zettabyte Era: Trends and Analysis. San Jose, CA, USA: Cisco Systems, Inc.; 2016.

[2]   Huawei. Global Connectivity Index 2016. Huawei Technologies Co., Ltd.; 2016.

[3]   Sterbenz JP, Çetinkaya EK, Hameed MA, *et al.* Evaluation of network resilience, survivability, and disruption tolerance: analysis, topology generation, simulation, and experimentation. Telecommunication Systems. 2013;52(2):705–736.

[4]   Gregori E, Improta A, Lenzini L, *et al.* Discovering the geographic properties of the Internet AS-level topology. Networking Science. 2013; 3(1):34–42.

[5]   Sun L, Song F, Yang D, *et al.* DHR-CCN, distributed hierarchical routing for content centric network. Journal of Internet Services and Information Security. 2013;3(1/2):71–82.

[6]   Wang Y, Li Z, Tyson G, *et al.* Optimal cache allocation for content-centric networking. In: Proceedings of International Conference on Network Protocols (ICNP). IEEE; 2013. p. 1–10.

[7]   Oliveira R, Pei D, Willinger W, *et al.* The (in) completeness of the observed internet AS-level structure. IEEE/ACM Transactions on Networking (ToN). 2010;18(1):109–122.

[8]   Roughan M, Willinger W, Maennel O, *et al.* 10 Lessons from 10 years of measuring and modeling the internet's autonomous systems. IEEE Journal on Selected Areas in Communications. 2011;29(9):1810–1821.

[9]   Donnet B, Friedman T. Internet topology discovery: a survey. IEEE Communications Surveys & Tutorials. 2007;9(4):56–69.

[10]   Motamedi R, Rejaie R, Willinger W. A survey of techniques for Internet topology discovery. IEEE Communications Surveys & Tutorials. 2015;17(2): 1044–1065.

[11]   Haddadi H, Rio M, Iannaccone G, *et al.* Network topologies: inference, modeling, and generation. IEEE Communications Surveys Tutorials. 2008 Second;10(2):48–69.

[12]   Zegura EW, Calvert KL, Donahoo MJ. A quantitative comparison of graph-based models for internet topology. IEEE/ACM Transactions on Networking. 1997;5(6):770–783.

[13]   Floyd S, Paxson V. Difficulties in simulating the Internet. IEEE/ACM Transactions on Networking (ToN). 2001;9(4):392–403.

[14]   Augustin B, Cuvellier X, Orgogozo B, *et al.* Avoiding traceroute anomalies with Paris traceroute. In: Proceedings of the 6th ACM SIGCOMM Conference on Internet Measurement. IMC '06. New York, NY: ACM; 2006. p. 153–158.

[15]   Roughan M, Willinger W, Maennel O, *et al.* 10 lessons from 10 years of measuring and modeling the internet's autonomous systems. IEEE Journal on Selected Areas in Communications. 2011;29(9):1810–1821.

[16] Pansiot JJ, Mérindol P, Donnet B, *et al.* Extracting intra-domain topology from MRINFO Probing. In: PAM. Springer; 2010. p. 81–90.

[17] Keys K. Internet-scale IP alias resolution techniques. SIGCOMM Computer Communication Review. 2010;40(1):50–55.

[18] Pastor-Satorras R, Vespignani A. Evolution and structure of the Internet: a statistical physics approach. Cambridge, UK: Cambridge University Press; 2007.

[19] Erdos P, Rényi A. On the evolution of random graphs. Publication of the Mathematical Institute of the Hungarian Academy of Sciences. 1960;5(1):17–60.

[20] Watts DJ, Strogatz SH. Collective dynamics of 'small-world' networks. Nature. 1998;393(6684):440.

[21] Faloutsos M, Faloutsos P, Faloutsos C. On power-law relationships of the internet topology. In: Proceedings of the Conference on Applications, Technologies, Architectures, and Protocols for Computer Communication. SIGCOMM '99. New York, NY: ACM; 1999. p. 251–262.

[22] Barabási AL. Network Science. Cambridge, UK: Cambridge University Press; 2016.

[23] Bauckhage C, Kersting K, Rastegarpanah B. The Weibull as a model of shortest path distributions in random networks. In: Proceedings of Int. Workshop on Mining and Learning with Graphs, Chicago, IL; 2013.

[24] Vazquez A. Polynomial growth in branching processes with diverging reproductive number. Physical Review Letters. 2006;96(3):038702.

[25] Kalisky T, Cohen R, Mokryn O, *et al.* Tomography of scale-free networks and shortest path trees. Physical Review E. 2006;74(6):066108.

[26] Bauckhage C, Kersting K, Hadiji F. Parameterizing the distance distribution of undirected networks. In: UAI; 2015. p. 121–130.

[27] Medina A, Lakhina A, Matta I, *et al.* BRITE: an approach to universal topology generation. In: Proceedings of Ninth International Symposium on Modeling, Analysis and Simulation of Computer and Telecommunication Systems. IEEE; 2001. p. 346–353.

[28] Baumann A, Fabian B. How robust is the internet?—Insights from graph analysis. In: Proceedings of International Conference on Risks and Security of Internet and Systems. Springer; 2014. p. 247–254.

[29] Bollobás B. Modern Graph Theory. vol. 184. Springer Science & Business Media; 2013.

[30] Calvert KL, Doar MB, Zegura EW. Modeling internet topology. New York, USA: IEEE Communications Magazine. 1997;35(6):160–163.

[31] Zegura EW, Calvert KL, Donahoo MJ. A quantitative comparison of graph-based models for Internet topology. IEEE/ACM Transactions on Networking (TON). 1997;5(6):770–783.

[32] John W, Tafvelin S, Olovsson T. Passive internet measurement: overview and guidelines based on experiences. Computer Communications. 2010;33(5):533–550.

[33] Kurose JF, Ross KW. Computer Networking: A Top-Down Approach. vol. 5. Boston, MA, USA: Addison-Wesley Reading; 2010.

[34]   Jacobson V. Traceroute Software. Berkeley, CA, USA: Lawrence Berkeley Laboratories; 1988.

[35]   Postel J. Internet Control Message Protocol. RFC Editor; 1981.

[36]   Rekhter Y, Li T. Internet Control Message Protocol (ICMPv6) for the Internet Protocol Version 6 (IPv6) Specification. RFC Editor; 1998.

[37]   Spring N, Mahajan R, Wetherall D, et al. Measuring ISP topologies with rocketfuel. IEEE/ACM Transactions on Networking. 2004;12(1):2–16.

[38]   Hawkinson J, Bates T. Guidelines for Creation, Selection, and Registration of an Autonomous System (AS). RFC Editor; 1996.

[39]   Vohra Q, Chens E. BGP Support for Four-Octet AS Number Space. RFC Editor; 2007.

[40]   Rekhter Y, Li T, Hares S. A Border Gateway Protocol 4 (BGP-4). RFC Editor; 2006.

[41]   Battista GD, Refice T, Rimondini M. How to extract BGP peering information from the internet routing registry. In: Proceedings of the 2006 SIGCOMM Workshop on Mining Network Data. MineNet '06. New York, NY: ACM; 2006. p. 317–322.

[42]   University of Oregon Route Views Project;. Available online at: http://www.routeviews.org.

[43]   RIPE Routing Information Service (RIS);. Available online at: http://www.ripe.net/ris.

[44]   Zhang B, Liu R, Massey D, et al. Collecting the internet AS-level topology. SIGCOMM Computer Communication Review. 2005;35(1):53–61.

[45]   Ng TE, Zhang H. Predicting Internet network distance with coordinates-based approaches. In: INFOCOM 2002. Twenty-First Annual Joint Conference of the IEEE Computer and Communications Societies. Proceedings. IEEE. vol. 1. IEEE; 2002. p. 170–179.

[46]   Poese I, Uhlig S, Kaafar MA, et al. IP geolocation databases: unreliable? SIGCOMM Computer Communication Review. 2011;41(2):53–56.

[47]   Center for Applied Internet Data Analysis (CAIDA);. Available online at: http://www.caida.org/.

[48]   Knight S, Nguyen HX, Falkner N, et al. The internet topology zoo. IEEE Journal on Selected Areas in Communications. 2011;29(9):1765–1775.

[49]   Mahadevan P, Krioukov D, Fomenkov M, et al. The Internet AS-level topology: three data sources and one definitive metric. ACM SIGCOMM Computer Communication Review. 2006;36(1):17–26.

[50]   Geant2 Looking glass. Available online at: http://stats.geant2.net/lg/.

[51]   Mills DL, Braun H. The NSFNET backbone network. In: Proceedings of the ACM Workshop on Frontiers in Computer Communications Technology. SIGCOMM '87. New York, NY: ACM; 1988. p. 191–196.

[52]   Newman ME, Watts DJ. Scaling and percolation in the small-world network model. Physical Review E. 1999;60(6):7332.

[53]   Yook SH, Jeong H, Barabási AL. Modeling the Internet's large-scale topology. Proceedings of the National Academy of Sciences. 2002;99(21):13382–13386.

[54] Bu T, Towsley D. On distinguishing between Internet power law topology generators. In: Proceedings of Twenty-First Annual Joint Conference of the IEEE Computer and Communications Societies. vol. 2. IEEE; 2002. p. 638–647.

[55] Chen Q, Chang H, Govindan R, *et al.* The origin of power laws in Internet topologies revisited. In: Proceedings of Twenty-First Annual Joint Conference of the IEEE Computer and Communications Societies. vol. 2. IEEE; 2002. p. 608–617.

[56] Barabási AL, Albert R. Emergence of scaling in random networks. Science. 1999;286(5439):509–512.

[57] Zhou S, Mondragón RJ. The rich-club phenomenon in the Internet topology. IEEE Communications Letters. 2004;8(3):180–182.

[58] Shi D. Critical thinking of scale-free networks: similarities and differences in power-law random graphs. National Science Review. 2014;1(3):337–338.

[59] Waxman BM. Routing of multipoint connections. IEEE Journal on Selected Areas in Communications. 1988;6(9):1617–1622.

[60] Jin C, Chen Q, Jamin S. Inet: Internet Topology Generator; University of Michigan, Ann Arbor, MS: Tech. Rep. CSE-TR-456-02; 2000.

[61] Aiello W, Chung F, Lu L. A random graph model for massive graphs. In: Proceedings of the Thirty-Second Annual ACM Symposium on Theory of Computing. STOC '00. New York, NY: ACM; 2000. p. 171–180.

[62] Medina A, Matta I, Byers J. On the origin of power laws in Internet topologies. ACM SIGCOMM Computer Communication Review. 2000;30(2):18–28.

[63] Zegura E. GT-ITM: Georgia Tech Internetwork Topology Models (Software). Georgia Tech.; 1996.

[64] Calvert K, Eagan J, Merugu S, *et al.* Extending and enhancing GT-ITM. In: Proceedings of the ACM SIGCOMM Workshop on Models, Methods and Tools for Reproducible Network Research. ACM; 2003. p. 23–27.

[65] Spring N, Mahajan R, Anderson T. The causes of path inflation. In: Proceedings of the 2003 Conference on Applications, Technologies, Architectures, and Protocols for Computer Communications. SIGCOMM '03. New York, NY: ACM; 2003. p. 113–124.

[66] Mühlbauer W, Uhlig S, Feldmann A, *et al.* Impact of routing parameters on route diversity and path inflation. Computer Networks. 2010;54(14): 2506–2518.

[67] Leguay J, Latapy M, Friedman T, *et al.* Describing and simulating internet routes. In: Networking. Springer; 2005. p. 659–670.

[68] De Haan L, Ferreira A. Extreme Value Theory: An Introduction. Berlin, Germany: Springer Science & Business Media; 2007.

[69] Fisher RA, Tippett LHC. Limiting forms of the frequency distribution of the largest or smallest member of a sample. In: Mathematical Proceedings of the Cambridge Philosophical Society. vol. 24. Cambridge University Press; 1928. p. 180–190.

[70] Massey Jr FJ. The Kolmogorov–Smirnov test for goodness of fit. Journal of the American statistical Association. 1951;46(253):68–78.

[71] Jin C, Wang H, Shin KG. Hop-count filtering: an effective defense against spoofed DDoS traffic. In: Proceedings of the 10th ACM Conference on Computer and Communications Security. CCS '03. New York, NY: ACM; 2003. p. 30–41.

[72] Wang H, Jin C, Shin KG. Defense against spoofed IP traffic using hop-count filtering. IEEE/ACM Transactions on Networking (ToN). 2007;15(1): 40–53.

[73] Iannelli F, Koher A, Brockmann D, *et al.* Effective distances for epidemics spreading on complex networks. Physical Review E. 2017;95(1):012313.

[74] Hunkeler U, Truong HL, Stanford-Clark A. MQTT-S A publish/subscribe protocol for wireless sensor networks. In: Proceedings of Int. Conf. on Communication Systems Software and Middleware. IEEE; 2008. p. 791–798.

[75] Xylomenos G, Ververidis CN, Siris VA, *et al.* A survey of information-centric networking research. IEEE Communications Surveys & Tutorials. 2014;16(2):1024–1049.

[76] Piro G, Amadeo M, Boggia G, *et al.* Gazing into the crystal ball: when the Future Internet meets the Mobile Clouds. IEEE Transactions on Cloud Computing. 2016.

[77] Sanguankotchakorn T, Jaiton P. Effect of triangular routing in mixed IPv4/IPv6 networks. In: Networking, 2008. ICN 2008. Seventh International Conference on. IEEE; 2008. p. 357–362.

Part III

Case studies and more

Chapter 7

Accurate modeling of VoIP traffic in modern communication

Homero Toral-Cruz[1], Al-Sakib Khan Pathan[2], and Julio C. Ramírez Pacheco[3]

7.1 Introduction

In the recent years, voice has become one of the most attractive and important services in telecommunication networks which can be transmitted via circuit-switched and packet-switched networks. The most common examples of circuit-switched and packet-switched networks are the public switched telephone network (PSTN) and Internet, respectively [1]. Compared to traditional resource-dedicated PSTN, the Internet is resource-shared. Therefore, the conditions in the PSTN are totally different from those in the Internet. There are several advantages in the case of voice transmission using internet protocol (IP) technology, also called voice over IP (VoIP) [2]: the reduced communication cost, the use of joined IP infrastructure, the use in multimedia applications, etc. It is also an interesting fact that sending wireless phone calls over IP networks is considerably less expensive than that of sending over cellular voice networks. However, such types of communications must ensure good performance and quality of voice transmissions. Faster wireless networking technologies and more powerful mobile telephones promise to help solve these problems [3,4]. With the currently available technologies, VoIP over mobile networks has not yet got that much popularity. However, the advantage of such technology is that keeping the wireless network as it is, with a higher quality of service (QoS) facility, the existing infrastructure of IP networks could be used for serving the users. If a company owns the Wi-Fi connection, such VoIP communication could be provided at a very low cost. Either Wi-Fi or WiMAX network or even if any of these is unavailable, cellular technologies could be used to pass the voice traffic to the Internet. This is known as mobile VoIP (mVoIP). mVoIP via cellular services could achieve QoS by prioritizing voice packets over those used for data and other traffic types [2].

[1]NETCOM Laboratory, Department of Sciences and Engineering, University of Quintana Roo, Mexico
[2]Department of Computer Science and Engineering, Southeast University, Bangladesh
[3]Department of Basic Sciences and Engineering, Universidad Del Caribe, Mexico

The emergence of new real-time services, such as VoIP and mVoIP, has allowed the growth and evolution of the Internet as a modern communication network, characterized by its complex nature. The main reason for its complex nature is the result of the convergence of information and media transmission (voice, video, and data) through the same communication channel. As there are a very high (and increasing) number of nodes (i.e., devices) connected to the network, and these are being added in a random and decentralized manner, the network has the property of *scale-free*, meaning that the node degree distribution follows a power-law distribution [5]. With the increase in number of nodes and connections, the traffic load in the network increases and congestions occur accordingly.

Congestion is a cause of the QoS impairment, which consists of delay issues (i.e., delay and jitter) and packet loss. For real-time communications, such as VoIP and mVoIP, jitter and packet loss can have high impact on the QoS. To achieve a satisfactory level of voice quality, the VoIP networks must be designed by using correct traffic models [2]. Traffic modeling in classical and modern communication networks mainly comprise the following steps [6]: (1) Selection of one or more models that may provide a good description of the statistical properties of packet traffic. In order to select an adequate traffic model, it is necessary to study the traffic characteristics. The main characteristics of a traffic source are its average data rate, burstiness, and correlation. The average data rate gives an indication of the expected traffic volume for a given period of time. Burstiness describes the tendency of traffic to occur in clusters. Data burstiness is manifested by the correlation function which describes the relationship between packet arrivals at different times and is an important factor in packet losses due to buffer and bandwidth limitations. (2) Estimation of parameters for the selected model. Parameter estimation is based on a set of statistics (e.g., mean, variance, density function or autocovariance function (ACV), and multifractal characteristics) that are measured or calculated from observed data traces. The set of statistics used in the inference process depends on the impact they may have in the main QoS parameters of interest. (3) Statistical testing for election of one of the considered models and analysis of its suitability to describe the traffic type under analysis.

Previously, the statistical properties of the IP traffic were described by Poisson model, where the autocorrelation decays exponentially fast; this is because in the early stages, the size of the Internet was small and its structure was simple (simple packet network). However, the explosive growth in the number of users and the growing diversity of real-time traffic have allowed to discover complex behaviors in the IP traffic, where the Poisson model is not able to capture all their statistical properties [5,7]. Such behaviors can have significant impact on network performance and can be well described by long-range dependence (LRD), self-similarity, and multifractality [6]. The LRD behavior manifests itself along a communication channel as a bursty activity in the packet rate, which produces a wide range of traffic volume away from the average rate and persists strongly on all relevant time scales [5]. In the LRD traffic, the autocorrelation decays slowly as a power-law function. This great variation in the traffic volume leads to buffer overflow and network congestion that result in packet loss and jitter, which directly impact the quality of VoIP applications.

This chapter presents the jitter and packet-loss modeling on VoIP traffic by means of network measurements. We basically modeled the main QoS parameters of VoIP traffic which could be related with regular VoIP communications over regular-wired networks as well as for mVoIP (which functions as an application that runs over any wireless network technology that provides data access to the Internet). Hence, our modeling work is basically relevant to both regular VoIP and mVoIP technologies.

7.2   Modern communication networks: from simple packet network to multiservice network

A simple packet network is an infrastructure consisting of a collection of network resources: terminals (hosts), nodes (routers), and links (copper wire), which are connected together to enable communication between users via their terminals, intended for transportation of some type of data (flow) in a way as to meet a supply and demand pattern between the nodes [1,8]. The hosts are connected together by a subnet. The subnet consists of two components [9]: communication links and routers, as is depicted in Figure 7.1.

The main function of the subnet is to transport data flows from host to host. The way that the network elements are connected to each other is known as topology. The supply and demand pattern, together with the topology type, induces data flows in the network, which in the case of packet network is known as traffic [8].

The communication networks have seen a tremendous development in the recent decades, and this historical development can be summarized in the following main events [10]: the introduction of automatic telephone exchange, the digitalization of telecommunications systems, the integration of circuit-switched technology and packet-switched technology, and the evolution of the mobile systems (1G, 2G, 3G, 4G, and beyond).

The development of network technology and the enhanced services that can be offered to the end users have been captured in the concept of a modern communication

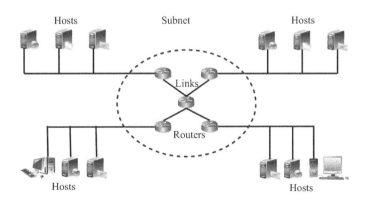

Figure 7.1   Simple packet network

network. This concept encapsulates the convergence of network processes where the convergent network is called multiservice network and is based on IP technology. In the multiservice network, packets should be transported transparently from host to host without excessive protocol conversion through an IP network core (Internet) [11]. However, with this convergence, a new technical challenge has emerged. The IP network core provides best-effort services in most of the cases and cannot guarantee the QoS of real-time multimedia applications, such as VoIP [12].

The efficient design of modern communication networks is a complex task and it involves complex mathematical topics. An efficient network design can be achieved by using accurate models and that ensures that the network has the necessary capabilities of providing services with a certain standard QoS [8].

In the abovementioned context, the connectivity of a communication network is modeled by using concepts from graph theory, i.e., a packet network can be modeled by means of a graph $G = (V, E)$, where V is the set of nodes, E is the set of links and the degree of a node is the number of nearest neighbors. The degree of a node is a local quantity; however, an interesting study is the node degree distribution of the entire network, because it gives important information about the global properties of a network and can be used to characterize different network topologies. From the perspective of node degree, the main topologies used in communication networks are the following [5]:

- *Regular-symmetric networks*: The regular-symmetric network has the same degree for all nodes, e.g., the ring network, rectangular toroidal network, triangular toroidal network, hexagonal toroidal network.
- *Random networks*: In the random network, the node degree distribution is well approximated by a binomial distribution.
- *Scale-free networks*: In the scale-free network, the node degree distribution is described by a power-law.

Recent studies have shown that the real topology of the modern communication networks is neither completely random nor completely regular-symmetric but it basically shows a prominent characteristic of self-organization, with which the node degree distribution is described by a power-law [5]. This scale-free network model was proposed by Barabási and Albert in [13] and it involves one of the following three operations for each time step: add new links between the existing nodes, rewire links, and add new nodes.

Often, we have a communication network with a particular topology and we seek traffic models to measure or predict the network performance as a function of some QoS parameters, such as delay and packet loss. This activity mainly involves the probabilistic relations between traffic, network resources, and QoS and it is better known as traffic theory [8].

At the early stage of the packet networks, the traffic and congestions were rather sparse, and one of the models which were widely applied for the traffic modeling was the Poisson model. However, with the evolution from the simple packet network to multiservice network, the classical Poisson model fails to model the new aggregated traffic. Besides, recent studies in modern communication networks have shown that

IP traffic exhibits self-similar, LRD, and multifractal behaviors. The abovementioned behaviors are presented in modern communication networks due to the following issues [8]:

- They transport a mixture of multiple traffic types, such as voice, video, and data traffic.
- They are growing in size continuously (becoming larger), and various transport technologies are becoming available.
- They are increasingly dynamic, i.e., new services are frequently introduced which change the network behavior.

7.3 Voice over IP (VoIP) and quality of service (QoS)

7.3.1 Basic structure of a VoIP system

VoIP is the real-time transmission of voice between two or more parties, by using IP technologies. A basic VoIP system mainly consists of three parts: the source terminal (sender), the multiservice network, and the destination terminal (receiver) [1].

Figure 7.2 represents the *source terminal* diagram from a typical VoIP system [14]. The shown blocks indicate the positions of measurement points and system components.

The A/D converter of the source terminal transforms the voice waveform into its digital version. In order to reduce the quantization error, a nonuniform quantification (either A-law or μ-law) is applied in its internal quantization block. The source encoding consists of compressing the digital signal in order to reduce redundancy and improve bandwidth utilization. The channel coding is a viable method to reduce information rate by increasing reliability. This goal is achieved by adding redundancy

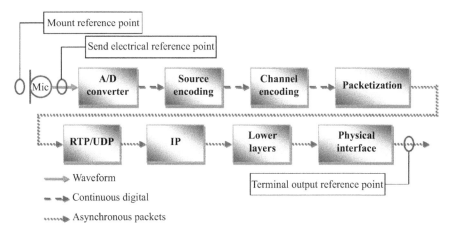

Figure 7.2   Source terminal diagram from a VoIP system. Adopted, with permission, from Reference [14]

to the information so that it is more distinguishable at the output of the channel (receiver).

The voice packetization protocols use a sequence number field in the transmit packet stream to maintain temporal integrity of voice, allowing the receiver to detect lost packets and to properly reproduce silence intervals during play out. The packetized voice stream traverses to lower layers, i.e., RTP/UDP (Real-time Transport Protocol/User Datagram Protocol), IP, etc., and it is sent asynchronously into the network [15]. The main impairment for the source terminal is named source terminal delay, which is defined as the sum of all delays of the system components belonging to the source, i.e., those between the mouth reference point and the terminal output reference point; more specifically, it is defined as the interval defined by the time that a signal enters the mouth reference point and the time that the first bit of the correspondence-encoded, packetized signal exits the terminal output reference point [15].

Then, the voice packets are introduced to the *multiservice network* with the aim of being routed to the destination terminal. In this network, numerous types of access networks with different types of terminals are integrated, e.g., they must interwork with legacy networks through gateways (GWs). Access networks can use a variety of Layer 1 and 2 protocols. In the multiservice transport network, packets should be transported transparently between endpoints without excessive protocol conversion and adaption. Also, all terminals should use the same network layer protocol to give a uniform end-to-end routing method [1]. However, the multiservice network is based on IP technology. As the current IP technology offers the best effort service and does not guarantee the QoS, voice traffic over IP networks is susceptible to suffer packet loss and delays, which results in voice-quality degradations.

Figure 7.3 represents the *destination terminal* diagram of a typical VoIP system [14]. The components and the positions of measurement points are indicated. The input bit stream that contains the asynchronous voice packets arrives at

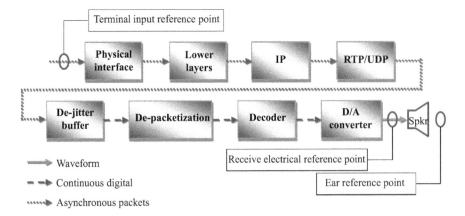

Figure 7.3   Destination terminal diagram from a VoIP system. Adopted, with permission, from Reference [14]

the terminal input reference point. Then it passes through the lower layers and then through the IP and RTP/UDP blocks. After that, the sequence of voice datagrams is passed to the de-jitter buffer, which performs very important tasks, in the sense of QoS because it is used to compensate the network jitter at the cost of further delay (buffer delay) and loss (late arrival loss). Therefore, the de-jitter buffer defines the relationship between jitter and loss on the receiver side.

At this point, voice stream is already impaired by the delay and loss due to the traversed network. Additional impairments occur due to the additional delay of the remaining blocks and the additional loss caused by the discarding of highly delayed packets, compared to de-jitter buffer size which is desirably optimized [16]. Therefore, an important design parameter at the receiver side is the de-jitter buffer size, because this parameter becomes the essential descriptor of intrinsic quality that supplants jitter. The packet headers are stripped off and voice samples are extracted in the de-packetization block. Finally, the de-jittered voice samples are decoded to recover the original voice signal.

7.3.2   VoIP frameworks: H.323 and SIP

Current implementations of VoIP have two main types of architectures, which are based on H.323 and session initiation protocol (SIP) frameworks, respectively [17]. H.323, which was ratified by International Telecommunication Union (ITU), is a set of protocols for voice, video, and data conferencing over packet-based network [18]. SIP, which is defined in Request For Comments 2543 (RFC 2543) of the Multi-party Multimedia Session Control working group of Internet Engineering Task Force (IETF), is an application-layer control signaling protocol for creating, modifying, and terminating sessions with one or more participants [19,20]. Regardless of their differences, the fundamental architectures of these two implementations are the same. They consist of three main logical components: terminal, signaling server, and GW. They differ in specific definitions of voice coding, transport protocols, control signaling, GW control, and call management.

7.3.2.1   H.323

ITU-T (ITU Telecommunication Standardization Sector) H.323 is a set of protocols of voice, video, and data conferencing over packet-switched networks such as Ethernet Local Area Networks (LANs) and the Internet that do not provide a guaranteed QoS [18,17]. The H.323 protocol stack is designed to operate above the transport layer of the underlying network. H.323 was originally developed as one of the several video-conferencing recommendations issued by the ITU-T. The H.323 standard is designed to allow clients on H.323 networks to communicate with clients on other videocon-ferencing networks. The first version of H.323 was issued in 1996, designed for use with Ethernet LANs and borrowed much of its multimedia conferencing aspects from other H.32.x series recommendations. H.323 is part of a large series of communica-tion standards that enable videoconferencing across a range of networks. This series also includes H.320 and H.324, which address the ISDN and PSTN communications, respectively. H.323 is known as a broad and flexible recommendation. Although H.323

specifies protocols for real-time point-to-point communication between two terminals on a packet-switched network, it also includes support for multipoint conferencing among terminals that support not only voice but also video and data communications. This recommendation describes the components of H.323 architecture. This includes terminals, GWs, Gatekeepers (GK), Multipoint Control Units (MCU), Multipoint Controller (MC), and Multipoint Processors (MP).

- *Terminal:* An H.323 terminal is an endpoint on the network that provides for real-time, two-way communications with another H.323 terminal, GW, or MCU. This communication consists of control, indications, audio, moving color video pictures, and/or data between the two terminals. A terminal may provide speech only; speech and data; speech and video; or speech, data, and video.
- *GW:* The GW is an H.323 entity on the network which allows intercommunication between IP networks and legacy circuit-switched networks, such as ISDN and PSTN. They provide signaling mapping as well as transcoding facilities. Transcoding is basically the ability to adapt digital files so that content can be viewed on different playback devices. For example, GWs receive an H.320 stream from an ISDN line and convert it to an H.323 stream and then send it to the IP network.
- *GK:* The GK is an H.323 entity on the network which performs the role of the central manager of VoIP services to the endpoints. This entity provides address translation and control access to the network for H.323 terminals, GWs, and MCUs. The GK may also provide other services to the terminals, GWs, and MCUs such as bandwidth management and locating GWs.
- *MCU:* The MCU is an H.323 entity on the network which provides the capability for three or more terminals and GW to participate in a multipoint conference. It may also connect two terminals in a point-to-point conference which may later develop into a multipoint conference. The MCU consists of two parts: a mandatory MC and an optional MP. In the simplest case, an MCU may consist only of an MC with no MPs. An MCU may also be brought into a conference by the GK without being explicitly called by one of the endpoints. The MC is an H.323 entity on the network which provides for the control of three or more terminals participating in a multipoint conference. It may also connect two terminals in a point-to-point conference which may later develop into a multipoint conference. The MC provides for capability negotiation with all terminals to achieve common levels of communications. It may also control conference resources such as who is multicasting video. The MC does not perform mixing or switching of audio, video, and data. The MP is an H.323 entity on the network which provides for the centralized processing of audio, video, and/or data streams in a multipoint conference. The MP provides for the mixing, switching, or other processing of media streams under the control of the MC. The MP may process a single media stream or multiple media streams depending on the type of conference supported.

The H.323 architecture is partitioned into zones. Each zone comprises the collection of all terminals, GW and MCU managed by a single GK. H.323 is an umbrella recommendation which depends on several other standards and recommendations to enable real-time multimedia communications. The main ones are

- *Call Signaling and Control:* Call control protocol (H.225), media control protocol (H.245), security (H.235), digital subscriber signaling (Q.931), generic functional protocol for the support of supplementary services in H.323 (H.450.1), supplemental features (H.450.2–H.450.11).
- *H.323 Annexes:* Real-time facsimile over H.323 (Annex D), framework and wire-protocol for multiplexed call signaling transport (Annex E), simple endpoint types—SET (Annex F), text conversation and Text SET (Annex G), Security for annex F (Annex J), hypertext transfer protocol (HTTP)-based service control transport channel (Annex K), stimulus control protocol (Annex L), and tunneling of signaling protocols (Annex M).
- *Audio CODECs:* Pulse Code Modulation (PCM) audio CODEC 56/64 kbps (G.711), audio CODEC for 7 kHz at 48/56/64 kbps (G.722), speech CODEC for 5.3 and 6.4 kbps (G.723), speech CODEC for 16 kbps (G.728), and speech CODEC for 8/13 kbps (G.729).
- *Video CODECs:* Video CODEC for ≥ 64 kbps (H.261) and video CODEC for ≤ 64 kbps (H.263).

The H.323 protocol is implemented by exchanging messages between the protocol endpoints and intervening entities. These messages are all encoded using Abstract Syntax Notation (ASN.1) which is a binary format for defining the syntax of information data. Even though an H.323 is a packet-based protocol, it remains tightly coupled with most traditional communication standards. As such, it defines various communication channels, each using its own subprotocol for the communication between the various H.323 entities. Specifically, the protocol distinguishes between three communication channels:

- *The RAS channel:* Is an unreliable channel between the H.323 endpoint and the GK. It is used to exchange registration, admission, bandwidth change, and status messages. The messages exchanged in this channel follow the recommendation in H.225.
- *The call signaling channel:* Is a reliable channel between H.323 endpoints (direct or routed through GKs). It is used to perform the call setup and teardown phases. The messages exchanged in this channel follow the recommendation in H.225.
- *The H.245 control channel:* Is a reliable channel between H.323 endpoints (direct or routed through GKs). It is used to exchange the H.245 call-control messages. The messages exchanged in this channel follow the recommendation in H.245.

The signaling of the whole call session is performed through messages exchanged on these three channels. H.323 defines one additional type of channel, the logical media channel. This channel carries the media content (voice, video, and/or data) and each session can have one or more channels established through the H.245 control channel. In H.323, a call session is perceived as consisting of five phases: call setup, initial communication and capability exchange, establishment of audio–visual communication, call services, and call termination.

7.3.2.2   Session initiation protocol

SIP was developed by IETF in reaction to the ITU-T H.323 recommendation. The IETF believed that H.323 was inadequate for evolving IP telephony, because its command structure is complex and its architecture is centralized and monolithic. SIP is an application layer–control protocol that can establish, modify, and terminate multimedia sessions or calls [19,20]. SIP transparently supports name mapping and redirection services, allowing the implementation of ISDN and intelligent network telephony subscriber services. The early implementations of SIP have been in network carrier IP-Centrex trials. SIP was designed as part of the overall IETF multimedia data and control architecture that supports protocols such as resource reservation protocol (RSVP), RTP, real-time streaming protocol (RTSP), session announcement protocol (SAP), and session description protocol (SDP). SIP establishes, modifies, and terminates multimedia sessions. It can be used to invite new members to an existing session or to create new sessions. The two major components in a SIP network are user agents (UAs) and network servers (registrar server, location server, proxy server, and redirect server).

- *UAs:* Is an application that interacts with the user and contains both a UA client (UAC) and UA server (UAS). A UAC initiates SIP requests, and a UAS receives SIP requests and returns responses on the user's behalf.
- *Registrar server:* Is a SIP server that accepts only registration requests issued by UAs for the purpose of updating a location database with the contact information of the user specified in the request.
- *Proxy server:* Is an intermediary entity that acts both as a server to UAs by forwarding SIP requests and as a client to other SIP servers by submitting the forwarded requests to them on behalf of UAs or proxy servers.
- *Redirect server:* Is a SIP server that helps locate UA by providing alternative locations where the user can be reachable, i.e., provides address-mapping services. It responds to a SIP request destined to an address with a list of new addresses. A redirect server does not accept call, does not forward request, and does not initiate any of its own.

The SIP protocol follows a web-based approach for call signaling, contrary to traditional communication protocols. It resembles a client/server model where SIP clients issue requests and SIP servers return one or more responses. The signaling protocol is built on this exchange of requests and responses, which are grouped into transactions. All the messages of a transaction share a common unique identifier and traverse the same set of hosts. There are two types of messages in SIP—requests and responses. Both of them use the textual representation of the ISO 10646 character set with UTF-8 encoding. The message syntax follows HTTP/1.1, but it should be noted that SIP is not an extension to HTTP.

- *SIP responses:* Upon reception of a request, a server issues one or several responses. Every response has a code that indicates the status of the transaction. Status codes are integers ranging from 100 to 699 and are grouped into six classes. A response can be either final or provisional. A response with a status code from 100 to 199 is considered provisional. Responses from 200 to 699 are final responses.

- 1xx informational: Request received, continuing to process request. The client should wait for further responses from the server.
- 2xx success: The action was successfully received, understood, and accepted. The client must terminate any search.
- 3xx redirection: Further action must be taken in order to complete the request. The client must terminate any existing search but may initiate a new one.
- 4xx client error: The request contains bad syntax or cannot be fulfilled at this server. The client should try another server or alter the request and retry with the same server.
- 5xx server error: The request cannot be fulfilled at this server because of server error. The client should try with another server.
- 6xx global failure: The request is invalid at any server. The client must abandon search.

The first digit of the status code defines the class of response. The last two digits do not have any categorization role. For this reason, any response with a status code between 100 and 199 is referred to as a "1xx response," any response with a status code between 200 and 299 as a "2xx response," and so on.

- *SIP requests:* The core SIP specification defines six types of SIP requests, each of them with a different purpose. Every SIP request contains a field, called a method, which denotes its purpose.
    - INVITE: INVITE requests invite users to participate in a session. The body of INVITE requests contains the description of the session. Significantly, SIP only handles the invitation to the user and the user's acceptance of the invitation. All of the session particulars are handled by the SDP used. Thus, with a different session description, SIP can invite users to any type of session.
    - ACK: ACK requests are used to acknowledge the reception of a final response to an INVITE. Thus, a client originating an INVITE request issues an ACK request when it receives a final response for the INVITE.
    - CANCEL: CANCEL requests cancel pending transactions. If a SIP server has received an INVITE but not yet returned a final response, it will stop processing the INVITE upon receipt of a CANCEL request. If, however, it has already returned a final response for the INVITE, the CANCEL request will have no effect on the transaction.
    - BYE: BYE requests are used to abandon sessions. In two-party sessions, abandonment by one of the parties implies that the session is terminated.
    - REGISTER: Users send REGISTER requests to inform a server (in this case, referred to as a registrar server) about their current location.
    - OPTIONS: OPTIONS requests query a server about its capabilities, including which methods and which SDPs it supports.

SIP is independent of the type of multimedia session handled and of the mechanism used to describe the session. Sessions consisting of RTP streams carrying audio and video are usually described using SDP, but some types of sessions can be described with other description protocols. In short, SIP is used to distribute session descriptions among potential participants. Once the session description is distributed, SIP can be used to negotiate and modify the parameters of the session and terminate the session.

7.3.3   Basic concepts of QoS

VoIP is one of the most sensitive services to ensure good QoS, and to compete with the traditional voice service like PSTN, we need a specific level of QoS. The QoS in VoIP applications depends on many parameters, such as bandwidth, one way delay (OWD), jitter, packet-loss rate (PLR), CODEC type, voice data length, and de-jitter buffer size. Particularly, jitter and PLR have an important impact on the voice quality. QoS is an important subject that takes a central place in the IP technology; it is a complex subject and its analysis involves mathematical disciplines such as probability theory and stochastic processes [21].

QoS can be defined from three different points of views [1]: QoS experienced by the end user, the QoS from the point of view of the application, and from the network.

From the end user's perspective, QoS is the end user's perception of the quality that he receives from the VoIP services provider. The end-user perception of the voice quality is determined by subjective (MOS—mean opinion score)/objective (E-Model) testing as a function of the some impairments (OWD, PLR, CODEC type).

From the point of view of the application, the QoS refers to the application's capabilities to reconfigure some parameters (voice data length, CODEC type, size of FEC (forward error correction) redundancy, de-jitter buffer size, etc.) to values based on network conditions in order to meet good levels of voice quality.

From the network's perspective, the term QoS refers to the network's capabilities to provide the QoS perceived by the end user as defined above. A QoS mechanism has the capability to provide resource assurance and service differentiation in the IP network. Without a QoS mechanism, an IP network provides the *best effort* service. The IETF has proposed many QoS mechanisms, such as [22] integrated services (IntServ) [23,24], differentiated services (DiffServ) [25], multiprotocol label switching [26,27], and traffic engineering [27].

7.3.4   QoS assessment

One of the most employed metrics for the QoS assessment is the MOS [28], which can be estimated using the E-Model's R factor (elaborated later) [29,30].

The MOS method is derived from the absolute category rating method for assessing voice-transmission systems. The MOS requires that some people (listeners) appraise the general quality of voice samples submitted to vocoders. This numerical method is based on averaging the scores provided by many listeners where the scores 1, 2, 3, 4, and 5 imply "bad," "poor," "fair," "good," and "excellent" ratings, respectively. Although the method seems very clear, the MOS grading is a hard procedure to be performed. Objective methods (such as the E-model) have been developed and implemented to estimate this punctuation in VoIP systems.

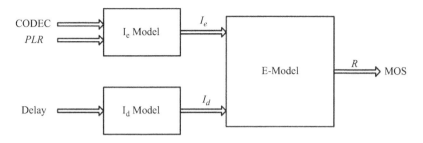

Figure 7.4   Prediction of voice quality by the E-Model

The E-Model is a computational model designed to produce the MOS without conducting subjective testing. The subjective testing is impractical because it is time-consuming and can be substituted for the E-Model.

In the E-Model, the effects of delay, packet loss, and other relevant impairments are combined into a total value called "Transmission Rating Factor" (R factor) that varies from 0 (worst case) to 100 (excellent). The factor is expressed as the sum of four terms as

$$R = R_0 - I_s - I_d - I_e + A \tag{7.1}$$

where R_0 represents the basic signal-to-noise ratio, I_s represents the combination of all impairments which occur (more or less) simultaneously with the voice signal, I_d represents the impairments caused by delay, I_e represents impairments caused by low bit rate CODECs, and A is the advantage factor that corresponds to the user allowance due to the convenience in using a given technology. Once the used CODEC is well-known, only network statistics (i.e. delay and loss) is needed for estimating the voice quality by means of the reduced R factor expression [29]:

$$R = 93.2 - I_d(\text{T}) - I_e(\text{CODEC}, \text{PLR}) \tag{7.2}$$

$$I_d = 0.024(T) + 0.11(T - 177.3)y(T - 177.3)$$

$$\text{T} = \text{OWD} = \frac{1}{2}\text{RTT}, \quad y(x) = \begin{cases} 0, \ x < 0 \\ 1, \ x > 0 \end{cases} \tag{7.3}$$

$$I_e(\text{G.711}) \sim 0 + 30\ln(1 + 15\text{PLR})$$
$$I_e(\text{G.729}) \sim 11 + 40\ln(1 + 10\text{PLR}) \tag{7.4}$$

where I_d is a function of the delay $_T$ and I_e is a function of the used CODEC type (G.711 or G.729) and $_{PLR}$. The delay components within the function I_d are OWD and round trip time delay. In order to simplify the expression for I_d, we used (7.3). Figure 7.4 illustrates how the E-Model may be used to predict the voice quality in VoIP applications.

Table 7.1   R-Factor, quality ratings, and the associated MOS

R-Factor	MOS	User's satisfaction
$90 \leq R < 100$	4.34–4.50	Very satisfied
$80 \leq R < 90$	4.03–4.34	Satisfied
$70 \leq R < 80$	3.60–4.03	Some users are dissatisfied
$60 \leq R < 70$	3.10–3.60	Many users are dissatisfied
$0 \leq R < 60$	1.00–3.10	Nearly all users are dissatisfied

Besides, the relationship between the R factor and MOS is given by the next expression:

$$
\begin{aligned}
\text{MOS} &= 1; & R &< 0 \\
\text{MOS} &= 1 + 0.035R + 7 \cdot 10^{-6} R(R - 60)(100 - R); & 0 &\leq R \leq 100 \\
\text{MOS} &= 4.5; & R &> 100
\end{aligned}
\tag{7.5}
$$

Typically, the R factor values are categorized as shown in Table 7.1 [29]:

7.3.5   Oneway delay

The delay experienced by a packet across a path consists of several components: propagation, processing, transmission, and queuing delays [31]. The OWD [32] is the time needed for a packet to traverse the network from source to destination terminal and is described analytically by the following equation:

$$
D(K) = \delta + \sigma + \sum_{h=1}^{s} \left(\frac{L}{C_h} + X_h(t) \right)
\tag{7.6}
$$

where $D(K)$ is the OWD of a packet K of size L δ represents the propagation delay, σ is the processing delay, s is the number of hops, L/C_h is the transmission delay and $X_h(t)$ is the queuing delay of a packet K of size L at hop $h(h = 1, \ldots, s)$ with capacity C_h [33].

7.3.6   Jitter

When voice packets are transmitted from source terminal (sender) to destination terminal (receiver) over IP networks, packets may experience variable delay, called jitter. The packet inter-arrival time (IAT) on the receiver side is not constant even if the packet inter-departure time (IDT) on the sender side is constant. As a result, packets arrive at the destination terminal with varying delays (between packets) referred to as jitter [34]. The difference between arrival times of successive voice packets that arrive on the receiver side is measured according to RFC 3550 [35]—this is illustrated in Figure 7.5. This figure shows the jitter measurement between the sending packets and the receiving packets.

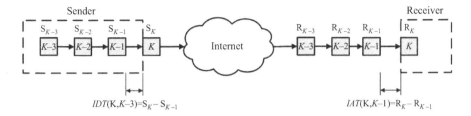

Figure 7.5 Jitter experienced across IP network

Let S_K be the RTP timestamp and R_K be the arrival time in RTP timestamp units for packet K. Then, for two packets K and $K - 1$, the OWD difference between two successive packets, K and $K - 1$ is given by the following equation [34]:

$$J(K) = (R_K - S_K) - (R_{K-1} - S_{K-1}) = (R_K - R_{K-1}) - (S_K - S_{K-1})$$

$$= \mathrm{IAT}(K) - \mathrm{IDT}(K) \tag{7.7}$$

$$\mathrm{IAT}(K) = J(K) + \mathrm{IDT}(K) \tag{7.8}$$

where $\mathrm{IDT}(K, K - 1) = (S_K - S_{K-1})$ is the IDT and $\mathrm{IAT}(K, K - 1) = (R_K - R_{K-1})$ is the IAT or arrival jitter for the packets K and $K - 1$. In the current context, $\mathrm{IAT}(K, K - 1)$ is referred to as jitter [34]

On the other hand, when voice packets are transported over IP networks, they may experience delay variations and packet loss. From (7.8), a relationship between jitter and packet loss can be established using the following equations [34]:

If packet $K - 1$ is lost,

$$\mathrm{IAT}(K) = J(K) + (2) \cdot \mathrm{IDT}(K) \tag{7.9}$$

Therefore, if n consecutive packets are lost,

$$\mathrm{IAT}(K) = J(K) + (n + 1) \cdot \mathrm{IDT}(K) \tag{7.10}$$

Therefore, (7.10) describes the packetloss effects in the VoIP jitter.

7.3.7 Packetloss rate

There are two main transport protocols used on IP networks, UDP and transmission control protocol (TCP). While UDP protocol does not allow any recovery of transmission errors, TCP includes an error recovery process. However, the voice transmission over TCP connections is not very realistic. This is due to the requirement for real-time (or near real-time) operations in most voicerelated applications. As a result, the choice is limited to the use of UDP which involves packetloss problems. Packet loss can occur in the network or at the receiver side, for example, due to excessive network delay in the case of network congestion [2].

Owing to the dynamic, timevarying behavior of packet networks, packet loss can show a variety of distributions. The loss distribution most often studied in speech

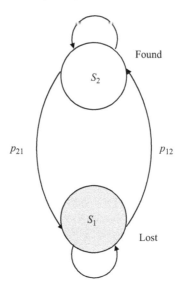

Figure 7.6   Two-state Markov chain

quality tests is random or Bernoulli-like packet loss. Random loss here means independent loss, implying that the loss of a particular packet is independent of whether or not previous packets were lost. However, random loss does not represent the loss distributions typically encountered in real networks. For example, losses are often related to periods of network congestion. Hence, losses may extend over several packets, showing a dependency between individual loss events. In this chapter, dependent packet loss is often referred to as bursty. The packet loss is bursty in nature and exhibits temporal dependency [36]. As noted earlier in the introduction section, if packet n is lost then normally there is a high probability that packet $n + 1$ will also be lost. Consequently, there is a strong correlation between consecutive packet losses, resulting in a bursty packetloss behavior. Hence, this temporal dependency can be effectively modeled by a finite Markov chain [3637].

Let $S = S_1, S_2, \ldots, S_m$ be the m states of an m-state Markov chain and let p_{ij} be the probability of the chain to pass from state S_i to the state S_j. The probabilities of transitions between states can be represented by the transition matrix P[2]:

$$P = \begin{bmatrix} S_1 & S_2 & \cdots & S_m \\ S_1 & S_2 & \cdots & S_m \\ \vdots & \vdots & \ddots & \vdots \\ S_1 & S_2 & \cdots & S_m \end{bmatrix} \tag{7.11}$$

such that $S_1 + S_2 + \cdots + S_m = 1$

In the twostate Markov chain (see Figure 7.6), one of the states (S_1) represents a packet loss and the other state (S_2) represents the case where packets are correctly

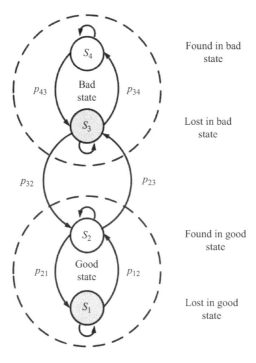

Found in bad
state

Lost in bad
state

Found in good
state

Lost in good
state

Figure 7.7   Four-state Markov chain

transmitted or received. The transition probabilities in this model are represented by p_{21} and p_{12}. In other words, p_{21} is the probability of going from S_1 to S_2, and p_{12} is the probability of going from S_1 to S_2. Different values of p_{21} and p_{12} define different packetloss conditions that can occur on the Internet [2].

The steady-state probability of the chain to be in the state S_1, namely the PLR, is given by the following equation [30]:

$$S_1 = \frac{p_{21}}{p_{21} + p_{12}} \tag{7.12}$$

and clearly $S_2 = 1 - S_1$

The collected data traces in real IP networks can be modeled accurately with a higher number of states, i.e., n-state Markov chains. However, for network planning, a trade-off is desirable between very accurate modeling of data traces and a low number of model input parameters in order to yield a model still usable for network planners with reasonable effort. Therefore, we used a simplification of an n-state chain, i.e., the four-state Markov chain. Figure 7.7 shows the state diagram of this four-state Markov chain. In this model, a "*good*" and a "*bad*" state are distinguished, which represent periods of lower and higher packet loss, respectively. Both for the "*bad*" and the "*good*" state, an individual two-state Markov chain represents the dependency between consecutively lost or found packets [2]

The two two-state chains can be described by four independent transition probabilities (two for each one). Two further probabilities characterize the transitions between the two two-state chains leading to a total of six independent parameters for this particular four-state Markov chain [2].

In the four-state Markov chain, states S_1 and S_3 represent packets lost, S_2 and S_4 packets found, and six parameters $(p_{21}, p_{12}, p_{43}, p_{34}, p_{23}, p_{32} \in (0, 1))$ are necessary to define all the transition probabilities. The four steady-state probabilities of this chain are [38]

$$S_1 = \frac{1}{1 + (p_{12}/p_{21}) + (p_{12}p_{23}/p_{21}p_{32}) + (p_{12}p_{23}p_{34}/p_{21}p_{32}p_{43})} \tag{7.13}$$

$$S_2 = \frac{1}{1 + (p_{21}/p_{12}) + (p_{23}/p_{32}) + (p_{23}p_{34}/p_{32}p_{43})} \tag{7.14}$$

$$S_3 = \frac{1}{1 + (p_{34}/p_{43}) + (p_{32}/p_{23}) + (p_{21}p_{32}/p_{12}p_{23})} \tag{7.15}$$

$$S_4 = \frac{1}{1 + (p_{43}/p_{34}) + (p_{32}p_{43}/p_{23}p_{34}) + (p_{21}p_{32}p_{43}/p_{12}p_{23}p_{34})} \tag{7.16}$$

The probability of the chain to be either in S_1 or in S_3, which corresponds to PLR, is then $r = S_1 + S_3$ [38].

7.4 Self-similarity processes in modern communication networks

Modern investigations of traffic measurements suggest that the IP traffic is self-similar [7,39–41]. The fact that IP traffic often has self-similar behavior means that it shows noticeable bursts at a wide range of time scales. One characteristic of this self-similar traffic is that it has LRD, i.e. the traffic is strongly correlated at all time scales of engineering interest [5]. Some consequences of LRD traffic are that it increases queue lengths and latency dramatically [5].

7.4.1 Self-similar processes

Certain processes are said to be self-similar if certain property of the processes is preserved with respect to scaling in space and/or time. An attractive property of the self-similar processes for modeling a time series of IP traffic is the degree of self-similarity, which is expressed with a single parameter called Hurst parameter. This parameter expresses the speed of decay of the autocorrelation function of the time series. In this section, a brief overview of self-similar processes is given from [42–44].

Continuous self-similarity: A real-valued continuous time series $\{X(t), t \in \Re\}$ is self-similar with the exponent $0 < H < 1$ if, for any $a > 0$, the finite-dimensional distributions of $\{X(at), t \in \Re\}$ are identical to the finite-dimensional distributions of $\{a^H X(t), t \in \Re\}$, i.e., $\{X(at), t \in \Re\} \overset{d}{=} \{a^H X(t), t \in \Re\}$, where $\overset{d}{=}$ denotes equality in distribution.

Discrete self-similarity: Let $X_t = (X_t; t \in N)$ denote a discrete time series with mean μ_X, variance σ_X^2, autocorrelation function $r(k)$, and ACV $\gamma(k)$, $k \geq 0$, where X_t can be interpreted as the jitter, at time instance t.

When considering discrete time series, the definition of self-similarity is given in terms of the aggregated processes, as following:

$$X_k^{(m)} = (X_k^{(m)}; k \in N) \tag{7.17}$$

where m represents the aggregation level and $X_k^{(m)}$ is obtained by averaging the original series X_t over nonoverlapping blocks of size m, and each term $X_k^{(m)}$ is given by

$$X_k^{(m)} = \frac{1}{m} \sum_{i=(k-1)m+1}^{km} X_i; \quad k = 1, 2, 3, \ldots \tag{7.18}$$

Then it is said that X_t is self-similar $(H - ss)$ with self-similarity parameter $(0 < H < 1)$ if

$$X_k^{(m)} \overset{d}{=} m^{H-1} X_t \tag{7.19}$$

Second-order discrete self-similarity: X_t is called exactly second order self-similar with Hurst parameter H if the variance and covariance of the aggregated time series are defined by (7.20) and (7.21), respectively:

$$var\left(X_k^{(m)}\right) = \sigma_X^2 \cdot m^{2H-2} \tag{7.20}$$

$$\gamma_X^m(k) = \frac{\sigma_X^2}{2}((k+1)^{2H} - 2k^{2H} + (k-1)^{2H})k \geq 1 \tag{7.21}$$

The time series X_t is called asymptotically second-order self-similar if

$$\lim_{m \to \infty} \gamma^m(k) = \frac{\sigma_X^2}{2}((k+1)^{2H} - 2k^{2H} + (k-1)^{2H}) \tag{7.22}$$

Second-order self-similarity (in the exact or asymptotic sense) has been a dominant framework for modeling IP traffic.

So far, the role of second-order self-similarity has been discussed but not much has been mentioned about the role of H and limiting values. The definition of LRD and its interconnection with the correlation factor $r(k)$ will now be discussed.

Let $r(k) = \gamma(k)/\sigma_X^2$ be the autocorrelation function of X_t with self-similarity parameter $0 < H < 1, H \neq 1/2$, then the asymptotic behavior of $r(k)$ is given by the following equation:

$$r(k) \sim H(2H-1)k^{2H-2}k \to \infty \tag{7.23}$$

In particular, if $1/2 < H < 1, r(k)$ asymptotically behaves as $ck^{-\eta}$ for $0 < \eta < 1$, where $c > 0$ is a constant, $\eta = 2 - 2H$ and this also means that the correlations are nonsummable: $\sum_{k=-\infty}^{\infty} r(k) = \infty$. That is, the autocorrelation function decays slowly. When $r(k)$ obeys a power-law, the corresponding stationary process X_t is called long-range dependent. On the other hand, X_t is short-range dependent if the sum $\sum_{k=-\infty}^{\infty} r(k) < \infty$ does not diverge.

Generally speaking, time series with LRD has a Hurst parameter $0.5 < H < 1$; on the other hand, time series with shortrange dependence (SRD) has a Hurst parameter $0 < H < 0.5$

Following are some simple facts regarding the value of H and its impact on $\gamma(k)$ [45].

1. $\gamma(k) = \begin{cases} 1, & k = 0 \\ 0, & k \neq 0 \end{cases}$ for $H = 0.5$. This is the well-known property of white Gaussian noise.

2. $\gamma(1) < 0$ for $0 < H < 0.5$

3. $\gamma(1) > 0$ for $0.5 < H < 1$

7.4.2 Haar wavelet-based decomposition and Hurst index estimation

The time series X_t can be decomposed into a set of time series [24], each defined by

$$C_{X,t}^{2,i} = X_t^{(2^{i-1}E)} - X_t^{(2^i E)} i \in \mathbb{N} \tag{7.24}$$

where $X_t^{(2^i E)}$ is the time series X_t after two operations, which are

Aggregation at level 2^i, as defined by (7.17) and (7.18), i.e., $m = 2^i$

Expansion of level 2^i, which consists in "repeat" each element of a time series 2^i times.

That is, $X_j^{(2^i E)} = X_k^{(2^i)}$ for $k = 1 + \lfloor (j-1)/2^i \rfloor$ and $j \in \mathbb{N}$

These zero mean components $C_{X,t}^{2,i}$ have three important properties:

They synthesize the original time series without loss, i.e.,

$$X_t = \sum_i C_{X,t}^{2,i} \tag{7.25}$$

They are pair-wise orthogonal:

$$\langle C_{X,t}^{2,i_1}, C_{X,t}^{2,i_2} \rangle = 0; \quad \text{for } i_1 \neq i_2 \tag{7.26}$$

If X_t is exactly self-similar, then the variance of the components comply:

$$\text{var}(C_{X,t}^{2,i}) = 2^{2H-2} \cdot \text{var}(C_{X,t}^{2,i-1}) \tag{7.27}$$

Properties 1, 2, and 3 imply the following equations:

$$\sigma_X^2 = \sum_i \text{var}(C_{X,t}^{2,i}) \tag{7.28}$$

$$\sigma_X^2 = \frac{1}{1 - 2^{2H-2}} \cdot \text{var}(C_{X,t}^{2,1}) \tag{7.29}$$

Then, the variance of the ith component is related to the variance of X_t as follows:

$$\text{var}(C_{X,t}^{2,i}) = (1 - r) \cdot r^{i-1} \cdot \sigma_X^2 \tag{7.30}$$

where

$$r = 2^{2H-2} \tag{7.31}$$

The plot $\log_2 [\text{var}(C_{X,t}^{2,i})]$ vs. i is equivalent to the wavelet-based diagram proposed in [46], the logscale diagram (LD); i.e., $\text{var}(C_{X,t}^{2,i}) = (E|d_X(j,\cdot)|^2/2^j)$ when using the Haar family of wavelet basis functions $\psi_{j,k}(t) = 2^{-j/2}\psi_0(2^{-j}t - k)$ (see [47]) where:

$$\psi_0(t) = \begin{cases} +1 & 0 \le t < 1/2 \\ -1 & 1/2 \le t < 1 \\ 0 & \text{otherwise} \end{cases} \tag{7.32}$$

For a finite-length time series with "L" octaves, the number of octaves (j) of the LD is related to index i of (7.24), according to (7.33).

$$j = i \tag{7.33}$$

The LD of an exactly self-similar time series is a straight line. Hence, a linear regression can be applied in order to estimate the Hurst index.

7.5   QoS parameters modeling on VoIP traffic

This section presents a case study to illustrate the jitter and packet-loss modeling on VoIP traffic by means of network measurements. In order to accomplish the above-mentioned study, extensive jitter and packetloss measurements were collected as follows [2]:

- Test calls were established by a VoIP application called "Alliance Foreign eXchange Station" [48].
- The jitter and packet loss were measured by Wireshark [49] to obtain a set of data traces.
- The measurement scenario was based on a typical H.323 architecture [18] see Figure 7.8.
- The parameter configuration employed in the test calls is shown in Table 7.2:
  - Four simultaneous test calls were established between A1/B1, A2/B2, A3/B3 and A4/B4 endpoints, see Figure 7.8 and Table 7.2.
  - The configurations used in the test calls are based on two parameters: CODEC type (G.711 and G.729) and voice data length (10, 20, 40 and 60 ms), see Table 7.2.
  - The measurement periods were 1 h for each test call (call duration time).
  - For each measurement period (1 h), four jitter and packetloss data traces were obtained.
  - The four configuration sets contain more than 113 million voice packets corresponding to 710 jitter and 710 packetloss data traces, measured during typical working hours [50].

7.5.1   *Jitter modeling by self-similar and multifractal processes*

Time-dependent statistics (e.g., correlation) is important for performance evaluation of IP networks. The statistics can be used to measure the impact of specific

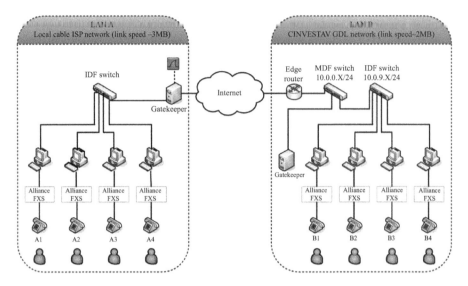

Figure 7.8   Measurement scenario [2]

Table 7.2   Parameter configuration employed in the test calls [2]

Set	A1/B1 (ms)	A2/B2 (ms)	A3/B3 (ms)	A4/B4 (ms)
1	G.711-10	G.711-20	G.711-40	G.711-60
2	G.729-10	G.729-20	G.729-40	G.729-60
3	G.711-10	G.711-20	G.729-10	G.729-20
4	G.711-40	G.711-60	G.729-40	G.729-60

impairments. Several studies have found that self-similarity and LRD can have a negative impact on the IP traffic, because they give rise to great losses and/or delays. For this reason, it is important to analyze the correlation structures (SRD and LRD) of the VoIP traffic. The Hurst parameter is used to measure the degree of self-similarity and LRD. Generally speaking, time series with LRD has a Hurst parameter of $0.5 < H < 1$; on the other hand, time series with SRD has a Hurst parameter of $0 < H < 0.5$

Unlike other statistics, the Hurst parameter, although mathematically well defined, cannot be estimated unambiguously from real-world samples. Therefore, several methods have been developed in order to estimate the Hurst parameter. Examples of classical estimators are those based on the R/S statistics [51] (and its unbiased version [52]), detrended fluctuation analysis [52,53], maximum likelihood (ML) [54], aggregated variance (VAR) [51], wavelet analysis [46], etc. In [55], Clegg developed an empirical comparison of estimators for data in raw form and corrupted in various ways. An important observation is that the estimation of the Hurst parameter may

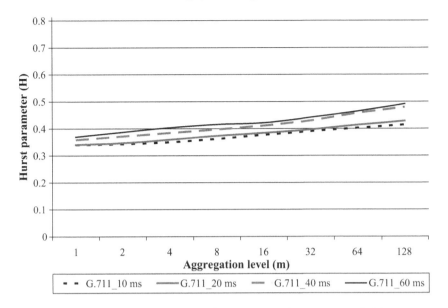

Figure 7.9   Hurst parameter for VoIP jitter data traces with SRD

differ from one estimator to another, which makes the selection of the most adequate estimator a difficult task. It seems to depend on how well the data sample meets the assumptions the estimator is based on. However, through analytical and empirical studies, it has been discovered that the estimators, that have the best performance in bias and standard deviation, and, consequently, in mean squared error (MSE), are the Whittle ML and the wavelet-based estimator proposed by Veitch and Abry in [46]. From these two estimators, the wavelet-based one is computationally simpler and faster [46,51].

Motivated by the above inferences, and following the methodology proposed in [56] to find correlations and LRD, the Hurst parameter is estimated by the wavelet-based estimator [46] of jitter data traces as a function of the aggregation level m ($m = \{1, 2, 4, 8, 16, 32, 64, 128\}$). Figures 7.9 and 7.10 show the Hurst parameter of representative jitter data traces to different aggregation levels m. Generally, Hurst parameters larger than 0.5 for all aggregation levels are a strong indication of LRD. It can be observed from Figure 7.9 that a set of jitter data traces has Hurst parameters larger than 0.5 for all aggregation levels. This indicates a high degree of LRD. In contrast, the other sets of jitter data traces shown in Figure 7.10 have Hurst parameters lower than 0.5. These results are thus not a strong indication of LRD. This indicates that the ACVs decay quickly to zero, indicating no memory property (SRD).

These results show that VoIP jitter exhibits self-similar characteristics with SRD or LRD; therefore, a self-similar process can be used to model the jitter behavior.

The discovery of LRD and weak self-similarity in the VoIP jitter data traces was followed by a further work that shows the evidence for multifractal behavior.

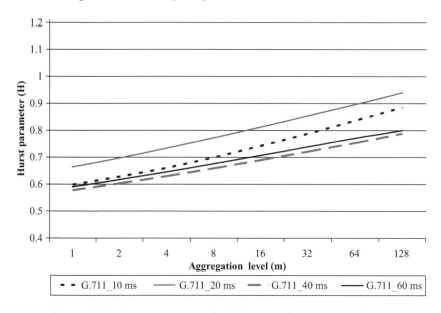

Figure 7.10   Hurst parameter for VoIP jitter data traces with LRD

The discovery of evidence for multifractal behavior is a richer form of scaling behavior associated with nonuniform local variability, which could lead to a complete and robust model of VoIP traffic over all time scales of engineering interest.

In order to accomplish this analysis, we decomposed the time series of VoIP jitter into a set of time series or components $C_{X,t}^{2,i}$ as it is defined in (7.24). The behavior of these components is used to determine the kind of asymptotic fractal scaling. If the variance of the components of a time series is modeled by a straight line, the time series exhibits monofractal behavior. Then, a linear regression can be applied in order to estimate the Hurst parameter. On the other hand, if the variance of the components cannot be adequately modeled with a linear model, the scaling behavior should be described with more than one scaling parameter, i.e., the time series exhibits multifractal behavior [57]. In Figures 7.11 and 7.12, we show the component behavior of the collected VoIP jitter data traces.

Figure 7.11 shows the component behavior of VoIP jitter data traces that belong to the data sets with SRD. It is observed that the variance of the components of this time series is modeled by a straight line; therefore, the time series exhibits monofractal behavior.

Figure 7.12 shows the component behavior of VoIP jitter data traces that belong to the data sets with LRD. It is observed that the variance of the components of this time series cannot be adequately modeled with a linear model, and the scaling behavior should be described with multiple scaling parameters (biscaling). Therefore, this time series exhibits multifractal behavior.

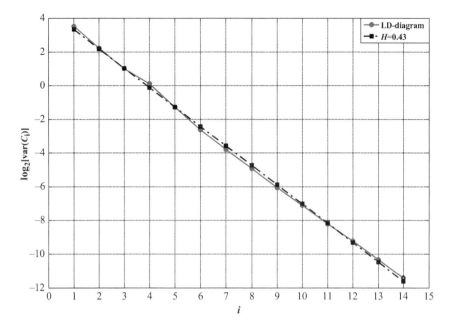

Figure 7.11   Component behavior of VoIP jitter data traces: monofractal behavior

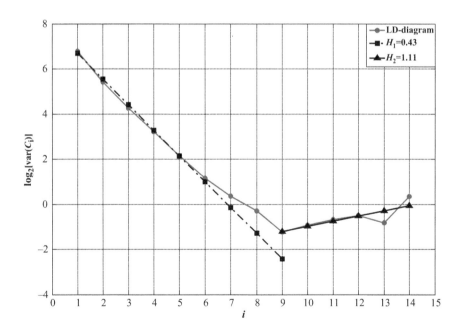

Figure 7.12   Component behavior of VoIP jitter data traces: multifractal behavior

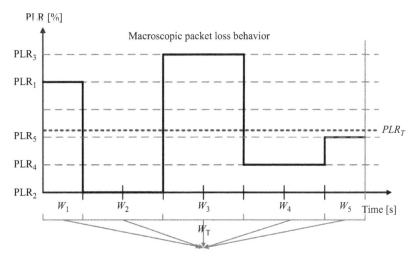

Figure 7.13   Microscopic and macroscopic behavior

These results show that VoIP jitter with SRD and LRD exhibit monofractal and multifractal behavior, respectively. This phenomenon explains the behavior of the data traces with SRD and high degree of self-similarity (scale invariance), because the self-similarity is defined for a single-scale parameter. On the other hand, the data traces with LRD exhibit weak self-similarity because of having associated nonuniform local variability (multifractal behavior).

7.5.2   Packet-loss modeling by Markov models

In this chapter, a description of VoIP packet loss based on narrow and wide time windows is used. The packet-loss behavior over a narrow time window is called microscopic, and the packet-loss behavior over wide time windows is called macroscopic [58]. Microscopic behavior refers to a packet-loss period observed on a "time window" W_1 of the packet-loss data trace where this packet-loss period has a specific PLR_1. On the other hand, macroscopic behavior refers to a set of microscopic periods $(W_1, W_2, W_3, \ldots, W_n)$ that are observed on all packet-loss data traces, where each microscopic period has a particular $(PLR_1, PLR_2, PLR_3, \ldots, PLR_n)$ as shown in Figure 7.13. Figure 7.13 illustrates different levels of PLR for each microscopic period. Therefore, the packet losses do not occur homogeneously. Instead, they are concentrated in some time intervals (i.e., the packet loss is bursty).

Microscopic behavior can be effectively modeled by a Markov chain with a low number of states [58]. On the other hand, macroscopic behavior can be effectively modeled by a Markov chain with a higher number of states [58]. Here, substates represent phases of a given microscopic behavior. Ideally, an *n*-state Markov chain

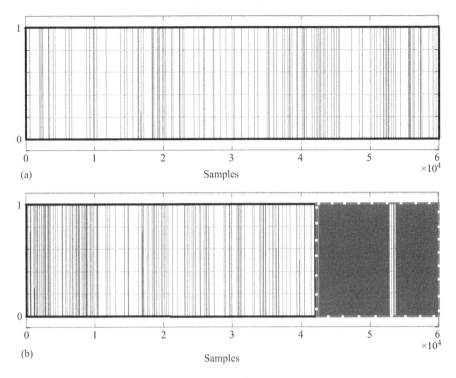

Figure 7.14   Packet-loss patterns from VoIP test calls: (a) homogeneous PLR and (b) nonhomogeneous PLR

is required to capture the macroscopic behavior. Especially for network planning, a trade-off is desirable between very accurate modeling of data traces and a low number of model input parameters in order to yield a model still usable for network planners with reasonable effort. Therefore, we used the two-state and four-state Markov chains proposed in Section 7.3.7 for modeling the microscopic and macroscopic periods, respectively.

Figure 7.14(a) and (b) shows some packet-loss patterns extracted from VoIP test calls. In Figure 7.14(a), we can see that packet-loss behavior is homogeneous, i.e., the packet-loss pattern is represented by a microscopic period with a low level of *PLR*. Therefore, this packet-loss pattern can be modeled by a two-state Markov chain.

On the other hand, in Figure 7.14(b), the packet loss is nonhomogeneous, i.e., the packet-loss pattern is represented by a concatenation of two microscopic periods.

In order to simplify this packet-loss description, the microscopic periods can be classified into two sets, one for low and one for high PLRs. Therefore, this packet-loss pattern can be modeled by a four-state Markov chain, where the microscopic period with lower PLR is represented by the good state (see Figure 7.6), and the microscopic period with higher PLR is represented by the bad state (see Figure 7.7).

The threshold used to delimit between a good state (low level of packet loss) or bad state (high level of packet loss) is a function of the perceived quality, good or poor, respectively, according to the computed MOS values.

In Figure 7.14(a) and (b), the microscopic period with lower PLR is delimited by the solid square, while the microscopic period with higher PLR is delimited by the dashed square.

7.5.3   Packet-loss simulation and proposed model

The current methodologies for simulating packet loss consist only of generating packet-loss patterns by Markov chains of different orders [58]. Therefore, the studies based on this methodology are limited, because the impact of this parameter is ana-lyzed separately from the others. A new methodology to simulate packet loss in two stages is proposed: first, by generating packet-loss pattern by a four-state Markov chain, and second, by applying this packet-loss pattern to a VoIP jitter data trace, i.e., the simulation of the effect of this packet-loss pattern in the VoIP jitter by the relationship between packet loss and jitter, as shown in (7.10).

In order to simplify this methodology and achieve the trade-off between very accurate modeling of data traces and a low number of model input parameters, we consider the following:

- The relationship between PLR and jitter, summarized by (7.10).
- The description of VoIP packet loss based on microscopic and macroscopic behaviors.
- A set of real jitter and packet-loss data traces, where the jitter data traces exhibit self-similar characteristics and the packet-loss data traces are homogeneous, as shown in Figure 7.14(a). Then, these packet-loss patterns can be represented by only one microscopic period with a low level of PLR, PLR_0 (near zero).
- In order to incorporate n microscopic packet-loss periods with high level of PLR, as it is shown in Figure 7.14(b), we generated packet-loss patterns by means of a four-state Markov chain.

Let $X = \{X_t : t = 1, \ldots, N\}$ be a VoIP jitter data trace with a length of N, self-similar (H parameter $0 < H_0 < 0.5$), and with a low PLR, PLR_0. The packet-loss patterns are generated by means of a four-state Markov chain and are represented as the binary sequences $P = \{P_t^\tau : t = 1, \ldots, N; \ \tau = 0, 1, 2, \ldots, T - 1\}$, where $P_t^\tau = 1$ means a packet loss, $P_t^\tau = 0$ means a success or received packet, N is the length of the packet-loss pattern, and T is the number of packet-loss patterns used. The relationship between jitter and packet loss from (7.10) is used to apply the packet-loss patterns to X_t by the algorithm shown in Table 7.3.

As a result of using the above algorithms, the new time series $\hat{X}_t^\tau$ was obtained, for $t = 1, \ldots, N$ and $\tau = 0, 1, 2, \ldots, T - 1$. For each $\hat{X}_t^\tau$, the H and parameter were calculated, and the functions $f(PLR_\tau, H_\tau)$ were generated.

The simulations are accomplished over the measured VoIP jitter data traces. Figure 7.15 illustrates the relationships between the PLR and the Hurst parameter.

Table 7.3   Algorithm for applying the
packet-loss patterns [2]

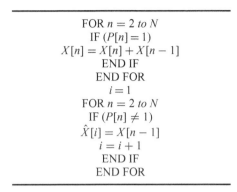

$$\text{FOR } n = 2 \text{ to } N$$
$$\text{IF } (P[n] = 1)$$
$$X[n] = X[n] + X[n-1]$$
$$\text{END IF}$$
$$\text{END FOR}$$
$$i = 1$$
$$\text{FOR } n = 2 \text{ to } N$$
$$\text{IF } (P[n] \neq 1)$$
$$\hat{X}[i] = X[n-1]$$
$$i = i + 1$$
$$\text{END IF}$$
$$\text{END FOR}$$

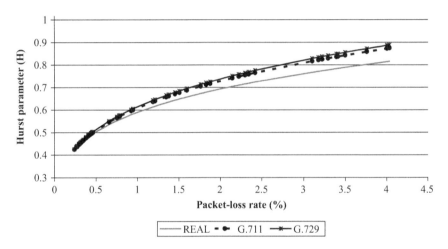

Figure 7.15   Relationship between PLR and H parameter: $f(\text{PLR}_\tau, H_\tau)$ *vs.*
$f_{\text{REAL}}(\text{PLR}_\varepsilon, H_\varepsilon)$

The figure shows the empirical functions $f(\text{PLR}_\tau, H_\tau)$ that were obtained from simulation results and the function $f_{\text{REAL}}(\text{PLR}_\varepsilon, H_\varepsilon)$.

The functions $f(\text{PLR}_\tau, H_\tau)$ resulted from applying T packet-loss patterns to representative VoIP jitter data traces X_t. In these functions, each point represents the PLR_τ and H_τ of a particular new time series $\hat{X}_t^\tau$.

The function $f_{\text{REAL}}(\text{PLR}_\varepsilon, H_\varepsilon)$ is generated by "E" jitter data traces. In this function, each point represents the PLR_ε and H_ε of a particular jitter data trace X_t^ε, where $t = 1, \ldots, N$, $\varepsilon = 1, 2, \ldots, E$, and "E" is the number of representative jitter data traces used.

Table 7.4   Fitted parameters of Figure 7.15 [2]

$f(\text{PLR}_\tau, \text{H}_\tau)$	$\hat{H}_0$	$\hat{a}$	$\hat{b}$	MSE
$G.711$	0.0428	0.5659	0.2760	0.001474
$G.729$	0.0430	0.5716	0.2805	0.002305
$f_{\text{REAL}}(\text{PLR}_\varepsilon, \text{H}_\varepsilon)$	0.0429	0.5471	0.2475	

The respective differences between the functions corresponding to simulation results $f(\text{PLR}_\tau, \text{H}_\tau)$ and the function $f_{\text{REAL}}(\text{PLR}_\varepsilon, \text{H}_\varepsilon)$ were quantified in terms of MSE.

Table 7.4 shows the fitted parameters and MSE between $f(\text{PLR}_\tau, \text{H}_\tau)$ and $f_{\text{REAL}}(\text{PLR}_\varepsilon, \text{H}_\varepsilon)$. The results are summarized in Figure 7.15 and Table 7.4. As a result of this analysis, we conclude that the presented methodology for simulating packet loss has achieved the trade-off between a very accurate modeling and a low number of model input parameters. Based on these simulation results, it is proposed that the relationships found between PLR and Hurst parameter can be modeled by a power-law function, characterized by three fitted parameters, as follows:

$$H_M = \hat{H}_0 + \hat{a}(\text{PLR})^{\hat{b}} \tag{7.34}$$

where H_M is the H parameter of the model found; $\hat{H}_0$, $\hat{a}$, and $\hat{b}$ are the fitted parameters; $\hat{H}_0$ is the H parameter when $\text{PLR} = 0$. The fitted parameters are estimated by linear regression. The strategy to find $\hat{H}_0$, $\hat{a}$, and $\hat{b}$ is such that it minimizes the MSE, i.e., $\text{MSE} = \int_r (\hat{H}_0 + \hat{a}r^{\hat{b}} - H_\tau)^2 dr$, and the validity of the proposed model corresponds to those ranges of $r = \text{PLR}$ (e.g., 0%–4%).

7.6   Conclusions

The fast development and evolution of communication networks and the emergence of enhanced services (e.g., VoIP) that can be offered to the end users have been captured in the concept of modern communication network. Modern communication networks are complex in nature and have the property of scale-free, meaning that the numbers of nodes and connections increases continuously and congestions occur accordingly. Congestion is a cause of impairment in real-time multimedia applications, such as VoIP. The QoS level for a VoIP application depends on many parameters; however, jitter and packet loss have an important impact on the QoS. However, an efficient network design can be achieved by using accurate models.

The current chapter presents the jitter and packet-loss modeling of VoIP traffic by means of network measurements and could be useful both for today's networks and future networks supporting VoIP/mVoIP technologies.

References

[1]   A.-S.K. Pathan, M.M. Monowar, and Z.M. Fadlullah, *Building Next-Generation Converged Networks: Theory and Practice* USA: CRC Press Taylor & Francis Group, 2013, 337–360.

[2]   H. Toral-Cruz, A.-S.K. Pathan, and J.C.R. Pacheco, Accurate modeling of VoIP traffic QoS parameters in current and future networks with multifractal and Markov models, *Mathematical and Computer Modelling Journal*, 57 (11–12) (2013): 2832–2845.

[3]   D. Geer, The future of mobile VoIP in the enterprise, *IEEE Computer*, 42 (6) (2009): 15–18.

[4]   S.K. Chui, O.-C. Yue, and W.C. Lau, impact of handoff control messages on VoIP over wireless LAN system capacity, *Proceedings of the 14th European Wireless Conference (EW)*, 22–25 June 2008, 1–5.

[5]   L. Kocarev and G. Vattay, *Complex Dynamics in Communication Networks*. Germany: Springer-Verlag Berlin Heidelberg, 2005.

[6]   A. Nogueira, P. Salvador, R. Valadass, and A. Pacheco, Modeling network traffic with multifractal behavior, *Telecommunication Systems*, 24 (2–4) (2003): 339–362.

[7]   O.I. Sheluhin, S.M. Smolskiy, and A.V. Osin, *Self-Similar Processes in Telecommunications*. England: John Wiley & Sons, Ltd, 2007.

[8]   C. Larsson, *Design of Modern Communication Networks: Methods and Applications*. The Netherlands: Academic Press, 2014.

[9]   J.F. Kurose and K. W. Ross, *Computer Networking: A Top-Down Approach*. USA: Addison-Wesley, 2010.

[10] T. Janevski, *Traffic Analysis and Design of Wireless IP Networks*. USA: Artech House, Inc., 2003.

[11] H. Hanrahan, *Network Convergence: Services, Applications, Transport, and Operations Support*. England: John Wiley & Sons, Ltd, 2007.

[12] J. Jo, G. Hwang and H. Yang, Characteristics of QoS Parameters for VoIP in the Short-Haul Internet, *Proc. International Conferences on Info-tech and Info-net (ICII), IEEE*, Beijing, China, 29 October–1 November 2001, pp. 498–502.

[13] R. Albert and A.-L. Barabási, Topology of Evolving Networks: Local Events and Universality, *Physical Review Letters*, 85 (24) (2000): 5234–5237.

[14] ITU-T Recommendation G.1020, Performance Parameter Definitions for Quality of Speech and Other Voiceband Applications Using IP Networks, 2006.

[15] L. Estrada-Vargas, Self-similar time series: analysis and modeling with applications to VoIP, Ph.D. Thesis, Electrical Engineering, Telecommunication Section, CINVESTAV, Guadalajara, Jalisco, Mexico, 2015.

[16] S. Madhani, S. Shah, A. Gutierrez, Optimized Adaptive Jitter Buffer Design for Wireless Internet Telephony, *Proc. Global Telecommunications Conference (GLOBECOM), IEEE*, Washington, D.C., USA, 26–30 November 2007, pp. 5248–5253.

[17]   A. Sulkin, PBX Systems for IP Telephony: Migrating Enterprise Communications. New York, NY: McGraw-Hill Professional, 2002.

[18]   ITU-T, H.323: Packet-Based Multimedia Communication Systems, Telecommunication Standardization Sector, Geneva, Switzerland, 2009.

[19]   G. Camarillo, SIP Demystified, USA: McGraw-Hill Professional, 2002.

[20]   J. Rosenberg, H. Schulzrinne, G. Camarillo, *et al.*, RFC 3261 – SIP: Session Initiation Protocol, Internet Engineering Task Force, June, 2002.

[21]   K.I. Park, *QoS in Packet Networks*. USA: Springer Science + Business Media, Inc, 2005.

[22]   R. Hunt, A review of quality of service mechanisms in IP-based networks: integrated and differentiated services, multi-layer switching, MPLS and traffic engineering, *Computer Communications*, 25 (1) (2002): 100–108.

[23]   R. Branden, D. Clark, S. Shenker, Integrated Services in the Internet Architecture: An Overview, RFC1633, 1994.

[24]   R. Braden, L. Zhang, S. Berson, S. Herzog, and S. Jamin, Resource ReSerVation Protocol (RSVP) – Version 1 Functional Specification, RFC 2205, 1997.

[25]   S. Blake, D. Black, M. Carlson, E. Davies, Z. Wang, and W. Weiss, An Architecture for Differentiated Services, RFC 2475, 1998.

[26]   C. Semeria, J. Stewart, Optimizing Routing Software for Reliable Internet Growth, Juniper Networks White Papers, www.juniper.net/techcenter/techpapers/200003.pdf, 2000.

[27]   D. Awduche, J. Malcolm, J. Agogbua, M. O'Dell, and J. McMaus, Requirements for Traffic Engineering Over MPLS, RFC 2702, 1999.

[28]   ITU-T, P. 800: Methods for Subjective Determination of Transmission Quality, Telecommunication Standardization Sector, Geneva, Switzerland, 1996.

[29]   ITU-T, G.107: The E-Model, A Computational Model for Use in Transmission Planning, Telecommunication Standardization Sector, Geneva, Switzerland, 2009.

[30]   ITU-T, G.108: Application of the E-Model: A Planning Guide, Telecommunication Standardization Sector, Geneva, Switzerland, 1999.

[31]   M.J. Karam and F.A. Tobagi, Analysis of the Delay and Jitter of Voice Traffic over the Internet, *Proc. Twentieth Annual Joint Conference of the IEEE Computer and Communications Societies (INFOCOM)*, IEEE, Anchorage, AK, USA, 22–26 April 2001, pp. 824–833.

[32]   ITU-T, G.114: One Way Transmission Time, Telecommunication Standardization Sector, Geneva, Switzerland, 2003.

[33]   R. Prasad, C. Dovrolis, M. Murray and K.C. Claffy, Bandwidth estimation: metrics, measurement techniques, and tools, *IEEE Network*, 17 (6) (2003): 27–35.

[34]   S. Kashihara, *VoIP Technologies*. Croatia: INTECH, 2011, 79–94.

[35]   H. Schulzrinne, S. Casner, R. Frederick and V. Jacobson, RFC 3550 – RTP: A Transport Protocol for Real-Time Applications, Internet Engineering Task Force, July, 2003.

[36] O. Hohlfeld, Stochastic Packet Loss Model to Evaluate QoE Impairments, *PIK Journal*, 1 (2009): 53–56.

[37] G. Haßlinger and O. Hohlfeld, The Gilbert-Elliott Model for Packet Loss in Real Time Services on the Internet, *Proc. 14th GI/ITG Conference on Measuring, Modelling and Evaluation of Computer and Communication Systems (MMB)*, 31 March 2008–2 April 2008, pp. 269–286.

[38] H. Toral-Cruz, D. Torres-Román, and L. Estrada-Vargas, "Analysis and Modeling of QoS Parameters in VoIP Traffic", Chapter 1 in Advancements in Distributed Computing and Internet Technologies: Trends and Issues. (A.-S.K. Pathan, M. Pathan and H. Y. Lee eds.), Hershey, PA, USA: IGI Global, 2011 pp. 1–22.

[39] T. Janevski, *Traffic Analysis and Design of Wireless IP Networks*. USA: Artech House Publishers, 2003.

[40] W.E. Leland, M.S. Taqqu, W. Willinger and D.V. Wilson, On the self-similar nature of Ethernet traffic (extended version), *IEEE/ACM Transactions on Networking (TON)*, 2 (1) (1994): 1–15.

[41] K. Park and W. Willinger, *Self-Similar Network Traffic and Performance Evaluation*. USA: John Wiley & Sons, Inc., 2000.

[42] K.M. Rezaul and V. Grout, A Survey of Performance Evaluation and Control for Self-Similar Network Traffic, *Proc. of the Second International Conference on Internet Technologies and Applications (ITA)*, 4–7 September 2007, pp. 514–524.

[43] G. Samorodnitsky, Long range dependence, *Foundations and Trends in Stochastic Systems*, 1 (3) (2007): 163–257.

[44] G. Zhang, G. Xie, J. Yang and D. Zhang, Self-Similar Characteristic of Traffic in Current Metro Area Network, *Proc. of the 15th IEEE Workshop on Local and Metropolitan Area Networks*, 10–13 June 2007, pp. 176–181.

[45] J. Gao, *Multiscale Analysis of Complex Time Series: Integration of Chaos and Random Fractal Theory, and Beyond*. USA: Wiley-Interscience, 2007.

[46] D. Veitch and P. Abry, A wavelet based joint estimator for the parameters of LRD, *IEEE Transactions on Information Theory*, 45 (3) (1999): 878–897.

[47] M.V. Wickerhauser, *Adapted Wavelet Analysis from Theory to Software*. USA: IEEE Press, 1994, 213–235.

[48] Advanced Information CTS (Centro de Tecnología de Semiconductores) Property, Alliance FXO/FXS/E1 VoIP System, www.cts-design.com

[49] Wireshark: A Network Protocol Analyzer, http://www.wireshark.org/

[50] H. Toral-Cruz, QoS Parameters Modeling of Self-similar VoIP Traffic and an Improvement to the E Model, PhD. Thesis, Electrical Engineering, Telecommunication Section, CINVESTAV, Guadalajara, Jalisco, Mexico, 2010.

[51] H.-D. J. Jeong, J.-S. R. Lee, and K. Pawlikowski, Comparison of various estimators in simulated FGN, *Simulation Modelling Practice and Theory*, 15 (9) (2007): 1173–1191.

[52] J. Mielniczuk and P. Wojdyllo, Estimation of Hurst exponent revisited, *Computational Statistics & Data Analysis*, 51 (9) (2007): 4510–4525.

[53]   P. Shang, Y. Lu and S. Kamae, Detecting long-range correlations of traffic time series with multifractal detrended fluctuation analysis, *Chaos, Solitons and Fractals*, 36 (2008): 82–90.

[54]   G. Horn, A. Kvalbein, J. Blomskold, and E. Nilsen, An empirical comparison of generators for self-similar simulated traffic, *Performance Evaluation*, 64 (2007): 162–190.

[55]   R. G. Clegg, A Practical Guide to Measuring the Hurst Parameter, *Proceedings of the 21st UK Performance Engineering Workshop, School of Computing Science, Technical Report Series*, 2006, pp. 43–55.

[56]   F.H.P. Fitzek and M. Reisslein, MPEG-4 and H. 263 video traces for network performance evaluation, *IEEE Network*, 15 (6) (2001): 40–54.

[57]   S. Stoev, M.S. Taqqu, C. Park and J.S. Marron, On the wavelet spectrum diagnostic for Hurst parameter estimation in the analysis of Internet traffic, *Computer Networks*, 48 (3) (2005): 423–445.

[58]   A. Raake, Short- and long-term packet loss behavior: towards speech quality prediction for arbitrary loss distributions, *IEEE Transactions on Audio, Speech, and Language Processing*, 14 (6) (2006): 957–1968.

Chapter 8

Exploratory and validated agent-based modeling levels case study: Internet of Things

Komal Batool¹ and Muaz A. Niazi²

This book chapter discusses two of the agent-based modeling (ABM) levels, i.e. exploratory agent-based modeling (EABM) and validated agent-based modeling (VABM). In first part of this chapter, we shall briefly explain EABM with the help of a case study of 5G networks modeled in an agent-based simulator called NetLogo [1] of the use cases of 5G networks is Internet of Things (IoT). We designed, implemented and experimented this case study to explore the futuristic approaches to ease the implementation of this under-developing 5G network which still needs to be explored. Next, we discuss another important level of modeling, i.e., VABM. Since ABM approach has turned into an attractive and efficient way for displaying large-scale complex systems, verification and validation (V&V) of these models have become questionable. Here, we shall briefly explain VABM with the help of the same case study of 5G networks modeled as in EABM. Using VABM, we shall validate the case study if it is a credible solution.

8.1 Introduction

We have entered into a world of Big Data where several devices communicate with each other. They generate their own data in large quantity when communicating with each other. Today's devices generate and accumulate huge amounts of data to their databases. ABM is a powerful way to put that data to work. An agent-based model featuring individuals can use real properties and behaviors taken from their databases. The results deliver refined optimization by providing a precise, easy, and up-to-date way to model, forecast and compare scenarios. One of uprising issues is modeling and understanding new network domains such as 5G networks. 5G networks is still a naive domain as it lacks standards and needs more exploration. Here, ABM can be of great help especially for complex networks where the dynamics of a system are nonlinear. In this chapter, first, we discuss ABM framework. ABM framework assists complex adaptive system (CAS) modeling in four levels. We discuss a brief definition

¹Riphah Institute of System Engineering, Riphah International University, Pakistan
²COMSATS Institute of IT, Pakistan

of each level. Next, we discuss one of the levels of ABM: EABM in detail. We evaluate EABM on one of the case studies of 5G networks, i.e., IoT using ABM simulator. In the next section, we highlight the importance of modeling in agent-based simulator. We also provide a brief comparison between different simulators commonly used by researchers. For modeling our 5G case study of IoTs, we have used NetLogo [1]. We evaluate different research queries to provide a proof of concept. Next, we discuss the case study in depth, its modeling approach, design and implementation. We shall then discuss the results. Lastly, we conclude this section of EABM.

8.1.1   Agent-based modeling framework

Recent work on the modeling and simulation of CASs has demonstrated the need for combining agent-based and complex network–based models [2], [3] and so. [4] describes a framework consisting of four levels of developing models of CASs. Reference [4] describes these levels using several example multidisciplinary case studies. These modeling levels are briefly discussed as follows:

1. Complex network modeling (CoNeM) level is for developing models using interactive data of various system components. This level of the framework involves the use of complex networks to model, visualize, simulate and analyze any CAS.
2. EABM level is for developing agent-based models for assessing the feasibility to model for further research. This can, e.g., be useful for developing proof-of-concept models such as for funding applications without requiring an extensive learning curve for the researchers. The EABM modeling paradigm allows researchers to experiment and develop proof-of-concept models of CAS with the goal of performing experimentation for improving understanding about a particular real-world complex system. Section 8.1.1.1 discusses EABM in detail.
3. Descriptive agent-based modeling (DREAM) is for developing descriptions of agent-based models by means of using templates and complex network-based models. Building DREAM models allows model comparison across scientific disciplines. DREAM allows researchers to develop semi formal, formal or pseudo-code-based specifications coupled with complex network representations of agent-based models allowing models to be better described for communication across disciplines without requiring the same usage of terminology.
4. VABM using virtual overlay multi-agent system (VOMAS) is for developing a verified and validated model in a formal manner. VOMAS involves the creation of a VOMAS for ensuring the validity of the simulation model by checking its conformance with the real world. Section 8.1.5 discusses VABM and VOMAS in detail.

8.1.1.1   Exploratory agent-based level

EABM uses some simple rules and play them out to see what patterns develop. Unlike traditional modeling schemes and paradigms, EABM helps in exploring a model possibilities for future research, its applicability, validity and parameters. Hence, providing a proof of concept for developing a model more dynamically. Figure 8.1

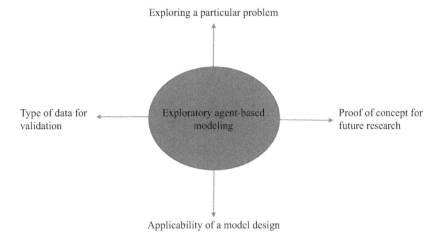

Figure 8.1   Exploratory agent-based modeling features

gives a brief overview of EABM features. Most agent-based models start out as exploratory in nature. Some examples of EABM studies include work by Palmer *et al.* for modeling artificial economic life [5], Becu *et al.* for modeling catchment water management [6], work by Holland [7] and by Premo for ethnoarchaeology [8] to work using ABM for modeling AIDS spread [9]. Other examples of exploratory agent-based modeling include a simulation of how research is considered as an emergent phenomenon as presented earlier in [10]. EABM explores wide range of possible explanations. It uses an understanding of pattern found in a case study data, collected from control, repeatable experiments that can be tested. With EABM, more testing and experiments can be done to explore the behavior of a system. A nonlinear system dynamic is studied and analyzed to evolve a model created. This will help in making new discoveries.

8.1.2 *Agent-based simulator*

Modeling and simulation is done before the actual implementation of large and complex systems. We design a model of an actual or theoretical physical system to analyze its execution. Simulation is a cost-effective solution as we can easily visualize the complete structure of a model in it. There are several network simulation tools for the demanding applications, e.g., NetLogo, NS2, OmNeT++, Opnet, NetLogo, Matlab, Tossim, J-Sim, Prowler, etc. [11]. But for evaluating exploratory agent-based model, we have chosen the NetLogo (v6.0.2 / August 10, 2017) [1] tool to implement the algorithm on complex networks for this reason being that it has proven its value in the modeling and simulation field of complex systems.

Table 8.1   Simulators comparison

Features	NetLogo	NS2/NS3	Omnet++	Opnet
License	GPL	GPL	Academic commercial	Academic commercial
Installation	Easy	Hard	Fair	Fair
Learning	Good	Good	Fair	Fair
Scalability $n > 100$	Excellent	fair	Good	Excellent
Real time execution modeling	Yes	No	No	No
3D radio modeling	Yes	No	No	No

8.1.2.1   Simulator: NetLogo

NetLogo [1] is a multi-agent-based programming language which has an organized developed interface. Unlike conventional simulators, modeling wireless networks as complex systems, this agent-based simulator is used. It helps in simulating models for sensor and actuator networks as well as adhoc networks. When using netlogo, a programmer is more focused on real problems. Rather, he needs not to worry about the transmission or communication of messages. NetLogo has been designed particularly for the research and development of the complex systems. Since from the start of Net-Logo, it is been used for the modeling and simulation of complex systems, including multi-agent systems, social simulations and biological systems [12]. NetLogo has the capability to simulate the issues particularly on human abstractions rather than purely technical aspects. Moreover, this provides the facility of direct addressability to one or more types or agents of objects/entities/nodes of the system. For example, we can address agents or turtles (in NetLogo language) just by their names and can tell them to change their characteristics, such as colors, or create links with other agents or turtles without taking care of the low-level inside information about the animation or how actual structure is going to be done. Thus, NetLogo provides several strong features such as:

- Direct addressability of nodes
- Ease of implementation
- Evaluation of self-organization and bio inspired algorithms
- The capability of providing user friendly interface
- Ability to perform on human abstractions
- User friendly interface
- 3D view
- Supports scalability

Another important feature of NetLogo is that it is not limited to the simulation of networks alone but can also be used to model human users, intelligent agents, mobile robots interacting with the system or virtually any concept (**ABM tools**). Table 8.1 provides a brief comparison of NetLogo with other simulating tools.

8.1.2.2 Research questions

While evaluating EABM, we focus on the following research queries by analyzing the impact of implementing a proposed algorithm [4] on a case study of 5G networks, i.e., IoT.

- It is difficult to actualize and test all situations for large-scale and complex networks. Can EABM be used to analyze such networks with such a huge number of devices?
- How metrics impact such large-scaled networks especially on new domain networks?
- Can we use EABM for providing a proof of concept for such domains?
- How a particular problem can be addressed with such networks?
- How can we investigate the impact of an algorithm on communication performance?
- Can we implement EABM on more of such case studies?

8.1.3 Case study: 5G networks and Internet of Things

Fifth generation mobile networks or 5th generation wireless systems, abbreviated 5G, are the proposed next telecommunications standards beyond the current 4G/IMT-advanced standards. 5G [13] planning aims at higher capacity than current 4G, allowing a higher density of mobile broadband users and supporting device-to-device, more reliable and massive machine communications. 5G research and development aims to provide lower latency than 4G equipment and lower battery consumption. These features shall aid in better implementation of the IoT [14], [15], [16]. These next-generation networks and standards will need to solve a more complex challenge of combining communications and computing together so that intelligence is at your fingertips [17] and readily available to machines that make up the IoT. With 5G, we will see computing capabilities getting fused with communications everywhere, so trillions of things like wearable devices do not have to worry about computing power because network can do any processing needed. The IoT is changing the world. This is transforming how organizations, governments, buyers and sellers connect with the world. The global IoT market shall grow to $457B by year 2020 [18], and the proliferation of connected devices as shown in Figure 8.2 and massive increase in data has already started an analytical revolution. These IoT devices will generate a massive amount of data that can be used to provide better insight for the given application, as well as drive tremendous opportunities for data centers and network operators to create new value and new revenue sources. 5G will deliver new capabilities and efficiencies not possible with today's wireless networks. Furthermore, 5G is being designed to be future-proof, providing forward compatibility with services and devices that are not yet known as today. Connecting the IoT will be an integral part of 5G, and we are excited by the boundless potential it will bring in taking us one step closer to the vision of a totally interconnected world.

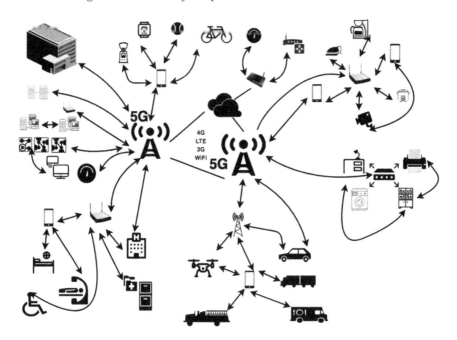

Figure 8.2   5G network: Internet of Things

8.1.3.1   Modeling approach and design

IoT is about connecting different devices within their communication range via a gateway. To demonstrate and address our research queries. we shall use EABM approach to provide a proof-of-concept for future research; for this use case of 5G networks, i.e., IoT. We design our model to implement a proposed algorithm by [4]. This algorithm is called self-advertising content sources (SACS). This algorithm works to locate content sources by putting flags in a network. These content source sends message to other devices present in their communication range. This process is called gradient establishment process as shown in Figure 8.3. Here, a node with red flag represents a content source. This can be any device such as cell phone, laptop, tablet or any other device. The other squares represent other computing devices. Once the message reaches these devices, being the first in the communication range, these devices keep the counter of the message and are shown by a 1 in the Figure 8.3. These devices next search for other devices and send the messages but with a lower SACS value. This process is continued until message time-to-live (TTL) expires or there is no device in their communication range. TTL is to avoid the redundancy of messages in the network.

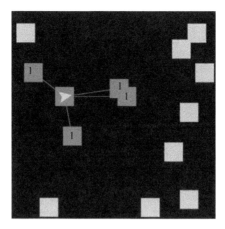

Figure 8.3   Gradient establishment process

As mentioned in [4], this simulation exploration has two types of agents:

1. Computing devices.
2. Message agents ((a) query message agents and (b) SACS setup message agents).

Computing device agents
The detailed design description of these agents is given below: Computing device agents. These agents are designed to simulate various types of computing devices. The devices can range from mobile phones, handheld games as well as laptop computers to stationary computers. As such, there are the basic design requirements for these agents that can be listed as follows:

1. The agents must have a certain amount of memory to hold certain information.
2. In addition, these agents must be able to communicate with other agents in a certain communication radius.
3. The agents must be able to perform basic computations (such as retrieval of data from storage or using data structures).

The idea of gateway agents is the intermediate devices to connect one device to other if that device is out of its communication range. The second type of computing device agents is the SACS source agents. The key idea is that each content source self-organizes its surroundings by putting flags for enticing query agents. The way it actually would be implemented in physical devices can be described as follows:

1. First, the device sends a message within its communication radius to see devices which are up and ready to respond.
2. Devices in range send back their IDs which are stored by the sending device.
3. Next, the sender device broadcasts the SACS setup message agent after assigning a TTL value to it.

```
Move-to device in-radius R
If { Data = 1   \\ data is present then 1 and
                \\ data is not there then 0
Go Back}
Else { if TTL>0
Move-to next device { else Die} }
```

Figure 8.4   Query algorithm

4.   On receiving the message agent, the neighbor devices can subsequently forward the message based on the policy of the message agent, which will be described below. Next, we discuss the design of the message agents.

Message agents

Message agents are of two types:

1.   SACS setup message agents
2.   Query message agents

Message agents forward messages to other devices. We describe both types of message agents as the following: (1) They help in establishing gradient value. SACS setup agents have a simple task to perform. Once they are on a specific device, they use the algorithm described in Figure 8.4 to hop from one device to the next. This is done by means of counting the distance (radius) by means of a hop count. As a result when the hop value reaches the end of the SACS radius, it implies the end of the line for the SACS setup process. However, during each hop, the agents also communicate their respective hop count value to the computing device agent, on which they are currently located. As such, the computing device agent stores this information locally for future queries. It is important to note here that SACS gradient can be established from a smaller radius to a large radius, where the radius here would be used in terms of communication hop counts. In the case of smaller radii, there are chances that subsequently when queries are looking for a particular content source, they might spend a long time hopping randomly before they can locate a particular SACS gradient "scent," i.e., the hop value flags left over by the SACS setup message agent in the initial phase. (2) The second type of message agents is the query agent type. The goals of the query message agents are different from the SACS setup message agents. The query agents are initiated by a user logged on to one of the computing devices. As such, the goal of the query agent is to look for information. Now, the interesting thing to note here is that according to the small world complex network theory, a large number of real-world networks have nodes which can locate most other nodes using a small number of hops. As such, when an owner of a set of devices is looking for specific information, chances are that the query would be able to locate this content on one of the computers locally or within a few hops away from the person. Query agents are thus responsible for content location discovery based on the "smell" of the

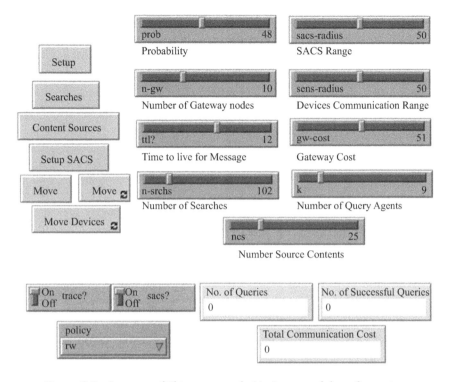

Figure 8.5   Internet of Things network: NetLogo model configurations

content discovered by means of the flag values dropped by the previous SACS setup messages during the initial setup phase.

8.1.3.2   Implementation

We simulate this model in NetLogo. Below, we describe all the functions and components of the implemented model. For implementing this model, we need to declare (1) global variables that is metrics declared for all the procedures and can be called by any procedure or breed. Then we define (2) breeds, i.e., agentsets, which can be assigned procedures, e.g., in our models, it can be devices, messages. Next we shall explain the (3) procedures or functions assigned to the defined breeds.

Global variables
We have several globally declared variables which include many input (from sliders) and output variables such as counters as shown in Figure 8.5.

Breeds
We have created two main breeds, computing devices (i.e., breed devices device) and message agents (i.e., breed messages message).

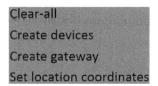

Figure 8.6   Setup algorithm

Procedures

1. Setup: Setup procedure first clears the older simulation. Creates new and gateway devices and adjust the location of the devices. Setup algorithm is shown in Figure 8.6.
2. Setup-devices: This function invokes the patches in the simulation. First of all, each patch creates a random number between 0 and 100. Next, this number is compared with the probability assigned via a global input variable. Based on a comparison of these two numbers, the patch might sprout a device at this location. Next, the device is initialized with certain values. Initially, the agent is given an unexplored status. The SACS distance is assigned equal to the SACS radius input global variable. Initially, all nodes are given "false" as the Boolean value for both the start as well as the goal and the gateway variables. The shapes of the computing devices are next adjusted to be random (one-of shapes), and they are slightly randomly moved to ensure that most devices will not overlap previous devices.
3. Setup-gateway: This function is concerned with creating the gateway nodes from the previously created computing device node agents. The working is based on a random selection of agents from these devices. The number of agents which are to be created as the gateway nodes is based on a global input variable. After making the node as gateway, it changes its color. To show it here, we have encircled the gateways in Figure 8.7.
4. No-overlapping: This function adjusts each device location so that they do not overlap each other.
5. Searches: This function calls Do-Search function and the input is taken from the slider "Number of Searches."
6. Do-Search: This function selects a device and checks it should not be a gateway. Then it calls "Create-Search-Node" function.
7. Create-Search-Node: This function creates a search device. It sets the Boolean variable to 1 which means start. This device becomes the starting point of the query. It is then set to blue color. the location is saved in a temporary value. It then executes the hatching of k-number of query agents taken as input from slider. For each of the query agents, it calls a "Setup Query" function.
8. Setup query: This assigns values to the query agents. Setups the query agents shape to circle and color green. This increments the total query count global variable.
9. Content sources: This function creates the content sources, taking the number from the slider "Number of Content Sources."

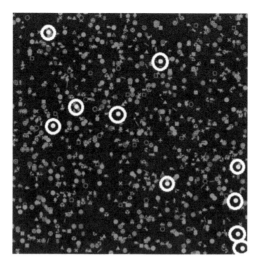

Figure 8.7   IoT Model: Initial Setup

10. Create content sources: This function selects a device node that is neither a gateway nor has a query, then it calls "Create Content Source Node" function.

11. Create content source node: Selected device node is set to true and color is set to red.

    After calling setup, searches and goals functions, the screen now looks as shown in Figure 8.7.

12. Setup SACS: It locates all the content sources nodes. Next it calls "Setup-SACS-d" function.

13. Setup-SACS-d: This function is repeatedly called on agents until the gradient is established. It stops if there are no devices in its range. Otherwise, it will also stop if the calling argument "d" is greater than the global variable "SACS-radius." If the SACS-distance is non-zero, it changes the color of the node to gray as shown in Figure 8.8. To highlight the connections, we show it with links. This way, the content sources are the only ones which will visually stand out from the other nodes. After setting these basic attributes, it compares the current argument "d" with the SACS-radius global variable. If d is less than or equal to this value, it creates a new agent set. This agent set is formed of all other nodes in a certain communication radius but is based on a condition of being unexplored till now. The communication radius is again a configurable value "sens-radius." Now, this agent set is not guaranteed to be non-empty so it is tested for emptiness. If non-empty, each of these agents is asked to execute the same function again but with an incremented "d" value. Thus, this process can continue till the SACS gradient is properly setup around all SACS content sources reflecting how the content sources can be self-advertising their contents. Afterwards, the simulation screen can be observed to reflect this setup as shown in Figure 8.9.

Figure 8.8   SACS Distance

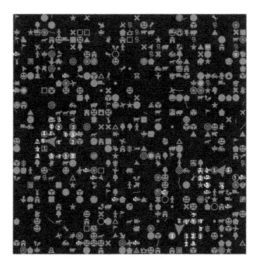

Figure 8.9   After Simulation

14. Move and move (forever): If any queries are there, it will continue by asking them to execute move-RW function, otherwise it will terminate. Move is for one tick and move (forever) is for continuous ticks.
15. Move-RW: The function executes by first creating a list of all nodes in the given sensing radius. This radius is basically equal in physical terms to the communication radius. In case there are no other devices in range, the query agent will simply die. If there are other devices, the next step is choosing one of the nodes as the next location. This is performed differently based on what is the selected

mode of movement of queries from the user interface. Based on the Boolean variable "SACS?", the next node is either a random node in the case of a false value for this variable, or else one of the nodes with minimum of "SACS-dist" variable. Note that this represents the gradient previously established by the SACS nodes. As such, once this node is selected, the query agent can first move to the selected node. Next, it can update its internal variables such as TTL value by decrementing it and also storing the tagged location node inside the "loc" variable. After moving and updating these values, it needs to calculate the cost associated with this move. If the location is a gateway node, then the move will incur cost equal to "gw-cost," a user input configurable variable, otherwise the cost will simply be incremented by 1. Finally, before this function terminates, a call is made to the CheckCS function, which is explained next.

16. CheckCS: This function is primarily for goal verification for queries. In this function, the query checks whether it has reached either one of a gateway node or else a content source. If the current location agent satisfies either of these conditions, then the overall number of successful queries is incremented and then the query agent terminates itself. If it has not reached the goal nodes, then again it checks its TTL value. In case, the TTL value has reached zero, then again it dies.

17. Move devices: This function is to test the effects of mobility of devices on SACS. It randomly moves a percentage of the devices over time, and when the devices are out of range, it calls setup-SACS again to reset the SACS setup messages.

8.1.4   Results and discussion

We experimented the IoT network for the explained algorithm proposed by [4]. We used behavior space of NetLogo to perform experiments by changing the given input metrics. This shall give us a proof of concept for our designed model for how it behaves under different parameters.

8.1.4.1   Simulation parameters

The **input metrics** to the simulation are as follows:

1. The number of devices used in the simulation
2. The number of content sources, NCS
3. The number of gateway nodes, N-GW
4. The number of query source nodes, NSRC
5. Number of random walkers per source node, KRW
6. Cost in terms of hop counts for using a gateway node for locating a content source: CGW.

Output metrics: The output parameters that are of interest in the evaluation of SACS are as follows:

1. Number of successful queries
2. Total communication cost
3. Total number of queries

Table 8.2   Simulation experiments parameters

Parameters	Experiment 1	Experiment 2	Experiment 3	Experiment 4
Gateway cost, GW-cost	20	20	20	20
Number of searches, N-SRCHS	200	200	200	200
Device communication range, sens-radius	4	4	4	4
Number of gateway, N-GW	20	20	20	20
Number of content sources, NCS	20	20	20	20
Time to live, TTL	20	[10 10 40]	20	[10 10 40]
SACS range, SACS-radius	[0 10 40]	[0 10 40]	[0 10 40]	[0 10 40]
Number of query agents, k	5	5	5	5
	Static	Static	Mobility	Mobility

Table 8.3   Experiment 1 statistics

Parameters	N	Min	Max	Mean	Std. deviation
No. of runs	250	1	250	125.50	72.313
TTL	250	20	20	20	0
SACS-radius	250	0	40	20	11.54
No. of successful queries	250	32	798	439.02	221.27
Total communication cost	250	1,000	1,000	1,000	.000
No. of queries	250	6,388	10,110	7,941.54	1,119.19

However, here the total number of queries will be constant and instead the cost of communication as well as the number of successful queries will be evaluated in depth.

8.1.4.2   Behavior space experiments

We have formulated our experiments as shown in Table 8.2. We conducted four experiments. We shall vary two parameters: (1) TTL and (2) SACS-radius while other remain constant. Also we conduct these experiments to evaluate the mobility of devices.

8.1.4.3   Descriptive statistics

Experiment 1

We vary the SACS value while TTL is kept constant. We repeat these simulation for 250 while repeated each simulation for 50 times. Table 8.3 shows the statistics.

Table 8.4   Experiment 2 statistics

Parameters	N	Min	Max	Mean	Std. deviation
No. of runs	1,000	1	1,000	500.50	288.819
TTL value	1,000	10	40	25	8.66
sacs-radius	1,000	0	40	20	11.54
No. of successful queries	1,000	32	798	439.02	221.27
Total communication cost	1,000	1,000	1,000	1,000	0
No. of queries	1,000	6,388	10,110	7,941.54	1,119.19

Table 8.5   Experiment 3 statistics

Parameters	N	Min	Max	Mean	Std. deviation
No. of runs	250	1	250	125.50	72.313
TTL	250	20	20	20	0
SACS-radius	250	0	40	20	11.54
No. of successful queries	250	16	698	357	196.87
Total communication cost	250	1,000	1,000	1,000	0
No. of queries	250	6,738	10,152	8,445	985.53

Table 8.6   Experiment statistics

Parameters	N	Min	Max	Mean	Std. deviation
No. of runs	1,000	1	1,000	500.50	288.819
TTL value	1,000	10	40	25	8.66
SACS-radius	1,000	0	40	20	11.54
No. of successful queries	1,000	12	860	436	424
Total communication cost	1,000	1,000	1,000	1,000	0
No. of queries	1,000	3,912	19,766	11,839	7,927

Experiment 2
We vary the both SACS value and TTL value. We repeat these simulation for 1,000 while repeated each simulation for 50 times. Table 8.4 shows the statistics.

Experiment 3
We vary the SACS value while TTL is kept constant. We repeat these simulation for 250 while repeated each simulation for 50 times. Table 8.5 shows the statistics.

Experiment 4
We vary the both SACS value and TTL value. We repeat these simulation for 1,000 while repeated each simulation for 50 times. Table 8.6 shows the statistics.

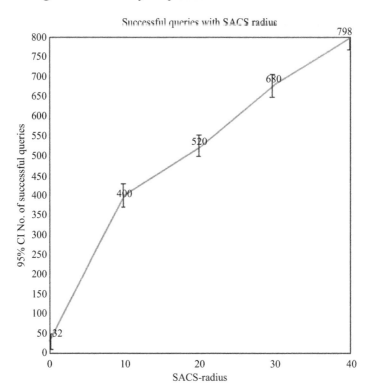

Figure 8.10   Successful queries vs SACS radius for static network

8.1.4.4   Discussion

The first set of results is the details of variation of SACS-radius and its effects on the successful queries. In Figure 8.10, there are five values for the SACS-radius ranging from 0 to 40. Here, the value 0 implies that the SACS is not used. As SACS radius is 20, the number of successful queries increase drastically and keeps on increasing as the SACS radius increases.

In next experiment, the results are covered in Figure 8.11. Here, we increase SACS radius and see the effect on cost of execution of the queries. Number of queries is extremely too much when SACS is not used but decreases as we start using it and keeps on decreasing as the SACS radius increases.

Next, in Figure 8.12, we experimented TTL value against Number of queries cost. We experiment to see if TTL causes the rise in the number. of successful queries or not and also we want to see the effect of this. The data is color coded based on the SACS radius. We see the number of queries/cost is high when no SACS is used. Higher TTL values give more SACS radius values.

We plot a graph Figure 8.13 using mean successful queries vs. TTL value. Now, we can note here that for TTL = 10, SACS appears to have the best possible successful query values. The differences between all SACS values are initially very small for

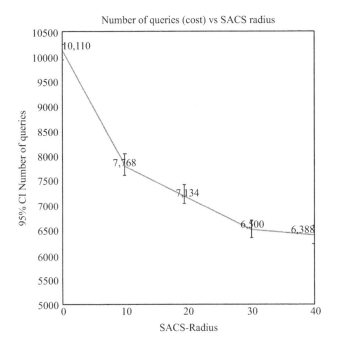

Figure 8.11   No. of Queries vs SACS Radius for Static Network

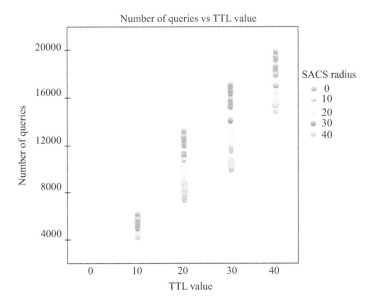

Figure 8.12   Number of queries vs TTL value for static network

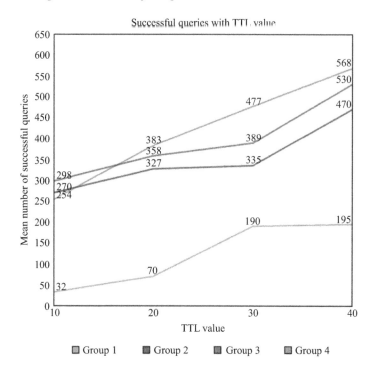

Figure 8.13   Number of queries vs TTL value for static network

TTL of 10. Subsequently, however, for an increase in SACS-radius values, there is a corresponding increase in the successful queries.

The first plot using mobility is for the evaluation of the effects of changing the SACS radius by measuring the number of successful queries. Here, in Figure 8.14, a drastic change is seen when using SACS, otherwise number of successful queries is low.

In next experiment, the results are covered in Figure 8.15. Here we increase SACS radius and see the effect on cost of execution of the queries. Number of queries is extremely high when SACS is not used but decreases as we start using it and keeps on decreasing as the SACS radius increases.

Next, in Figure 8.16, we experimented TTL value against number of queries cost. We experiment to see if TTL causes the rise in number. of successful queries or not and also we want to see the effect of this. The data is color coded based on the SACS radius. We see the number of queries/cost is high when no SACS is used. Higher TTL values give more SACS radius values.

We plot a graph Figure 8.17 using mean successful queries vs TTL value. Now, we can note here that for TTL = 10, SACS appears to have the best possible successful query values. The differences between all SACS values are initially very small for TTL of 10. Subsequently, however, for an increase in SACS-radius values, there is a corresponding increase in the successful queries.

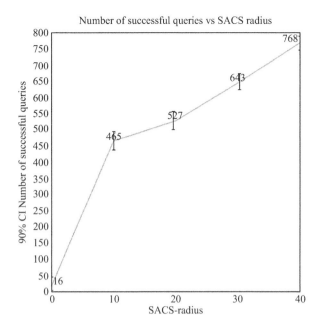

Figure 8.14   Number of queries vs TTL value for static network

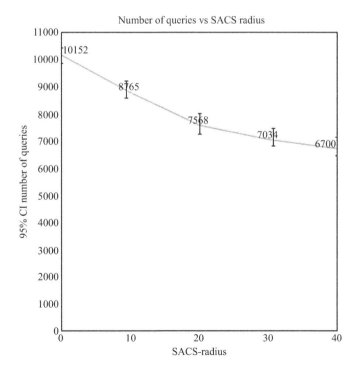

Figure 8.15   Number of queries vs SACS radius for mobility

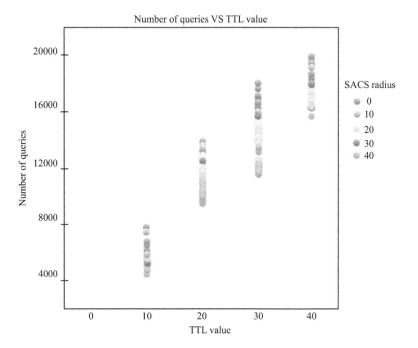

Figure 8.16   Number of queries vs TTL value for mobility

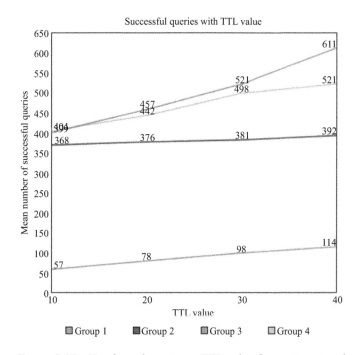

Figure 8.17   Number of queries vs TTL value for static network

8.1.5 Conclusion

This section gives an in-depth detail of exploratory agent-based model with a case study of 5G network modeled over ABM simulating tool: NetLogo. We discussed exploratory ABM and defined how it can be used for exploring new domains. Using EABM, we provided a proof of concept, identified problems and analyzed on agent-based model. We used IoT 5G network as a use case. We used this network case study which to the best of our knowledge cannot otherwise be modeled using any other simulator easily because this particular domain is still unexplored in various ways. We saw it is easy to actualize and test all situations for large-scale and complex networks using ABM. Our EABM of the SACS algorithm clearly demonstrates how different CAS researchers can use ABM to perform extensive simulation experiments using parameter sweeping to come up with comprehensive hypotheses. Our results demonstrated the effective testing and validation of various exploratory hypotheses. We demonstrated how one hypothesis can lead to the next and how simulation experiments can be designed to test these hypotheses.

8.2 Validated agent-based modeling level case study: Internet of Things

We next discuss another important level of modeling framework, i.e., VABM. For EABM, we designed, implemented and experimented a case study of 5G networks to explore the futuristic approaches and its ease to implement this under developing network. Since ABM approach has turned into an attractive and efficient way for displaying large-scale complex systems, V&V of these models have become questionable. Here, we shall briefly explain VABM with the help of a case study of 5G networks modeled in an agent-based simulator called NetLogo [1]. We have already discussed NetLogo simulator in Section 8.1.2.1. One of the use cases of 5G networks is IoT, see section 8.1.3 for further details. Using Validated ABM, we shall validate a modeled case study if it is a credible solution or not. Simulation provides lots advantages for modeling complex systems but once they are simulated, question arises if these simulated models can be trusted. A reliable simulated model which has a random behavior must correlate with the real world or what it is expected to behave. Therefore, a simulated modeled is first required to be verified if it is working as expected and then it is required to be validated according to the requirements of assigned parameters.

8.2.1 Introduction

ABM has become a significant and efficient way of modeling systems especially complex systems. Modeling explains and makes it easy to understand real world networks which are costly to explain or larger in scale. A successful simulation that is able to produce a sufficiently credible solution can be used for prediction. Since it is not possible in terms of execution concern and unnecessary (counting components that do not have much impact on the framework) to construct a simulation model that

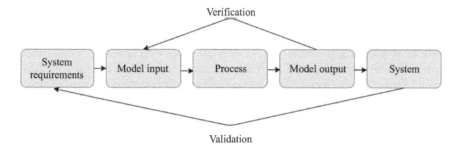

Figure 8.18   Validation and verification of a system

caters all the detail and behavior of the real system, a few presumptions must be made about the framework to build a simulation model. Therefore, a simulation model is an abstract representation of a physical system and intended to enhance our ability to understand, predict or control the behavior of the system. However, the abstractions and assumptions introduce inaccuracies to the simulation model. One of the important tasks after simulating a system is determining how accurate a simulation model is with respect to the real system. There are no set rules or models to validate a simulated system although researchers and industrial people use various techniques to verify and validate their simulated systems. In the previous section, EABM, we discussed ABM frameworks levels. We discussed EABM in depth and explored underdeveloping network case study, i.e., IoT of 5G networks. Also, we highlighted the importance and purpose of using simulator, NetLogo and provided a brief comparison between different ABM simulators. We used NetLogo to simulate the IoT case study and evaluated different research queries to provide a proof of concept. In this section, we shall validate the similar simulated 5G network and see if the model is feasible or not. We shall validate one of the use cases of 5G networks, i.e., IoT. Next, we discuss the case study in depth, its modeling approach and design. We shall then discuss results. Lastly, we conclude this section of VABM.

8.2.2   Validated agent-based level

Model verification involves verifying if a model is being building correctly or determines whether the programming implementation of the abstract or conceptual model is correct. This process includes debugging model, finding incorrect conceptual implementations, and checking calculations. Where as validation process validates a model's credibility. As shown in Figure 8.18, the process of validation of a model could be defined as the process of verifying that the model produces outputs in line with the design objectives and also with empirical observations. Through the validation against empirical data, it is possible to verify if the stylized representation of the real-world implemented in the model is accurate enough to generate emergent behaviors that are similar to the real world system. The ABM outputs have been validated against the observed SOCA algorithm implementation. To validate the ABM, a simple strategy was followed. The set of parameters in the model for which empirical data are not known was selected as elements of what is called the parameters. Parameter were changed and observed over networks. Then simulated. There are many principles and techniques of model V&V that have been presented, e.g., in [20]. However, it is difficult and time-consuming to use all possible techniques for validating every model that is developed. Modelers depend upon choosing the appropriate techniques. Choosing an appropriate techniques assures the acceptable accuracy and credibility of their model. We shall use VOMAS as proposed by [4]. This scheme is designed in a manner which can be implemented on any kind of agent-based model. VOMAS has the capability to monitor position-based (spatial) as well as non-position-based agents. Subject matter experts (SMEs), are individuals who can provide validity to a model. SME is who can give the accurate specification as well as analyze the outputs and simulations runs of a model. VOMAS approach allows experts to be involved in the design of the agent-based model as well as the custom-built VOMAS from scratch. By involving SMEs from the start of the project, which are essentially equivalent to clients in the software engineering domain, VOMAS approach allows the simulation study to be a stronger candidate for success. "The Virtual Overlay Multi-agent System is created for each simulation model separately by a discussion between the simulation specialist as well as the SMEs. When the actual simulation is executed, the VOMAS agents perform monitoring as well as logging tasks and can even validate constraints given by the system designer at design time. VOMAS has been designed to cater for both face validity as well as model assumptions and IO-transformations. Model assumptions are ensured by the use of invariants. Face validation is ensured by means of various techniques based on spatial and non-spatial validation and animation-based validation. IO-transformations are ensured by means of essential logging components. Thus, in other words VOMAS provides the complete validation package" [21]. It has the following features:

- Maximizes the value of a model
- Complete optimization of a system
- Reduce cost by enhancing employee efficiency and effectiveness
- Effective management
- Save money and time by discovering defects at early stage.
- Reduces risk, legal liability, not regulatory, is often the most important reason to perform validation.
- Promotes continual process improvement

8.2.2.1 Validation techniques

There are diverse measurements in which a model might be assessed. One way researchers check models is to utilize greatest probability estimation. Since more complicated models with more degrees of freedom result in better fits, some researchers use a penalty for the complexity of the model. Especially, the use of minimum description length might be a promising tool. Increasingly experimental researchers and ABM start to be combined. Traditionally, experiments where used to challenge the standard model of rational choice; nowadays alternative models are tested in more complicated dynamic and spatial settings, in the lab and in the field. This may prompt better tried elective models that one can use in ABMS. Different approaches to assess models are the participatory model advancement with partners, or the utilization of Turing tests. Here [22] presents several unique validation and verification techniques that are being widely used in industrial and research models of manufacturing, engineering, and business processes. These models are mostly discrete event simulation models and are aimed to lower the cost of the system, its process and efficiency of working. Face validity is asking the concerning experts if the models output is accurate to the system. [4] explains validation methods, taxonomy in detail. The VOMAS technique is considered to be an extension of the Companion Modeling [5] that involves both SME as well as Simulation Specialists in developing of an overlay multi-agent system for the purpose of validation. Any good modeled simulation should not be considered as a full depiction of a real system. There can be different reasons ranging from non-availability of complete data sets, weather conditions, complex and costly parameters, etc. which can only be tested in real situations. Some level of abstraction is always involved in simulating a model. There is no set or defined way of verifying and validating a model. Particular techniques may be chosen depending upon the pertaining paradigm of a model.

8.2.2.2 Virtual overlay multi-agent system

For V&V of this simulation model, we have followed a formal methodological approach based on Companion modeling, termed as VOMAS. This methodology has been described in detail in [21]. In a VOMAS approach, a set of simulation agents are built in the form of a multi-agent system as a companion to the agent-based simulation model. The idea of building a VOMAS is based on the need for a formal model of V&V of agent-based models. VOMAS uses an Object-Oriented software engineering approach. In a VOMAS, the system designers work on the design of a Virtual Multi-Agent System starting alongside the design of the original agent-based simulation model. In a VOMAS simulation model, agents act together in the form of a cooperative multi-agent system and assist in data collection (e.g., logs) in addition to being available as a formal framework for SME to provide feedback to the Simulation Engineers and vice versa. The VOMAS approach helps bridge the gap between simulations and real world by enforcing interactions between the SME and the Simulation Engineers from the start of the project. Virtual overlay agents are created for run-time communication with the agents in the agent-based model. The design of the two models is thus influenced by a discussion between the simulation specialists as well as the SME. This entire exercise results in two models; one model is the actual ABM. And the second one, is the VOMAS agents model which monitor the ABM and validate various design requirements. The main duties of the VOMAS-based agents can be summarized as following:

1. To monitor various parameters specified by the SME during the design process, during actual execution of the simulation experiments.
2. To report generation of any extraordinary values or violations of invariants again specified on the basis of interactions with the SME.

3. To log activities of agents during simulation experiments. These logs are provided to SMEs for post-simulation data analysis.

Thus, in other words, VOMAS is based on an interactive process going back and forth between the simulation experiments and analysis of logs by SMEs and simulation specialists.

8.2.2.3 Research questions

We focus on the following research queries by using VOMAS [4] validating technique for validating 5G networks.

- What process parameters and controls are necessary to be determined for validating a model?
- Can the process design reproducible?
- Determining the set of conditions encompassing upper and lower limits of a model for recognizing system behaviors in different situations including those parameters which can cause big change or failure of a system.
- Can this model remain in a state of control while performing other communication activities?
- Can validation process be used for qualitative variability of a process and its control?
- Can validation be used for scrutinizing a process performance for development and deployment of process controls for a system?
- Can a scientific study performed prior to implementing a change to a process can support the implementation of a change without revalidation?
- Can validation process cause reduction of cost?

8.2.3 Case study: 5G networks and Internet of Things

This section has already been explained in EABM (Section 8.1.3).

8.2.3.1 Modeling approach and design

5G networks are upcoming networks. The validity of these networks is very important due to various factors such as scalability, cost, effectiveness and usage in daily lives. In this section, we present the overview, experimental design and implementation details for 5G networks case study of IoT to demonstrate the generalized applicability of the proposed validation methodology and framework proposed in [4].

8.2.3.2 Basic simulation model

We start by developing a 5G network. In the real world, IoT is one of the case studies of this network. A message is spread in the network on different parameters such as:

1. The cause of message in network.
2. The message TTL, energy of device, communication power.
3. The amount of congestion of other messages in the past.
4. The devices mobility.
5. The current rate power consumption of devices to send and receive messages.

6. The particular type of devices.
7. The structural aspects of the IoT network.

To develop the basic IoT model, we thus need to develop several modules incrementally as given in the next subsections.

8.2.3.3   IoT creation module

In this module, our goal is to develop the IoT network. Our basic model extends upon the SACS implemented model created in exploratory agent-based model. We implemented several parameters for configuring a simulated network. We first develop an IoT network made up of devices covering a random set of locations. However, to ensure we can control number of devices, we set a slider for this. Based on slider value and location set, we can build an environment model with devices. Please see the model description in EABM chapter.

8.2.3.4   Basic IoT module

After the network has been created, we can develop a baseline model. The way the SACS module works is that at a given source node, message can start at a random device. Now, the message has three distinct states:

1. Started
2. Spreading
3. Dying

In the "started" state, message is initiated. After formation, the message next moves to the "Spreading" stage. In the spreading stage, message hatches to the next device/s in the communication range. TTL parameter is set, but as message hops to next, it decreases. TTL value is time to live. Thus depending upon the parameter, the message keeps going until it reaches the destination or ends its life which serves as a threshold to stopping the communication. The "Dying stage" gradually serves as a means of stopping the message. The effects of a communicating devices can be seen in Figure 8.19.

Up until now, we have developed the IoT model with different parameters and simulations modules. The goal of this section is to demonstrate how this model can subsequently be validated by means of building an associated VOMAS model. As described earlier in the VOMAS Methodology, we shall start by first writing our validation question in the model. Our key validation question can thus be stated as follows:

Question

How can we validate that the developed IoT model is representative of actual networks? We see a standard network. By analyzing the other factor affecting the network, it becomes clear that while there is message such as either the other messages, it is very difficult to get any message of the exact structural aspects of IoT over time. In other words, the message availability is limited to Geographical Information Systems based data visible via satellites. This data does not demonstrate the micro-dynamics of the IoT. As such, instead of this approach, which would not have guaranteed model

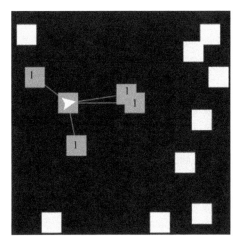

Figure 8.19   IoT initial setup for message

validity, we have to use a different approach. One possible approach in the absence of real data is thus to perform cross-model validation or "docking" of the models.

8.2.3.5   VOMAS agent design

A naïve way of designing a VOMAS could be to have each device calculate SACS value over time. We have previously described how the algorithm works.

8.2.4   Results and discussion

VOMAS forms a virtual overlay system over the agent-based model. In the previous section, we developed a 5G case study. Here, our goal is to present the results of the experimentation involved in the validation of the simulations. It is pertinent to note here that while we can perform considerably more experimentation here similar to previously done in earlier chapters in each of these case studies, we shall however only focus on V&V of the effects of the case study to demonstrate the effectiveness of the proposed generalized methodology by [4]. Our focus here will be on the different effects of the simulation execution as well as the resultant effects on the VOMAS validation. Here, we shall demonstrate how VOMAS-based validation builds on empirical validation methods while adding value addition in the form of being a customizable software engineering modeling scheme. Some of the parameters of the experiments are similar to the previous case study. While others are new, being related to the devices nodes are given below.

8.2.4.1   Simulation parameters

The **input metrics** to the simulation are as follows:

1.   The number of devices used in the simulation
2.   The number of content sources, NCS

3. The number of gateway nodes, N-GW
4. The number of query source nodes, NSRC
5. Number of random walkers per source node, KRW
6. Cost in terms of hop counts for using a gateway node for locating a content source: CGW.

Output metrics: The output parameters that are of interest in the evaluation of SACS are as follows:

1. Number of successful queries
2. Total communication cost
3. Total number of queries

However, here the total number of queries will be constant, and instead the cost of communication as well as the number of successful queries will be evaluated in depth.

8.2.5   Validation discussion

Here, our basic simulation model can be viewed in Figure 8.7. As can be noted from the simulation output, the sensors are all connected with each other; however, not every sensor will be directly accessible from each other sensor using a single hop. Thus, the information in this network needs to be aggregated and confirmed with local sensors before it is forwarded to the remote sink. When the simulation is actually executed, each of the nodes measure local parameters and display them. Thus, in this case study, we have demonstrated the effectiveness of using VOMAS to verify and validate the simulation model.

8.2.6   Conclusion

In this part of the chapter, we applied one of the techniques for validation of an agent-based model. We validated a case study of 5G networks using VOMAS methodology, i.e., VOMAS. This VABM level of the framework builds upon previous framework levels such as EABM allowing for VABM. As a means of unification of all ideas and concepts, the methods and case studies extensively involve both the use of agent-based models and complex network models and methods in different scientific disciplines. We presented a case study of IoT demonstrating the broad applicability of the proposed methods involving building customized validation schemes based on the particular case study. We presented a case study of IoT of 5G network which has a complex structure of communication. This network involves several devices communicating with each other while taking care of several other parameters. Unlike traditional validation exercises, VOMAS validation involves in-simulation agents, which observe and, if needed, interact with the simulation environment. The individual virtual agents validate the communicating devices as the messages proceeds in the network. Also, we verified how the network evolves when the message agents are given specific TTL or reach flagged devices.

References

[1] *2017 Roundup Of Internet Of Things Forecasts.* (last accessed on 27-May-2018). https://www.forbes.com/sites/louiscolumbus/2017/12/10/2017-roundup-of-internet-of-things-forecasts/#439ad5b41480/.

[2] *5G and Its Incredible Numbers that Will Rock the Global Economy.* (last accessed on 12-July-2018). https://ict.io/la-5g-et-ses-chiffres-incroyables-qui-vont-faire-vibrer-leconomie-mondiale/.

[3] Andrews, Jeffey G, Stefano Buzzi, Wan Choi, *et al.* What will 5g be? *IEEE Journal on Selected Areas in Communications* 2014; 32(6).1065–1082.

[4] Balci, Osman. Verification validation, and testing. Handbook of simulation 1998; 10.335–393.

[5] Becu, Nicolas, Pascal Perez, Andrew Walker, Olivier Barreteau & Christophe Le Page. Agent based simulation of a small catchment water management in Northern Thailand: description of the catchscape model. *Ecological Modelling* 2003; 170(2). 319–331.

[6] Gubbi, Jayavardhana, Rajkumar Buyya, Slaven Marusic & Marimuthu Palaniswami. Internet of things (IoT): a vision, architectural elements, and future directions. *Future Generation Computer Systems* 2013; 29(7). 1645–1660.

[7] Holland, John H. Studying complex adaptive systems. *Journal of Systems Science and Complexity* 2006; 19(1). 1–8.

[8] Korkalainen, Marko, Mikko Sallinen, Niilo Kärkkäinen & Pirkka Tukeva. Survey of wireless sensor networks simulation tools for demanding applications. In *Networking and services, 2009. ICNS'09. fifth international conference on*, 102–106. IEEE 2009.

[9] Laghari, Samreen & Muaz A Niazi. Modeling the internet of things, self-organizing and other complex adaptive communication networks: a cognitive agent-based computing approach. *PLoS One* 2016; 11(1). e0146760.

[10] Niazi, Muaz & Amir Hussain. Agent-based tools for modeling and simulation of self-organization in peer-to-peer, adhoc, and other complex networks. *IEEE Communications Magazine* 2009; 47(3). 166–173.

[11] Niazi, Muaz A. Emergence of a snake-like structure in mobile distributed agents: an exploratory agent-based modeling approach. *The Scientific World Journal* 2014.

[12] Niazi, Muaz A. Towards a novel unified framework for developing formal, network and validated agent-based simulation models of complex adaptive systems. *arXiv preprint arXiv:1708.02357.* 2017.

[13] Niazi, Muaz A & Amir Hussain. A novel agent-based simulation framework for sensing in complex adaptive environments. *IEEE Sensors Journal* 2009; 11(2). 404–412.

[14] Niazi, Muaz A, Qasim Siddique, Amir Hussain & Mario Kolberg. Verification & validation of an agent-based forest fire simulation model. In *Proceedings of the 2010 spring simulation multiconference*, 1. Society for Computer Simulation International. 2010.

[15] Niyato, Dusit, Marco Maso, Dong In Kim, Ariton Xhafa, Michele Zorzi & Ashutosh Dutta Practical perspectives on IoT in 5g networks: from theory to industrial challenges and business opportunities. *IEEE Communications Magazine* 2017; 55(2). 68–69.

[16] Palattella Maria Rita, Mischa Dohler, Alfredo Grieco, *et al.* Internet of things in the 5g era: enablers, architecture, and business models. *IEEE Journal on Selected Areas in Communications* 2016; 34(3). 510–527.

[17] Palmer, Richard G, W Brian Arthur, John H Holland, Blake Le Baron & Paul Tayler. Artificial economic life: a simple model of a stockmarket. *Physica D: Nonlinear Phenomena* 1994; 75(1–3). 264–274.

[18] Premo, Luke S. Exploratory agent-based models: towards an experimental ethnoarchaeology. In *Digital discovery: exploring new frontiers in human heritage. CAA.* 29–36. 2006.

[19] Saxena, Navrati, Abhishek Roy, Bharat JR Sahu & HanSeok Kim. Efficient IoT gateway over 5g wireless: a new design with prototype and implementation results. *IEEE Communications Magazine* 2017; 55(2). 97–105.

[20] Siddiqa A, Niazi M. A novel formal agent-based simulation modeling framework of an aids complex adaptive system. International Journal of Agent Technologies and Systems (IJATS). 2013 Jul 1;5(3). 33–53.

[21] Wilensky, Uri. Netlogo 1999.

[22] Xiang, Xiaorong, Ryan Kennedy, Gregory Madey & Steve Cabaniss. Verification and validation of agent-based scientific simulation models. *Agent-directed simulation conference,* 47–55. 2005. ISBN: 1-56555-291

Chapter 9

Descriptive agent-based modeling of the "Chord" P2P protocol

Hasina Attaullah[1], Urva Latif[1], and Kashif Ali[1]

9.1 Introduction

Chord [1,2] is a structured peer-to-peer (P2P) distributed protocol designed as a scalable lookup service to overcome the flaws of previously defined protocols like Gnutella, Napster [3], and Freenet [4,5]. Napster uses centralized directory to maintain indexes of files to be shared, so the main drawback of it is single point of failure and denial of service attack. Although Gnutella is decentralized, it uses flooding scheme to lookup for files that generate high volume of traffic in a network which causes scalability issue, and queries may not necessarily be resolved. Chord is scalable as it requires O(log n) lookup messages to resolve a query, and for its scalability, it is used in different approaches [6–8]. Chord is fault tolerance and it balances the load by exploiting functionality of distributed hash keys [9].

In Chord, thousands of peers linearly interact with each other to form a complex P2P network. In Chord, many peers independently self-organize and adapt itself as a client (downloading) or server (uploading). Like complex adaptive systems (CASs) [10] in Chord, downloading is the emergent phenomenon.

We have two different techniques for modeling of CAS. It is either agent-based modeling [11] or complex network–based modeling [12]. In the case of Chord, the number of query events increases with the increase of number of nodes, so it is difficult for discrete event simulations to analyze and handle such a large number of events. Agent-based modeling provides better understanding of CASs, so Chord being a complex in nature is also modeled in agent based model (ABM) to get the higher level of abstraction of a system.

In this chapter, we have modeled Chord in DescRiptivE Agent-based Models (DREAM) [13] for higher level of abstraction. For textual description of Chord, Overview, design concepts and details (ODD) [14] is used. To best of our knowledge, Chord is not yet modeled in DREAM and ODD in the literature.

The main contributions of our works are as follows:

- We have replicated the results of PeerSim-Chord [15] in NetLogo [16].
- We have modeled Chord in ODD for its textual description.

[1]Cosmose Research Group, COMSATS University, Pakistan

- For pseudo-code-based specification and centralities measure, we have modeled Chord in DREAM.
- We have compared and discussed the results of PeerSim and NetLogo.

Rest of the chapter is organized as follows. Second section covers background and literature review, and in the third section we have modeled Chord in ODD and DREAM. Fourth section discusses the results, and the last section concludes the chapter.

9.2 Background and literature review

9.2.1 CAS literature

The systems have multiple agents connected with each other and interact nonlinearly. Each agent has certain state variables associated with it. Change in state variables of agents effects the whole system. Such systems are referred as CASs [10]. CAS can be natural or artificial in nature; social network and growing plants are examples of natural CAS, while complex adaption communication networks, i.e., P2P, ad-hoc or wireless network are examples of artificial CAS [13]. Niazi *et al.* proposed a framework formal agent-based simulation framework that can be used for sensing complex wireless networks as well as an environment [17]. Later, they extended their work and proposed a model for sensing emergent behavior in complex networks [18].

9.2.2 Modeling and simulation of CACOONS

As the network size increases with advancement of technology, the nonlinear interactions of the agents increase complexity and the complex emergent behavior is observed [19]. Usefulness of the agent-based models to develop complex networks is demonstrated in literature. Agent-based modeling is useful for modeling of CASs . It is also useful for the P2P networks Complex Adaptive COmmunicatiOn Networks and eNvironmentS (CACOONS) [20]. Niazi *et al.* [21] evaluated and demonstrated the usefulness of agent-based tools for modeling of P2P and complex communication networks. Modeling of complex network of chord as agent based helps in simulating different scenarios of node adding, leaving and maintenance of network. It also helps to understand the problems of complex network when the system receives a lot of updates and congestion occurs. Deep insight of complex network scenario modeling can be achieved through it.

9.2.2.1 Agent-based modeling

Thomas Schelling's (1971) [22] pioneering work urban residential segregation model was a basic concept of ABMs. Thomas' model was not computational, but agents were interacting with environment. ABM share their origins with cellular automata (CA) in the work of Von Neumann, Ulam, and Conway. Agent-based models (ABM) emphasize on individual component. Agents can be humans, vehicles, products, equipment, or companies, whatever is appropriate for the specific system. It is a computational model for simulating the actions of agent, their interaction with environment, their behavior, and studying their behavior and influence of agent on the system as a whole. An ABM consists of agents and their specific model measures and types:

- decision-making rules in the real-world learning rules;
- a method for agent movement, e.g., move and interact; and
- an environment variable sets that can be impacted by the agents.

Agent-based modeling of network protocols can help understanding intricate scenarios of complex networks. Agent-based simulation tools, i.e., NetLogo can simulate P2P networks better than many network simulation tools [23]. Batool *et al.* [24] give guidelines for developing wireless sensor networks using agent-based modeling.

CABC framework levels

There are four levels of ABM framework depending upon the available data, interest of researcher and time spend on research. CAS research can be conducted by using one of the following level.

1. Complex network model is designed when the data of interaction is available and analysis is done to observe emergent pattern in CAS.
2. Exploratory agent-based modeling is done when interaction data is not available. It is used for exploring previous concepts, authenticate concept, and predict future research dimension.
3. Descriptive agent-based modeling to compare interdisciplinary models to learn and enhance knowledge.
4. Validation-based modeling to predict simulation results according to the case.

9.2.2.2   Complex network modeling

Complex networking model is adapted from graphs. It has nodes and links. Once the data for modeling is captured, there are many network tools that can be used for analyzing the network. Centrality measures can be used for quantitative analysis of network. Degree centrality simply defines number of connections of nodes with other nodes. For direction, there is in degree and out degree. High degree centrality value nodes will be more important, and any fault in its working will affect the whole network working. Eccentricity is the measure of distance from a node to all other nodes. Node that is near to other nodes will get high value. Betweenness is the measure of nodes ability to observe communication in network.

9.2.3   Chord P2P protocol

9.2.3.1   Architecture and working

Chord has different functions for node joining, leaving, mapping the keys and recovery from failures. Each of these is explained in detail in below sections.

Table 9.1   Hashes for nodes

Hash	Key
H(192.168.23.1)	N0
H(192.167.1.1)	N3
H(192.168.22.2)	N4
H(192.169.23.4)	N6
H(191.167.1.3)	N8

Table 9.2   Example hash keys for data items

Hash	Key
H(A)	K0
H(B)	K3
H(C)	K4
H(D)	K6
H(E)	K8

9.2.4   Hashing and key mapping

It uses the consistent hashing for assigning keys to nodes and objects. Hash function used in chord is SHA-1 [25]. SHA-1 uses IPs for generating keys for nodes and generated keys for data item as well. Tables 9.1 and 9.2 depict the example of hash table for lookup a node and for data item keys. After hash keys are generated, keys are assigned to their relevant nodes.

9.2.5   Node joining

Successor nodes are nodes in a clockwise direction and predecessor are in counter-clockwise direction. Predecessor and successor pointers are sustained by every node in a chord. When any node k wants to join a chord ring, it first calls join (n) function (n is any node in a ring). Node n finds the successor of k and returns it to k. The node k starts pointing to relevant node as successor. The rest of the chord nodes are not aware of the join () operation of k node.

9.2.6   Finger table

To accelerate the lookup process chord uses the finger table. Finger table maintains the information of some of nodes for efficient retrieval of information. Every node keeps a finger table with k records (where $N = 2^k$). First entry in a finger table is a start which is calculated at any node 'n' using $(n + 2^i - 1)$ mod N. Then second entry

Table 9.3   Finger table entries

Finger table			
Start $(n + 2^{i-1})$ mod N	Interval (finger[i].start, finger [i+1].start)	Successor Next node key in clockwise direction	Key

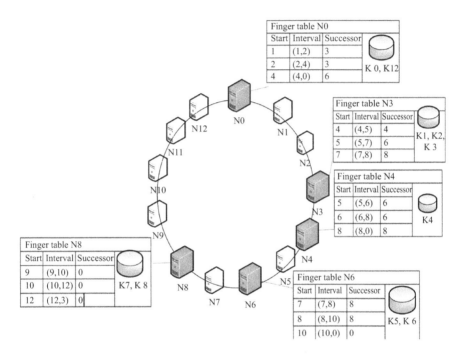

Figure 9.1   Finger table at active nodes. Adapted, with permission, from Reference [2]

is an interval $=$ (finger[i].start, finger [i + 1].start) and third entry in a finger table is successor of node n as shown in Table 9.3. Every node maintains its finger table and updates it regularly after any node join and leave. Figure 9.1 shows finger table for active nodes in a chord ring.

9.2.7   Stabilization

As nodes frequently join and leave a chord ring which makes chord network dynamic, the key task of chord is executing these operations to preserve the ability to find each key in the ring. In chord when node wants to join or leave the ring, the finger

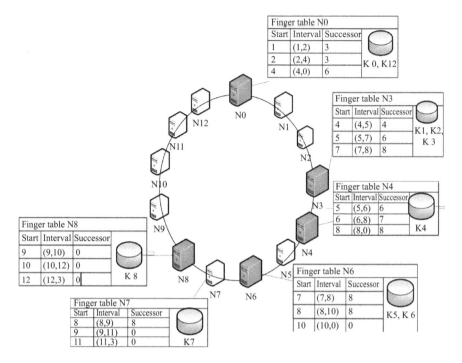

Figure 9.2   Finger table at nodes after stabilization. Adapted, with permission, from Reference [2]

table should be updated accordingly for up-to-date information. This will ensure the efficient retrieval of information. To attain this objective, Chord needs to maintain nodes successor correctly and key should be maintained by node successor. When joining or leaving a chord ring, node impact the ring by changing successor and predecessor of previously joined nodes, to maintain the entries correctly. Chord uses stabilize () function to update the changes in finger tables. And for fast lookup of entries, it is also necessary for finger table to be accurate.

- In stabilization at first join function is being called and then node's finger and predecessor are initialized. Suppose in Figure 9.2, N7 wants to join chord ring, initially successor of N6 is N8 and predecessor of N8 is N6. When N7 runs stabilize(), it asks N8 for its predecessor and resolves whether N8 should be N7 successor instead. Stabilize() then informs node N7 successor of N7 presence, giving the successor the chance to change its predecessor to N7 depicted in Figure 9.3.
- Existing nodes finger tables are updated to rationalize the changes. For example, in Figure 9.2, finger table N4 is updated as node N7 is added to the ring. Successor of node N6 is now updated as N7 in finger table of node 4.

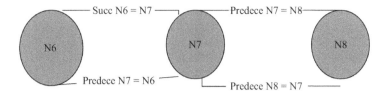

Figure 9.3   Successor and predecessor assignment after new node joining

- The last process is shifting of item keys to new joining node. In Figure 9.2, K7 is shifted from N8 to N7 when N7 joined the chord ring. And now N8 has only responsibility of K8 as K7 is moved to new added node.

9.2.8   Performance of chord

- Load balancing: Consistent hashing provides load balancing by distributing load on different nodes, and each node is accountable for around K/N keys (K keys, N nodes).
- Scalable: Chord is scalable as it requires $O(\log n)$ lookup messages.
- Fault tolerant: When nodes join or leave a network, only $O(\log^2 N)$ messages are necessary to rebuild routing invariants and finger tables.

9.2.9   PeerSim

9.2.10   Literature review

Many variants of chord are proposed in literature. Different Chord-based protocols are proposed in literature to improve and enhance the functionality of a Chord P2P protocol. Some of renowned work is investigated and presented below.

9.2.10.1   Security-based chord

Chord protocol is decentralized, scalable, and provides load balancing. Security is an important aspect in any P2P network. In this paper [26], author proposed distributed public key–based protocol which is built on Chord called as Chord-PKI. Chord-PKI exploits the functionality of a Chord lookup function and distributes storage, retrieval, certification and revocation functionalities between the nodes. In this way making the protocol self-sufficient and fully distributed. Chord-PKI assured security under different network conditions. Revocation and trust models are still needed to be improved for large-scale networks.

When we talk about security in P2P systems, malicious node attacks are the major concern. One of these attack is Eclipse [27] that adds malicious entries in finger table of a node which lead to false lookup or lookup latency increases. In this paper [28], two mitigation technique is proposed to overcome the Eclipse attack. Results show that it avoids attack, but the approach is not compared with other proposed approach for verification purpose.

File distributed P2P systems are vulnerable to attacks. As mostly cryptographic approaches does not protect these systems from denial of service attack or lookup query attack. In this paper [29], a layer LocationGuard is proposed which resides above the chord network layer. It protects files on any P2P systems from attacks by providing location key of random string, secure algorithm for file protection and location-inference guards. The results show that the proposed LocationGuard approach is effective against denial of service attack and peer compromise attack.

In this paper, Sybil attacks [30] are taking into care. Sybil attacks are most dangerous attacks for P2P systems. Basically, Sybil attack obtains multiple identities called "sybil" in order to perform different malicious activities in a network. In this paper [31], the authors proposed a way for attacker to figure out certain number of node IDs to attack a network. It shows that this technique limits Sybil attack on network. The limitation of this paper is that it does not consider other attacks if it happens, i.e., routing attack.

Security is a major issue in P2P systems. When Chord is making its DHT, it is not resilient to malicious nodes. In this paper [32], the authors proposed a CR-Chord in which a node maintains cycles in its finger table. This cycle is the path to finger and return back to original node. Results showed that CR-Chord outperforms chord in terms of lookup failure and lookup availability. The only limitation of CR-Chord is that it maintains extra entry in finger table which is overhead.

As a lot of nodes keep joining and leaving the network, so it is necessary to introduce a trust model to check the credibility of any node. Meng *et al.* [33] proposed guarantee-based trust model in chord. So each node that is providing any service will have a guarantee peer that will rate the service quality of that node. Service demanding node will check reputation of all the nodes that are providing that service and then avail service form the node that has best reputation of all. Results show that the proposed model is efficient than the previous model and reduces the risk of malicious attacks. Results of model in real world are not discussed.

9.2.10.2   Peer data management-based chord

In the chord protocol data, lookup is not very efficient, so there is need of mechanism that should be able to query the data as fast as possible. Some variant of chord based on peer-data management is explained below.

Peer-data management is a major issue in P2P systems. The main problem is how to efficiently utilize metadata on shared resources and to reduce the search query time. In ML-Chord [34], resources are distributed into multilayers based on ontology. Results shows that ML-Chords performs better in the case of failure, average query cost, node joining cost, and average maintenance cost as compared to Chord. A limitation of ML-Chord is with the increase of number of nodes, average query response time increases.

Peer-data management is a promising area in P2P system. In R-Chord [35], a semantic-based resource organization model is proposed. It integrates three different models DHT Chord protocol, Resource Space Model, and P2P Semantic Link model to retrieve resources effectively. It performs better in terms of average query response time as compared to chord. Limitation of this approach is that only average query response time is measured, other parameters like latency and failures are ignored.

The similarity data retrieval is a major issue in P2P systems. In M-Chord [36], generalized vector technique iDistance [37] is used to map a data into one-dimensional sphere. The data is divided among the nodes in a ring network. Similarity search algorithm for query processing is proposed. Results show that it performs better in query processing as compared to chord. Limitation of this work is while processing query, maximum hop-count increases with the increase of nodes in a network.

Search for data on node is successful only when exact keyword is given for search but mostly we only know some features of data. Moreover, the keyword distribution is mostly skewed which means some keywords occur often and other occur rarely. To deal with this problem, concept of multi-keyword queries is implemented and every node is assigned with a keyword to refer the objects that have that keyword [38]. Keyword feature vector, of size Q, of the data is formed and stored at different nodes in the Q dimensional feature space. Query efficiency of the system should be improved.

In chord with the passage of time, network gets large and time to query the data gets high. The existing data query approaches support the key-value based search which does not support the semantic-based search. A small-world-based overlay for P2P search is proposed [39] in which author introduced the design of an overlay network, named semantic small world (SSW). SSW integrates four ideas: semantic clustering, dimension reduction, the small world network, and the efficient search algorithm that boost the data search in semantic-based search in P2P system. The simulation results of the tests are quite reasonable than the traditional chord network. In the proposed approach, the information of physical network is missing because if the nodes are far away from each other than the result would be different.

9.2.10.3 Mobility-based chord

Mobile node and wireless-based Chord is presented in literature. Some of the work is studied and presented in below section.

Mostly P2P protocols does not consider the node's mobility. Chord also does not have any mechanism if mobile nodes join the chord ring. In this paper [40], mobile chord (M-Chord) is presented to accommodate the node's mobility. M-Chord implements hierarchical structure to establish network of mobile nodes and it presents registering mechanism on top of Chord. M-Chord performs well in terms of efficiency and robustness as compared to chord. M-chord limitations are there in extensive load on super node.

Mobility of the nodes effects the performance of the protocol. Mobile robust Chord (MR-Chord) has been introduced that uses modified distribution hash protocol [41]. The real-time fix scheme is used to fix fingers in the case of query failure. The detect fix scheme is used to fix fingers in the network initially. Weather the node is moving or not, MR-Chord performs better than original Chord procedure in terms of lookup success and delay time.

In [42], the authors of this paper have implemented the chord into wireless mesh networks. In this approach, they have proposed the visibility of wireless infrastructure, the 1-hop broadcast nature of wireless communication and MAC layer cross-layering.

Author's combined these protocol and make a MESH-CHORD to minimize the problem due to mobility and to achieve better results. Results show that message overhead reduces for about 40% in static environment and cross layering increases the successful operations to 94%, while in the case of original Chord, it will drop to 70%. If the traffic is high, MESH-CHORD slows down.

9.2.10.4   Hierarchy-based chord

Chord and other structured protocols considered peers at the same level, which means they have flat distributed hash table. Practically it is not possible for every peer to offer same functionalities. In this paper [43], author proposed hierarchical DHT system to solve the problem of flat DHT. Nodes are assigned different hierarchies based on calculated load factor of each node. Operation cost of every hierarchy is also calculated for optimal design. Evaluation results shows that hierarchical P2P systems performs better than flat DHT in terms of node's capability. Limitation of this approach is that it assumes equal group sizes in hierarchy of super-peers.

Chou *et al.* [44] address the problem of hot spots in networks. Hot spots are points which are accessed more frequently than other nodes. This results in congestion in that part of the network. They presented a model named SCALLOP, which is a structured scalable and loading balancing P2P network. Scalability is achieved using complete balancing tree such that if the load exceeds certain level, it will be equally distributed between neighboring sibling nodes. Performance results of SCALLOP are compared with Chord and Koorde [45] protocols. SCALLOP works fine for small networks, for extremely large networks an extension of SCALLOP Sparse SCALLOP [46] is introduced as updating of all lookup tables after adding and removing the node in an extensive network causing a lot of updating messages. This SCALLOP causes more bottle neck as compared to SCALLOP but performs better than Chord, so a trade-off lies here.

9.2.10.5   Routing and latency-based chord

Delay in routing and high latency is the major problem in Chord protocol. Ruben *et al.* [47] worked to improve Jain's Fairness Index from 0.6 to 0.9 by improving the finger selection algorithm. Different sizes of the zones between the node effect unfairness in routing. The new algorithm enhanced Chord removes the size of zone effect on routing. When a query is received for finger, one of the randomly chosen successor of the node is sent back to improve balancing and there is no extra overhead added to previous chord algorithm.

Routing in the Chord is inefficient, so in this paper [48] authors have designed a methodology to improve the routing capability in the Chord network. To make the routing efficient in the chord networks, they have combined the routing proximity with the traditional chord algorithm and named it P-Chord. The results show that P-Chord is 18.52% lower in the case of routing latency than the original Chord. In the proposed approach, mechanism of node joining and leaving is not discussed.

Due to mismatch between an overlay and its physical network, the routing latency in existing DHT in chord is relatively high. The ability of IPv6 hierarchical feature and standard address allocation scheme proposed [49] a method to get incorporate topology information (IPv6 address have these information) and use it in the construction of DHT system that is used in the Chord. It creates many small local chord using IPv6 hierarchical address prefix and embed them into Global Chord. This approach does not define node joining mechanism.

In the Chord, the peers are not aware of the underlaying physical path of other peers which cause end-to-end delay because of the largest path for routing. Apply anycast (one-to-nearest-one-of-many) [50] of IPv6 to tell the nodes of the Chord network more about their underlying network topology to achieve better routing efficiency. Chord know the physical topology of the network which is done by using any-cast of IPv6, so it is performed on the 1,740 nodes which join and leave the network randomly and used the relative plenty delayed and which show that to query data along the network is 28% efficient than the original chord. The problem they have is it will work only for ideal any-cast, if it is not the case, it will be failed which is quite hard to implement in today's network.

9.2.10.6   Load distribution and resource allocation based Chord

Imbalanced distribution of load along the network nodes is an immense challenge, but some authors consider it as an opportunity to improve system performance [51]. Skewed data distribution degrades the performance of the system. To tackle this problem acceleration, nodes are placed in network that are replica nodes of the popular nodes. A number of replica nodes depend upon the popularity of the nodes. As the skewness of data increases, system performance increases. There exists a trade-off between performance improvement and cost paid for the network.

In the original Chord P2P networks, assigning a resource to peers when new resource is published, this method is quite slow and does not needs to be carried out. In the following paper [52], the method Self-Chord is proposed. Self-chord decouples the meaning of resources and peers; this results in two sets of keys that can have different multiplicity. A key can be assigned to resources depending on the requirement of a specific domain on the desired roughness of desired resource categorization. Self-Chord does not place resource keys to specified hosts, as Chord does; this feature is actually unnecessary and limits the system flexibility. The tests have been performed on 256 nodes because their linear approach is very slow, the number of steps needed to reach the target peer is logarithmic with respect to the number of peers since each step allows the search space to be approximately halved, as in Chord. This paper has open research for P2P frameworks because it presents a P2P system that inherits the beneficial characteristics of structured systems self-organization, adaptivity, scalability, and fast recovery from external perturbations.

9.2.10.7   Other chord-based approaches

Gao *et al.* [53] discussed a model for content discovery system that helps to excess content that is present on other nodes along the network. Previous models like chord performance is improved by introducing scalability and more features. Scalability is enhanced by adding rendezvous points, to handle registration and queries, and load balancing matrix. Attribute value pairs are used for increasing the ability to search content without using its canonical name. If the registration rate or data-hosting rate of a node exceed specified threshold or the handling matrix is in dynamic state, the registration may fail.

Number of hops (path length) and actual time required for lookup are two parameters that play role in key lookup. Enhanced bidirectional chord and enhanced bidirectional chord with lookup-parasitic random sampling are proposed to reduce lookup path length and latency, respectively [54]. Two finger table approach used in these will increase robustness. Enhanced bidirectional chord protocol can reduce path length up to 36% and latency 36%. Other approach can reduce path length and latency up to 33% and 63%, respectively. System robustness for the proposed model is not discussed.

Tai *et al.* [55] proposed a LISP-PChord to increase the scalability and trustworthiness. The users keep on changing theirs IPs, and the routing tables of the system updating keeps going on. Pointer nodes are introduced and mapping is done with physical nodes using LP and genetic algorithm. There are three layers: pointer space, logical space, and physical space. Pointer space meets the routing fairness and desired load balance through divisibility.

The motivation behind this paper [56] is to propose a WILCO that is not only aware of the P2P availability but also peer's location and services as well. This paper proposes a wireless location-aware Chord-based overlay mechanism for WMNs (WILCO). The location awareness of the proposed mechanism is achieved through a novel geographical multilevel Chord-ID assignment to the MRs on grid WMNs. Also an improved finger table is proposed in the paper to make use of the geographical multilevel ID assignment to minimize the underlay hop count of overlay messages. The proposed scheme outperforms the original Chord and the state-of-the-art MeshChord in terms of lookup efficiency and it significantly reduces the overlay message overhead. Algorithms efficiently deploy P2P services, content replication, and update and make use of the location awareness to improve service quality.

In this paper [57], an improved version of Chord algorithm has been proposed. The way Chord is initially designed does not consider physical characteristics of a network. Also, while calculating the optimal path for data query, network latency is ignored. These issues have been fixed by first analyzing and identifying the redundant information in the Chord finger table. Then that redundant information is modified in a way so that it can be utilized in a more efficient manner. There was no runtime addition or deletion of nodes from the network, and their effects were not recorded.

9.3 ODD model of a "Chord"

The ODD protocol was first proposed by Grimm *et al.* [58] and later updated and standardized in [14]. ODD model is a textual description of any agent-based model and individual-based models. The basic principal of ODD was to make models more descriptive and understandable.

9.3.1 Purpose

The aim of this work is to design and implement agent-based model of P2P network protocol chord in NetLogo and compare its results with object-based model of chord in PeerSim. Chord can provide efficient search, data locater and authentication, etc. Agent-based model of chord helps us in acquiring deep understanding of chord agents and its procedures. Through this model, we can perform behavioral study for agents by initializing variables to different values and simulate results.

9.3.2 Entities, state variables, and scales

Following are the entities of chord model:

9.3.2.1 Agents/Individuals

There are four types of agents that are nodes, update-nodes, seeker nodes, and ping. Node is the most vital agent of the model as it depicts the node of the network. Basically, all the active nodes have finger tables and data stored in those tables depicts state variables of that node. State variables of nodes are

---

Hid: Hash id of the node calculated using HSA-I
Suc: Successor of the node
Pred: Predecessor of the node
Fingers: Finger table of node
In-ring: To depict if the node is part of ring
Next: To point to finger table entry

Update-nodes handle fingers table updates of the network. State variables of update nodes are

---

ConnectTo: The point to which finger is to connect
In-ring: To show if the node is part of ring
Dest: Destination node whose finger table is to be updated
Entry: The slot of finger table that is subject of update

Seeker-nodes are the nodes that are in search of successor and predecessor. State variables of seeker-nodes are

---

Sender: The node that is seeking for successor
Seeking: The node that is sending messages
Dest: The node to which message is sent
Entry: The slot for which message is sent
In-ring: To check if the node in part of ring

Pings are the messages that are sent by nodes to other nodes in network to find successor and to connect its fingers to nodes.

Sender: The sender of the message
Dest: The destination of ping message
In-ring: To check if the node is part of ring

9.3.2.2   Spatial units

Total number of nodes traversed for finding a key in the network represents the spatial units of the model. In our model, total number of links shows total hops.

9.3.2.3   Environment

Environment depicts the hardware on which this chord protocol is working. In our model, this protocol works on nodes. Nodes can be routers, hubs, or any gateways in real environment.

9.3.2.4   Collectives

All the nodes combine to form the shape of a ring using function in-ring. All the nodes in network define their successors, predecessors, and location of their fingers and links are formed. Any key is looked up in the network using links between the nodes.

9.3.3   Process overview and scheduling

First of all, setup procedure is called to set the environment for this protocol. Hash id is assigned to every node. Initially, nodes are defined, their predecessor is set to nobody and successor are set to itself. The node following the node under consideration on the identifier ring is the successor, while the previous node from the node under consideration on the identifier ring is its predecessor. Then for nodes which want to join their successor, predecessor and finger tables are defined. Successors of the node are found by calling find-successor procedure. New node is added to the ring and the maintenance procedure is called. Maintenance procedure stabilizes the network and fix fingers of nodes. Fingers are placed using formula $(n + 2k - 1) \bmod 2\,m$, $1 \leq k \leq m$, where n represents the node number and m represents the number of digits in node hash id. When a node joins or leaves the network, finger table entries of all nodes that are connected to it upgrade their finger table as a message is circulated by the successor of leaving node informing nodes about its departure.

9.3.4   Design concepts

9.3.4.1   Basic principles

To lookup a key in the network, we maintain finger table. Each node in the network maintains a finger table in which information about the nodes to which the node is connected through its fingers is stored. Where to put fingers is calculated using formula where n represents the node number and m represents the number of digits in node hash id. Complexity for key lookup is O(log N). Node joining or leaving the network will create $O(\log^2 N)$ messages to maintain the finger tables. Scope of basic principles is of sub-models level. The model will provide insight about the basic principles as through agents we can relate working of model in real word scenarios through the agents. The model use previously developed theory of agent-based traits from which system dynamics emerge. In this model, individuals are represented as network nodes and their behaviors are tracked with time.

9.3.4.2 Emergence

The nodes connect to form a ring. For look-up key is forwarded through the network using the links formed between nodes.

9.3.4.3 Adaptation

When a node leaves or joins the network, the finger tables of all nodes associated with that node gets updated. The updating process ensures successful lookup of keys, as to achieve success in key lookup, proper data storage at node finger table is necessary.

9.3.4.4 Objectives

The adaptive trait of updating finger tables increase the success of key lookup. The lookup success is measured on two basics. Number of hops and success in finding the particular key determines the success of key lookup.

9.3.4.5 Learning

When a node leaves a network, its predecessor and successor update their finger tables and inform other nodes about it.

9.3.4.6 Sensing

When a node wants to join the network, it send messages to the nearby nodes and then gets connected to the nearest node. Then its immediate successor is found on the circle, and rest of the network is updated through stabilization.

9.3.4.7 Stochasticity

The process of node joining or leaving are stochastic. There is no particular frequency for it.

9.3.4.8 Interaction

There are direct interactions between nodes. These interactions are represented through links. For looking up a key in the network, these direct links are used.

9.3.4.9 Collectives

All the nodes join to form a ring-shaped structure, and each node effects this collective work. Each node is connected to its successor, predecessor, and other nodes through its fingers.

9.3.4.10 Observation

All the nodes join to form a ring-shaped structure, and each node effect this collective work. Each node is connected to its successor, predecessor, and other nodes through its fingers.

9.3.5 Initialization

Hash degree, number of node and add nodes are the state variables. Initially, all nodes are defined, and but they are not connected to each other. The number of nodes are there initially and its value can be set through slider. Initialization can be varied among different simulations.

9.3.6 Input data

Input data is not given by any external source. The values of variables defined are used by the model.

9.3.7 Sub-models

Sub-models make it easy to understand a code.

9.3.7.1 Set-up

It is called to clear previous variables and set-up environment for new simulation.

9.3.7.2 Init-node

It initializes nodes and sets their state variable values.

9.3.7.3 Create-network

It creates a ring of some nodes, and rest of the nodes keep on joining the network.

9.3.7.4 Go

It is the main procedure that starts the simulation. All other main procedures like start messages, update messages, and maintenance are called from it.

9.4 DREAM model of a "Chord"

In this section, we will give detail DREAM model of a chord protocol.

9.4.1 Agent design

9.4.1.1 State charts (of agents)

In our system, if a node wants to join a network, it first sends ping message to the network; the node which sends ping then changes its state and becomes seeker-node. The seeker-node gets successors and predecessor response from a node in a network. Then other nodes in the network changes its state to update-node. The finger table is updated in update-node state. The state diagram of agents is shown in Figure 9.4.

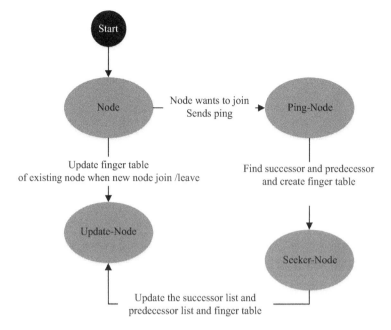

Figure 9.4   State diagrams of a node agent when it joins network

9.4.2   Activity diagrams

The overall flow activities of chord network implementation in NetLogo and some of the descriptions are as follows: procedures are the main cores of chord network of agent based in NetLogo. Setup activity is used to set default values which are necessary for the network creation, and this activity is dependent upon global agent which will provide the values for default parameters. After the network is setup, the next activity will be the go which will start the network and whole operations almost happened in this activity, but this activity is dependent upon the agent activity of seeker-node which is used to help get attributes for the new node which wants to join the network and the previous node which is the part of the network and the control will be shifted to go and this agents activity that is seeker-node and update-node will be calling by the go again and again. This flow of activity is shown in Figure 9.5 and Figure 9.6.

9.4.3   Flowchart

A flowchart is used to describe the algorithms and processes. A flowchart basically gives the step-by-step information that is needed for working of any process or algorithm. The flowchart shown in Figure 9.7 explains the steps of initial network formation in Chord algorithm. First, it initialize the values of network size, finger size, node that can join the network and leave the network. The hash function is used to get the hash value of node IP. After the hash IDs has been assigned to nodes, nodes try to

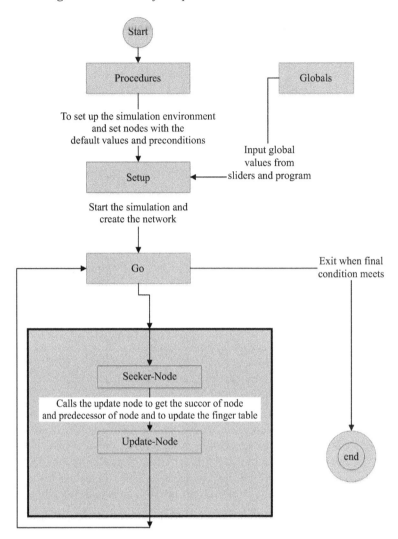

Figure 9.5 Activity diagram of Chord protocol

find successors using seeker messages. Fingers of nodes are placed at the respective nodes in the ring. When a node leaves or join, maintenance procedure is called to update finger table (Table 9.4).

9.4.4 Pseudo-code based specification

9.4.4.1 Agents and breed

Here we will describe agents and breed. For our model, we have four different types of agents: agent node, update-node agent, seeker-node agent, and ping-node agent.

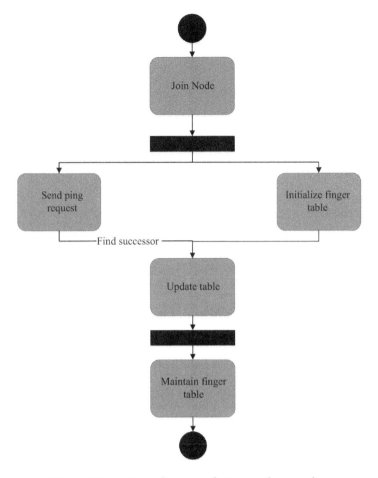

Figure 9.6   Activity diagram of a join node procedure

Node agent

Node agents are the main agents of a chord protocol as depicted in Table 9.5. They represent nodes in a network. They are placed in a ring network. They can communicate with each other using messages. Breed Node is the first one to explain in pseudo-code based specification. It has six variables which are hid, predecessor, successor, fingers, in-ring, and next. The variable "hid" is Hash id and is identification number assigned to nodes generated by a hash algorithm SHA [25]. Predecessor is the id of a previous node of any given node. Successor is the id of next node in a network. Next variable is fingers, it is fingers of nodes which are linked to any specific node at any given time in a finger table. The variable in-ring is used to create ring for nodes. The last variable "next" is next node entry in a finger table.

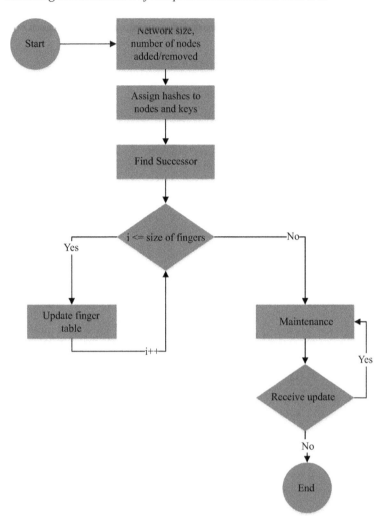

Figure 9.7   Flowchart of a Chord network

The next is "seeker-Node." seeker-Node shown in Table 9.6 is the one who wants to join ring network. It is used to find successor of a node. Seeker node seeks for a key in a network. It has five different variables. The "sender" variable is used for the node who wants to know its successor. The "seeking" is the data key which any node wants to find, "destination" variable is the successor of node sender. The variable entry is the entry in a finger table, and last variable "in-ring" is used to join ring.

The next breed is "update-Node" breed. "update-Node" breed is used to respond to "seeker-Node" breed. It responds to seeker node by sending message to seeker-Node. Message either contains desired successor address or a closest preceding node address if desired successor is not found. It has four variables, "connectTonode,"

*Table 9.4   Table showing nodes connection when
creating network in Pajek*

Node from	Node to
ABM	Agents
ABM	Globals
Globals	InputGlobals
Globals	OutputGlobals
Agents	Agent-Attributes
Agents	Agent-Breed
ABM	BS-Expts
ABM	Procedures
BS-Expts	network-size
Agent-Breed	Node
Agent-Breed	Update-Node
Agent-Breed	Seeker-Node
Agent-Breed	Ping-Node
InputGlobals	network-size
InputGlobals	stabilize-count
InputGlobals	Speed
Node	Node-Attributes
Update-Node	Update-Node-Attributes
Seeker-Node	Seeker-Node-Attributes
Node-Attributes	hid
Node-Attributes	predecessor
Node-Attributes	successor
Node-Attributes	fingers
Node-Attributes	in-ring
Node-Attributes	next
Update-Node-Attributes	ConnectToNode
Update-Node-Attributes	in-ring
Update-Node-Attributes	destination
Update-Node-Attributes	entry
Seeker-Node-Attributes	sender
Seeker-Node-Attributes	seeking
Seeker-Node-Attributes	in-ring
Seeker-Node-Attributes	destination
Seeker-Node-Attributes	entry
Procedure	setup
Procedure	init-node
Procedure	create-network
Procedure	start-messages
Procedure	lookup-messages
Procedure	maintenance
Procedure	find-successor
Procedure	join-node
Procedure	lookup
Procedure	stabilize
Procedure	notify
Procedure	lookup-ping
Procedure	fix-fingers
Procedure	report
Procedure	init-fingers
Procedure	receive-update

Table 9.5   Agent Breed Node

Breed **Node:** This breed represents the Node
Internal Variables: <hid predecessor successor fingers in-ring next > **hid:** hash ID of a given node **predecessor:** previous node in a network **successor:** next node of a given node **fingers:** fingers (links) of a given node **in-ring:** network ring which is made up of nodes **next:** next node entry in finger table

Table 9.6   Agent Breed seeker-Node

Breed **seeker-Node:** This breed represents the seeker Node
Internal Variables: <sender seeking destination entry in-ring > **sender:** node which wants to join a ring **seeking:** node seeking for a successor **destination:** destination node with which nodes want to connect **entry**: update entry in finger table **in-ring:** ring network

Table 9.7   Agent Breed update-Node

Breed **update-Node:** This breed represents the update Node
Internal Variables: <connectTonode in-ring destination entry > **connectTonode:** creating links which connects nodes. **in-ring:** network ring which is made up of nodes **destination:** destination of a node. **entry:** finger table entry

"in-ring," "destination," and "entry." The variable "connectTonode" is used to connect seeker node to its successor. The variable "in-ring" is used to join ring and variable "destination" is either address of successor or address of closest preceding node. The last variable "entry" is used to update entry in finger table. As depicted in Table 9.7

The agent "ping-Node" is used when a new node wants to join a network, it sends a ping message to network (Table 9.8).

9.4.4.2   Globals

Next we will explain globals used in our simulation model Chord. We have four global input variables described in Table 9.9. The variables "network-size,"

Table 9.8   Agent Breed ping-Node

Breed **ping-Node:** This breed represents the ping Node

Internal Variables: < sender destination in-ring >
sender: sender which sends ping.
in-ring: network ring which is made up of nodes
destination: node which receive pings

Table 9.9   Input Globals

Input Globals: <network-size,node-join,node-leave,update-frequency speed,stabilize-count,>

Sliders :
    network-size : it is used to give total number of nodes in a ring
    node-join : no. of nodes want to join the network
    node-leave : no. of nodes want to leave the network
    update-frequency : time after which stabilize is called to update the finger table.
Switch :
    speed : distance between two nodes
    stabilize-count : used to count stabilization call

"node-join," "node-leave," and "update-frequency" are taken from sliders and are provided by user. The other two variables "speed" and "stabilize-count" are declared globally in a program. The variable "stabilize-count" is to count how many times the procedure stabilize is called, and the last variable "speed" is used to calculate the distance between two nodes (Table 9.9).

9.4.4.3   Procedures

First procedure to be called is setup. Setup is the key procedure which should be called at the start of the simulation every time. Purpose of setup is to clear previously declared environment variables and to prepare the workspace for new simulation. It clears the output generated by previous simulation and reset ticks as well. The setup procedure in our case will call init-node procedure which initializes the nodes. After nodes are initializes, message is sent to each node to set their labels as hash ids and set their color to given color. Create network procedure is called to create empty chord network for new nodes. And at the end, nodes are arranged in a ring shape using layout-circle function (Table 9.10).

Init-node is a function which will create nodes for simulation. It is called from procedure setup. It creates nodes using global variable network-size from slider. It sets successor and predecessor to nobody. It sends message to node to set its shape to circle and calculate hash ids for nodes as their identification key. Hash ids are calculated using $2^{HashDegree}$, where hash degree is taken from slider as global variable. And it initializes finger table to 160 entries (Table 9.11).

Table 9.10   Procedure setup

Procedure **setup** : setting up the simulation

Input: It Uses global variables and parameters from user interface
Output: All nodes and environment is setup for simulation
Execution: Called at the start
Context: Observer
begin
    1. clear the screen
    2. clear all nodes
    3. clear output
    4. reset ticks
    5. create nodes
End

Table 9.11   Procedure init-node

Procedure **init-node** : To initialize nodes and generate hash keys

Input: Uses internal variables of nodes
Output: The nodes are configured according to variables
Execution: Called by setup
Context: Agent
begin
    1. initialize successor and predecessor
    2. initialize ring
    3. generate hash keys
    4. initialize finger table
End

The next procedure is "Create-network" procedure. It is called from setup. It creates an empty chord ring. It sets the ring for nodes to join. Successor and predecessor are initialized to nobody in this procedure. This procedure also initializes fingers of nodes by calling procedure "init-fingers" (Table 9.12).

Next we are going to explain procedure "go." The procedure "go" is for executing one step of the simulation. It uses global variables from interface. It is called repeatedly to run the simulation and to get the results. Other procedures called from "go" is start-messages, lookup messages, and ring maintenance. All these procedures are responsible for creating ring, updating finger tables and maintaining chord ring (Table 9.13).

Here we will explain specification of a procedure "start-messages." When any new node wants to join a chord ring, it will send messages to its neighboring node. It sends messages to agent seeker-node, update-node, and ping to find distance from

Table 9.12   Procedure create network

Procedure **create-network** : Creating network of initialize nodes

---

Input: Uses global variables from interface
Output: The network is configured according to initialize node variables
Execution: Called by setup
Context: Observer
begin
    1. set successor to self
    2. set predecessor to nobody
    3. initialize fingers
    4. create ring network
End

Table 9.13   Procedure go

Procedure **go** : for executing one step of the simulation

Input: Uses global variables from interface
Output: Equates functions and single step of each agent
Execution: Called repeatedly for execution of the simulation
Context: Observer
begin
    1. start-messages to nodes
    2. lookup messages
    3. maintenance of a ring
End

Table 9.14   Procedure start messages

Procedure **start-messages** : for joining the network node send messages

---

Input: Uses local variables and agents from interface
Output: Find destination node
Execution: Called from procedure go
Context: Observer
begin
    1. find distance from destination
    2. start sending messages
End

---

neighboring node and forward it accordingly to that specified speed. The procedure "start-messages" is called from procedure go (Table 9.14).

The procedure "lookup-messages" is called from procedure "go." It sends message to agent "Node" to get successor list and then sort it. After that it calls procedure

Table 9.15   Procedure maintenance

Procedure **maintenance** : for maintaining ring network

Input: Uses local variables and global variables from interface
Output: Finger table fixed and updated
Execution: Called from procedure go
Context: Observer
begin
    1. call of procedure stabilize
    2. call of procedure fix fingers
End

"receive-update" to update the finger table of a node and also ping all the nodes which are in successor list.

Procedure **lookup-messages** : to find successor list and ping them

---

Input: Uses local variables and agents from interface
Output: Successor list
Execution: Called from procedure go
Context: Observer
begin
    1. find successor list
    2. receive-update
    3. ping nodes in successor list
End

The procedure "maintenance" connects all new joining nodes in a ring. Then it calls procedure "stabilize" and "fix-finger" to update the finger table periodically (Table 9.15).

The specification of procedure "find-successor" is explained for a given node. In this procedure, if a given node is in the range of the requesting node, then it will return successor id to requesting node. If it is not in range, then it will return the id of closest preceding node (Table 9.16).

The "join-node" procedure is called from "lookup-ping" procedure. It initializes successor and predecessor to "nobody" and initializes fingers by procedure init-fingers. It generates messages from requesting nodes to destination node. It flooded messages until link is created between two nodes (Table 9.17).

Next procedure explained here is "init-fingers." In this procedure, variable next of agent "Node" is initialized to "0." It is called from procedure "create-network" and "join-node" (Table 9.18).

The procedure "stabilize" updates the finger table. It is called periodically for up-to-date information. Whenever a new node joins a network, the node calls a "stabilize"

Table 9.16   Procedure find successor

Procedure **find-successor** : to find successor of a particular node

---

Input: Uses local variables and messages
Output: Successor node found
Execution: Called from procedure go
Context: Observer
begin
    1. **If** successor is in range or node itself is a successor
     2. return successor id
     3. **else**
     4. return id of closest preceding node
End

Table 9.17   Procedure Join node

Procedure **Join-node** : procedure to join a node in the ring

---

Input: Uses local variables
Output: Node joined the ring
Execution: Called from procedure lookup-ping
Context: Observer
begin
     1. initialize fingers in finger table
     2. initialize seeker node entries
End

Table 9.18   Procedure init-fingers

Procedure **init-fingers**: initialize fingers of any given node

---

Input: Uses local variables
Output: Fingers initialized
Execution: Called from procedure maintenance
Context: Observer
begin
     1. initialize fingers in finger table
End

---

procedure. After procedure is called, node sets its entry of successor node and notifies its successor node to acknowledge it as a predecessor. All the nodes, which are already connected in the ring, run "stabilize" periodically and it asks the joining node to set its predecessor to the nearest node (Tables 9.19–9.21).

Table 9.19   Procedure Stabilize

Procedure **Stabilize**: procedure runs periodically to update finger table

Input: Uses local variables
Output: Node joined the ring
Execution: Called from procedure maintenance
Context: Observer
begin
    1. newly joined node sets its successor
    2. newly joined node sets its predecessor
End

The next procedure is "notify." The procedure "notify" is called from procedure "stabilize." In "notify" sets its predecessor.

Table 9.20   Procedure notify

Procedure **notify**: procedure to join node in a ring
Input: Uses local variables *Output: Node joined the ring* *Execution: Called from procedure stabilize* *Context: Observer* **begin**   1. notify node to set its predecessor **End**

Table 9.21   Procedure fix fingers

Procedure **fix-fingers** : fix fingers in finger table
Input: Uses local variables *Output: Node joined the ring* *Execution: Called from procedure maintenance* *Context: Observer* **begin**   1. update finger table **End**

The procedure "fix-fingers" is called from procedure "maintenance." It periodically runs and update finger table. It generates messages to all those nodes which will be its fingers. The specification of "receive-update" is explained here. When a node joins or leaves the network, a "receive-update" message is generated.

Table 9.22   Procedure receive update

Procedure **receive-update** : update finger table
Input: Uses local variables and message *Output: Finger table updation* *Execution: Called from lookup messages* *Context: Agent* **begin**   1. update entries of a finger table Update network ring **End**

Table 9.23   Procedure lookup ping

Procedure **lookup-ping** : new node sends messages on joining, this procedure handles it

Input: Uses local variables and message
Output: Finger table updation
Execution: Called from lookup messages
Context: Agent
begin
    1. update entries of a finger table
    2. update network ring
End

The "receive-update" procedure is called from procedure "lookup-message." It updates finger table at the given entry (Table 9.22).

The specification of procedure "lookup-ping" is explained here. In this procedure, node sends ping messages to nodes which will be its fingers.

9.4.4.4   Experiments

In this section, we have explained the procedures related to experiments (Tables 9.23–9.25).

9.5   Results and discussion

In this section, we will explain the findings and result we have obtained using PeerSim and NetLogo.

9.5.1   Metrics (table and description)

The implementation of a Chord is replicated in NetLogo and is validated by matching our results with the results produced in PeerSim [15]. We have simulated a network in NetLogo comparable to PeerSim [15] to obtain analogous results.

Table 9.24   Experiment to Vary Nodes added/removed

Experiment **Vary Nodes added/removed** : Experiment for noting the effects
of changing the number of nodes added removed
Input: <network-size,node-join,node-leave>
Setup procedures: <setup>
Go procedures: <go>
Repetition: <40>

Inputs:
Nodes added/removed: [5,10,−5,−10]
network-size: [5,000]
Stop condition: Ticks = 500
Final commands: none

Table 9.25   Experiment to Vary Network size

Experiment **Vary network-size**: Experiment for noting the effects
of changing the network size
Input: <network-size,node-join,node-leave>
Setup procedures: <setup>
Go procedures: <go>
Repetition: <40>

Inputs:
Network-size: [500,1000,5000,10000]
nodes added/removed: [20,−5]
Stop condition: Ticks = 500
Final commands: none

The interface of a Chord, which is implemented in NetLogo, is shown in
Figure 9.8. The interface design consists of graphical window where the nodes interacts with each other. It also has different type of sliders to select for different input.
The first one on top left is to adjust the network size. Next one is to add further nodes
in a network. Two different buttons are also present on the graphical interface, one
is to set up a simulation and other one to run a simulation. At the left bottom of
the interface, number of links and number of stabilization are plotted. To properly
observe the behavior of Chord in NetLogo, we have used same parameters as given
in PeerSim implementation. Input parameters are

- Varying network size
- Adding and removing number of nodes

And for output

- we are counting number of links (hop count in PeerSim)
- Stabilization count (how many times procedure stabilization is called)

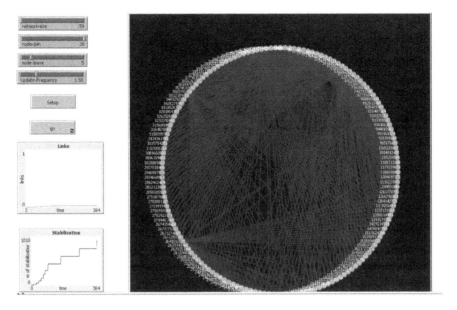

Figure 9.8   The graphical interface of Chord in NetLogo

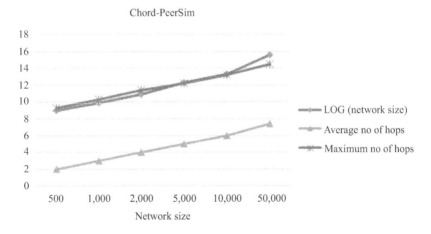

Figure 9.9   Plot showing maximum number of hops, average number of hops and log of network size in PeerSim

The results of PeerSim and NetLogo are explained in subsequent sections.

9.5.2   PeerSim results

Plots are presented for hop count, stabilization, and failures by varying network size, new node joining, and node leaving. In Figure 9.9, average number of hop count,

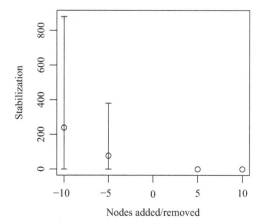

Figure 9.10   Plot showing stabilization of Chord network in PeerSim

maximum number of hop count, and log of network size is plotted. By varying network size to 500, 1,000, 5,000, 10,000, and 20,000, we have different values for hop count. We can see that average number of hop count increases with the increase of network size. The average hop count for network size 500 is 2, and it increases to 7.36 with the network size 50,000.

In Figure 9.10, the interval graph for stabilization count is plotted by varying nodes added and removed. The network size in this case is fixed, i.e., 5,000. The interval graph shows that when nodes are removed, stabilization is called more frequently. Stabilization of maximum value in case when 10 nodes are removed is 881, whereas mean is 240. But in case when no node is removed but new nodes are added in the network, there is not any stabilization call.

In Figure 9.11, we have seen that number of failures rises when more nodes are joining and leaving the network.

9.5.3   ABM results

In NetLogo, we have replicated the results of PeerSim. We have varied network size and then observed the hop count and stabilization calls. Then we have added and removed the nodes from the network and observed the results.

We can see in Figure 9.12 the average and maximum hop count with varying network size. In Figure 9.13, we have plotted interval graph for a number of stabilization calls verses nodes added and removed. So we have more stabilization calls when 10 nodes are removed from the network as compared to node added in the network. When there is not any node joining the network, we still have stabilization call as it is called after some specific interval.

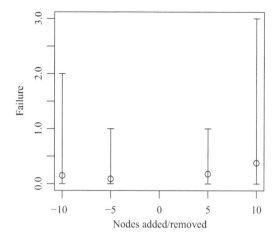

Figure 9.11   Plot showing failure of Chord network in PeerSim

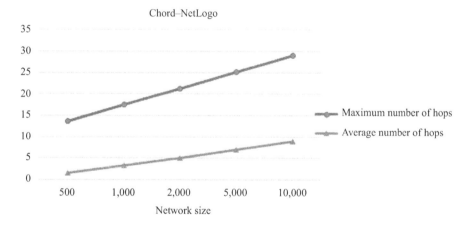

Figure 9.12   Plot showing maximum hop count and average hop count with varying network size in NetLogo

9.5.4   Comparison of PeerSim and ABM

In this section, we will compare PeerSim and ABM.

- Simple and easy: It is noticeable that the implementation of Chord in NetLogo is simple and easy as compared to PeerSim. In NetLogo, we have about 100 lines of code for Chord, whereas same protocol in PeerSim has several hundreds of lines of code. It shows that how suitable it is to select NetLogo as compared to other P2P simulators.
- Graphical user interface (GUI): Mostly P2P simulators are not graphical. To understand the behavior of network, some additional packages need to be added

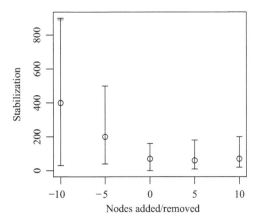

Figure 9.13   Plot showing stabilization of Chord network in NetLogo

in some simulators to interpret nodes interaction. PeerSim also does not have a GUI, but NetLogo GUI interface is user friendly, easy to understand and interpret. One can see the behavior of network and interaction of nodes easily. We can add sliders in the interface and add any input at any time. In Figure 9.8, we can see that we can add or remove nodes using sliders. We can give network size according to scenario requirement.

- Plots visualization: In PeerSim, we cannot visualize the plots while simulation is running. In the case of NetLogo, we have real-time monitors where we can plot the behavior of network.

- Results comparison: Comparing the results of PeerSim and NetLogo, we have observed the similar trends in results. We can see the average number of hop is greater in NetLogo as compared to PeerSim. Also, maximum hop count is greater in NetLogo as shown in Figure 9.9. Overall trend is same in both; with the increasing network size, number of hops increases. But the slight difference in values is because of different simulation environment. When we compare number of stabilization calls, in the case of PeerSim when a node joins, there is not any stabilization call, but in NetLogo, finger table is updated after stabilization call. So we have more stabilization calls in NetLogo as compared to PeerSim. When there is not any node joining the network, we still have stabilization call in NetLogo as it is called after some value.

9.5.5 DREAM network models

In this section, we have calculated centralities of a network which we have explained in Section 9.2. The three centralities, degree, betweenness and eccentricity, are measured which are shown in Table 9.26. After calculating degree centrality, the network graph in Figure 9.14 is manipulated and loaded with degree values, and new network graph is shown in Figure 9.15. We can see that size of some nodes increases. The size of node

Table 9.26 Table of eccentricity, betweenness and degree centrality measures of the Chord network

Id	Eccentricity (%)	Betweenness (%)	Degree (%)
ABM	0.660452	4	0
Agents	0.521469	3	0
Globals	0.238983	3	0
InputGlobals	0.191525	7	15.545455
OutputGlobals	0	1	0
Agent-Attributes	0	1	0
Agent-Breed	0.641243	5	11.075758
BS-Expts	0.098305	4	7
Procedures	0.489831	18	42.232323
node-join	0	1	0
node-leave	0	1	0
network-size	0	1	0
stabilize-count	0	1	0
Speed	0	1	0
update-frequency	0	1	0
network-sizeB	0	1	0
Node	0.209605	2	0
Update-Node	0.155367	2	0
Seeker-Node	0.183051	2	0
ping-Node	0.126554	2	0
Node-Attributes	0.191525	7	14.166667
Update-Node-Attributes	0.129944	5	10
Seeker-Node-Attributes	0.161017	6	12.090909
hid	0	1	0
predecessor	0	1	0
successor	0	1	0
fingers	0	1	0
in-ringn	0	1	0
next	0	1	0
ConnectToNode	0	1	0
in-ringu	0	1	0
destination	0	1	0
entry	0	1	0
sender	0	1	0
seeking	0	1	0
in-rings	0	1	0
destination-s	0	1	0
entry-s	0	1	0
setup	0	1	0
receive-update	0	1	0
init-node	0	1	0
create-network	0	1	0
start-messages	0	1	0
lookup-messages	0	1	0
maintenance	0	1	0
find-successor	0	1	0

(Continued)

Table 9.27 (Continued)

Id	Eccentricity (%)	Betweenness (%)	Degree (%)
join-node	0	1	0
lookup	0	1	0
stabilize	0	1	0
notify	0	1	0
lookup-ping	0	1	0
fix-fingers	0	1	0
report	0	1	0
leave-network	0	1	0
init-fingers	0	1	0
ping-Node-attributes	0.098305	4	7.888889
sender-p	0	1	0
destination-p	0	1	0
in-ringp	0	1	0
node-join-b	0	1	0
node-leave-b	0	1	0

Procedure is larger from all as node procedure has many others procedures attached with it. The size of node Input Globals is also larger as it has many input global variables attached with it. Other nodes which has larger size are agent attributes; every agent attribute has many attributes attached with it, so their size also increased.

9.5.5.1 Plots of centralities

In this section, we have plotted and analyzed the centralities which we have calculated in previous section.

In Figure 9.16, degree centrality is plotted . In this plot, we can analyze that the highest degree is of node Procedure. The second one which has highest value are agents attributes, i.e., node attributes and input globals. The nodes which do not have further nodes attached with them has degree value of 1.

Next betweenness centrality is shown in Figure 9.17. Here, we can see that highest centrality is for node ABM. ABM is the node with which all the main nodes are attached. Next highest centrality is for node Procedure and agent breed. Nodes which do not have further nodes attached with them have zero betweenness centrality.

Last centrality is eccentricity centrality. Eccentricity centrality is shown in Figure 9.18. We can note here that like degree centrality, node Procedure has the highest value. The second highest value is for input-globals followed by node attributes

9.5.5.2 Plots of centralities using power-law

Degree centralities are then plotted in R [59] after applying power-law on sorted values of centralities. In network, there are a few nodes that have many links and are playing the most important part in network, while there are many nodes with one or two links. When we apply power law on the sorted values of centralities, we get a

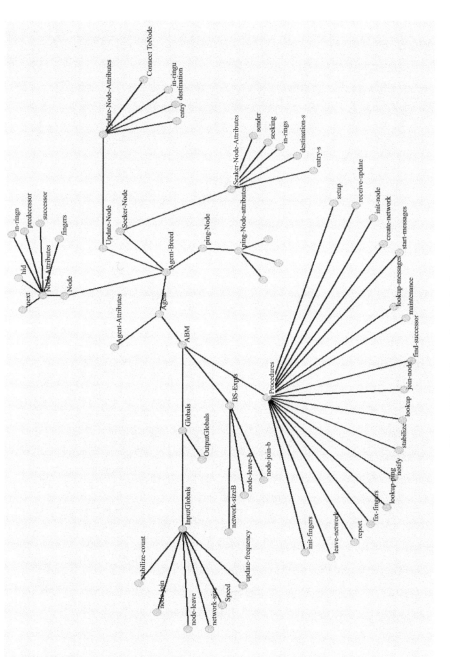

Figure 9.14   Network model of a chord

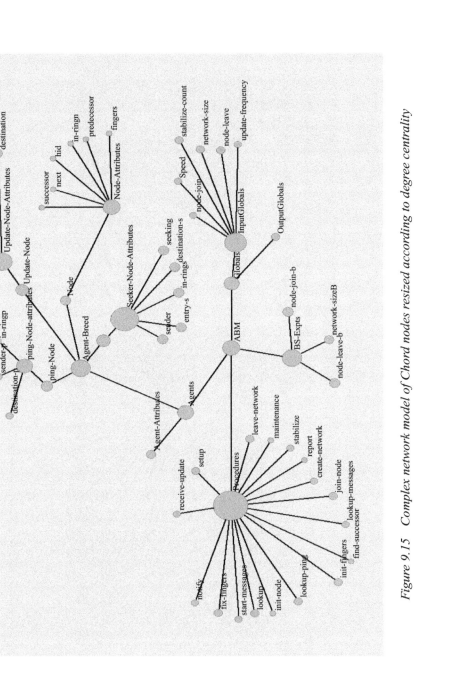

Figure 9.15 Complex network model of Chord nodes resized according to degree centrality

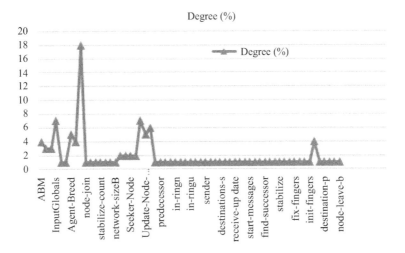

Figure 9.16   Plot showing the degree centrality

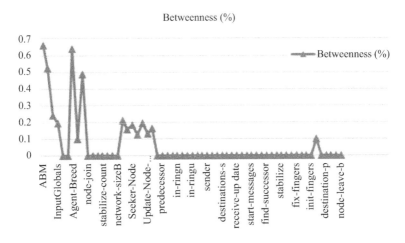

Figure 9.17   Plot showing the betweenness centrality

decaying curve that shows that there are a few nodes with more links as compared to other nodes.

In Figure 9.19, degree centrality is plotted which shows a cumulative curve in decreasing order. In Figures 9.20 and 9.21, eccentricity centrality and betweenness centrality are shown, both of which showing the same curve behavior like degree centrality.

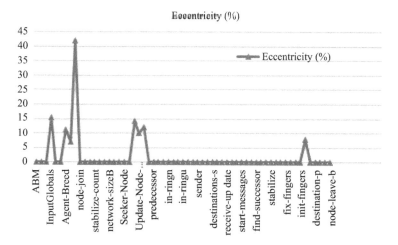

Figure 9.18   Plot showing the eccentricity centrality

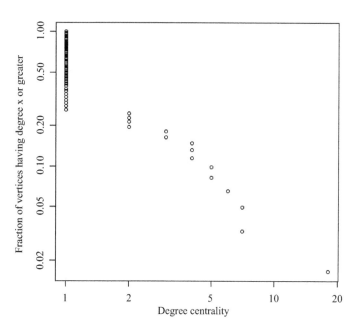

Figure 9.19   Plot showing the degree centrality

9.5.6   Discussion (ODD vs. DREAM pros and cons of both) and which is more useful for modeling the chosen P2P protocol

ODD protocols give the overview, design concepts, and details of the CHORD and explains all the main elements, i.e., emergence, adaptation, objectives, prediction,

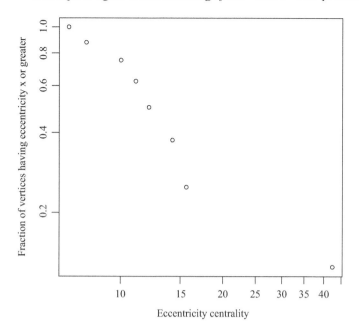

Figure 9.20   Plot showing the eccentricity centrality

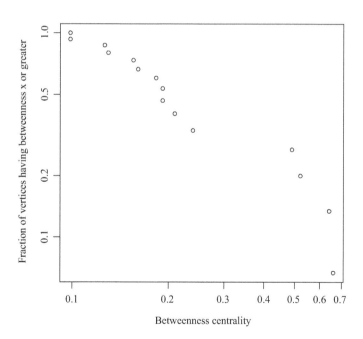

Figure 9.21   Plot showing the betweenness centrality

sensing, interaction, stochasticity, collectives and observation, required for modeling of Chord. Using ODD CHORD protocol, working can be implemented in any language, but its working may not be that much evident. DREAM explains CHORD protocol using its procedure pseudo code. It explains model working as well as its procedures in details. It shows the close correlation of all the sub-models. Using it CHORD can be easily understood, evaluated, compared and implemented in any language. So, for the modeling of CHORD protocol, DREAM model is the best.

9.5.7 Chord and theory of computation

In the theory of computation, we have three different branches, the complexity theory, the computability theory and the automata theory. Here we are relating the Chord protocol with the theory of complexity.

9.5.7.1 Complexity theory

The theory of complexity is the measure of an efficiency of a system in terms of its working. In a Chord protocol, we measure complexity based on its different functionalities. The measure of scalability (how scalable the Chord network is), to what extent it balances the load among different peers and if it is fault tolerant, then what will be its efficiency measure.

- Load balancing: Consistent hashing provides load balancing by distributing the load on different nodes and each node is accountable for around K/N keys (K keys, N nodes).
- Scalable: Chord is scalable as it requires $O(\log n)$ lookup messages.
- Fault tolerant: When nodes join or leave a network, only $O(log^2 N)$ messages are necessary to rebuild routing invariants and finger tables.

9.6 Conclusions and future work

In this chapter, we have modeled P2P protocol Chord in DREAM for descriptive agent-based modeling and in ODD for textual description. We have seen that using DREAM and ODD, now it is easy to use and understand Chord and it can be implemented in any simulator using these models. The protocol Chord is then implemented in NetLogo to replicate the results of PeerSim. We have shown that NetLogo provides user with better GUI, easy, and simple implementation and good visualization tools for analysis as compared to PeerSim. Future work may include implementations of Chord enhancements in NetLogo and to improve Chord in terms of lookup latency.

References

[1] Balakrishnan H. Chord: a scalable peer-to-peer lookup service for internet applications. In: ACM SIGCOMM. Citeseer; 2001.

[2] Stoica I, Morris R, Liben-Nowell D, *et al.* Chord: a scalable peer-to-peer lookup protocol for internet applications. IEEE/ACM Transactions on Networking (TON). 2003;11(1):17–32.

[3] Napster L. Napster. URL: http://www napster com. 2001.

[4] Clarke I. A distributed decentralised information storage and retrieval system. Doctoral dissertation, Master's thesis, University of Edinburgh; 1999.

[5] Clarke I, Sandberg O, Wiley B, *et al.* Freenet: a distributed anonymous information storage and retrieval system. Designing privacy enhancing technologies. Berlin, Heidelberg: Springer; 2001. p. 46–66.

[6] Dabek F, Brunskill E, Kaashoek MF, *et al.* Building peer-to-peer systems with Chord, a distributed lookup service. In: Hot Topics in Operating Systems, 2001. Proceedings of the Eighth Workshop on. IEEE; 2001. p. 81–86.

[7] Cai M, Frank M, Chen J, *et al.* MAAN: A multi-attribute addressable network for grid information services. Journal of Grid Computing. 2004;2(1):3–14.

[8] Balakrishnan H, Kaashoek MF, Karger D, *et al.* Looking up data in P2P systems. Communications of the ACM. 2003;46(2):43–48.

[9] Litwin W, Neimat MA, Schneider DA. LHa scalable, distributed data structure. ACM Transactions on Database Systems (TODS). 1996;21(4):480–525.

[10] Niazi MA, Hussain A. Cognitive agent-based computing-I: a unified framework for modeling complex adaptive systems using agent-based & complex network-based methods. Netherlands: Springer; 2012.

[11] Axelrod RM. The complexity of cooperation: agent-based models of competition and collaboration. Princeton, NJ: Princeton University Press; 1997.

[12] Newman ME. The structure and function of complex networks. SIAM Review. 2003;45(2):167–256.

[13] Niazi, MAK. Towards A Novel Unified Framework for Developing Formal, Network and Validated Agent-Based Simulation Models of Complex Adaptive Systems. Doctoral dissertation, University of Stirling; 2011.

[14] Grimm V, Berger U, DeAngelis DL, *et al.* The ODD protocol: a review and first update. Ecological Modelling. 2010;221(23):2760–2768.

[15] Montresor A, Jelasity M. PeerSim: a scalable P2P simulator. In: Proc. of the 9th Int. Conference on Peer-to-Peer (P2P'09). Seattle, WA; 2009. p. 99–100.

[16] Tisue S, Wilensky U. NetLogo: a simple environment for modeling complexity. In: International Conference on Complex Systems. vol. 21. Boston, MA; 2004. p. 16–21.

[17] Niazi MA, Hussain A. A novel agent-based simulation framework for sensing in complex adaptive environments. IEEE Sensors Journal. 2011;11(2):404–412.

[18] Niazi MA, Hussain A. Sensing emergence in complex systems. IEEE Sensors Journal. 2011;11(10):2479–2480.

[19] Niazi MA, Hussain A. Complex adaptive communication networks and environments: Part 1. SAGE Publications Sage UK: London, England; 2013.

[20] Niazi MA, Hussain A. Complex adaptive communication networks and environments: Part 2. SAGE Publications Sage UK: London, England; 2013.

[21]   Niazi M, Hussain A. Agent-based tools for modeling and simulation of self organization in peer-to-peer, ad hoc, and other complex networks. IEEE Communications Magazine. 2009;47(3):166–173.

[22]   Shoham Y. Agent-oriented programming. Artificial Intelligence. 1993;60(1): 51–92.

[23]   Alharbi H, Hussain A. An agent-based approach for modelling peer to peer networks. In: Modelling and Simulation (UKSim), 2015 17th UKSim-AMSS International Conference on. IEEE; 2015. p. 532–537.

[24]   Batool K, Niazi MA, Sadik S, *et al.* Towards modeling complex wireless sensor networks using agents and networks: a systematic approach. In: TENCON 2014-2014 IEEE Region 10 Conference. IEEE; 2014. p. 1–6.

[25]   Burrows JH. Secure hash standard. Department of Commerce: Washington DC; 1995.

[26]   Avramidis A, Kotzanikolaou P, Douligeris C, *et al.* Chord-PKI: a distributed trust infrastructure based on P2P networks. Computer Networks. 2012;56(1):378–398.

[27]   Sit E, Morris R. Security considerations for peer-to-peer distributed hash tables. In: International Workshop on Peer-to-Peer Systems. Springer; 2002. p. 261–269.

[28]   Rottondi C, Panzeri A, Yagne C, *et al.* Mitigation of the eclipse attack in Chord overlays. Procedia Computer Science. 2014;32:1115–1120.

[29]   Srivatsa M, Liu L. Mitigating denial-of-service attacks on the chord overlay network: a location hiding approach. IEEE Transactions on Parallel and Distributed Systems. 2009;20(4):512–527.

[30]   Douceur JR. The Sybil attack. In: International Workshop on Peer-to-Peer Systems. Berlin, Heidelberg: Springer; 2002. p. 251–260.

[31]   Uruena M, Cuevas R, Cuevas A, *et al.* A model to quantify the success of a Sybil attack targeting reload/chord resources. IEEE Communications Letters. 2013;17(2):428–431.

[32]   Nechaev B, Korzun D, Gurtov A. CR-Chord: improving lookup availability in the presence of malicious DHT nodes. Computer Networks. 2011;55(13):2914–2928.

[33]   Meng X, Liu D. GeTrust: a guarantee-based trust model in Chord-based P2P networks. IEEE Transactions on Dependable and Secure Computing. 2016;10(7):134–147.

[34]   Lu EJL, Huang YF, Lu SC. ML-Chord: a multi-layered P2P resource sharing model. Journal of Network and Computer Applications. 2009;32(3): 578–588.

[35]   Liu J, Zhuge H. A semantic-based P2P resource organization model R-Chord. Journal of Systems and Software. 2006;79(11):1619–1631.

[36]   Novak D, Zezula P. M-Chord: a scalable distributed similarity search structure. In: Proceedings of the 1st International Conference on Scalable Information Systems. ACM; 2006. p. 19.

[37]   Jagadish HV, Ooi BC, Tan K-L, Yu C, Zhang R. iDistance: An adaptive B+-tree based indexing method for nearest neighbor search. ACM Transactions on Database Systems (TODS 2005). p.364–397.

[38]   Joung YJ, Yang LW, Fang CT. Keyword search in DHT-based peer-to-peer networks. IEEE Journal on Selected Areas in Communications. 2007;25(1):46–61.

[39]   Li M, Lee WC, Sivasubramaniam A, *et al.* SSW: a small-world-based overlay for peer-to-peer search. IEEE Transactions on Parallel and Distributed Systems. 2008;19(6):735–749.

[40]   Li M, Chen E, Sheu PC. A chord-based novel mobile peer-to-peer file sharing protocol. In: Asia-Pacific Web Conference. Springer; 2006. p. 806–811.

[41]   Woungang I, Tseng FH, Lin YH, *et al.* MR-Chord: improved chord lookup performance in structured mobile P2P networks. IEEE Systems Journal. 2015;9(3):743–751.

[42]   Canali C, Renda ME, Santi P, *et al.* Enabling efficient peer-to-peer resource sharing in wireless mesh networks. IEEE Transactions on Mobile Computing. 2010;9(3):333–347.

[43]   Zoels S, Despotovic Z, Kellerer W. On hierarchical DHT systems—an analytical approach for optimal designs. Computer Communications. 2008;31(3):576–590.

[44]   Chou JC, Huang TY, Huang KL. SCALLOP: a scalable and load-balanced peer-to-peer lookup protocol for high-performance distributed systems. In: Cluster Computing and the Grid, 2004. CCGrid 2004. IEEE International Symposium on. IEEE; 2004. p. 19–26.

[45]   Kaashoek MF, Karger DR. Koorde: a simple degree-optimal distributed hash table. In: International Workshop on Peer-to-Peer Systems. Springer; 2003. p. 98–107.

[46]   Chou JY, Huang TY, Huang KL, *et al.* SCALLOP: a scalable and load-balanced peer-to-peer lookup protocol. IEEE Transactions on Parallel and Distributed Systems. 2006;17(5):419–433.

[47]   Cuevas R, Uruena M, Banchs A. Routing fairness in chord: analysis and enhancement. In: INFOCOM 2009, IEEE. IEEE; 2009. p. 1449–1457.

[48]   Hong F, Li M, Wu M, *et al.* PChord: improvement on Chord to achieve better routing efficiency by exploiting proximity. IEICE Transactions on Information and Systems. 2006;89(2):546–554.

[49]   Xiong J, Zhang Y, Hong P, *et al.* Reduce Chord routing latency issue in the context of IPv6. IEEE Communications Letters. 2006;10(1):62–64.

[50]   Dao LH, Kim J. A Chord: topology-aware Chord in anycast-enabled networks. In: Hybrid Information Technology, 2006. ICHIT'06. International Conference on. vol. 2. IEEE; 2006. p. 334–341.

[51]   Rao W, Chen L, Fu AWC, *et al.* Optimal resource placement in structured peer-to-peer networks. IEEE Transactions on Parallel and Distributed Systems. 2010;21(7):1011–1026.

[52]   Forestiero A, Leonardi E, Mastroianni C, *et al.* Self-chord: a bio-inspired P2P framework for self-organizing distributed systems. IEEE/ACM Transactions on Networking (TON). 2010;18(5):1651–1664.

[53]   Gao J, Steenkiste P. Design and evaluation of a distributed scalable content discovery system. IEEE Journal on Selected Areas in Communications. 2004;22(1):54–66.

[54] Wu YC, Liu CM, Wang JH. Enhancing the performance of locating data in chord-based P2P systems. In: Parallel and Distributed Systems, 2008. ICPADS'08. 14th IEEE International Conference on. IEEE; 2008. p. 841–846.

[55] Tai Z, Sheng W, Dan L. LISP-PCHORD: an enhanced pointer-based DHT to support LISP. China Communications. 2013;10(7):134–147.

[56] Le-Dang Q, McManis J, Muntean GM. Location-aware chord-based overlay for wireless mesh networks. IEEE Transactions on Vehicular Technology. 2014;63(3):1378–1387.

[57] Ding S, Zhao X. Analysis and improvement on Chord protocol for structured P2P. In: Communication Software and Networks (ICCSN), 2011 IEEE 3rd International Conference on. IEEE; 2011. p. 214–218.

[58] Grimm V, Berger U, Bastiansen F, *et al.* A standard protocol for describing individual-based and agent-based models. Ecological Modelling. 2006;198 (1–2):115–126.

[59] Team RC. R: A language and environment for statistical computing. 2013; 201.

Chapter 10

Descriptive agent-based modeling of Kademlia peer-to-peer protocol

Hammad-Ur-Rehman[1]* and
Muhammad Qasim Mehboob[1]*

10.1 Introduction

Kademlia, a peer-to-peer (P2P) protocol based on DHTs (Distributed Hash Tables), offers desirable features which are not offered by other protocols simultaneously. One notable feature is, in other protocols a large number of messages were needed to know about other nodes. Kademlia minimizes these number of messages [1]. While doing so, the key lookups configuration information automatically spreads among neighboring nodes. The nodes have all the desirable knowledge needed to route the specific query through the paths which have low latency. Another benefit of this protocol is the algorithm [1] which is used to find other node's existence and can also resist attacks by which basic service denial occurs.

Being a P2P system, Kademlia can be modeled in form of either complex network-based [2,3] or agent-based models. Models represent the complex system in term of its multiple components, behavior and communication among them for management and other tasks. Communication between nodes in a P2P system like Kademlia is complex in nature. Finding a node, Joining and Store adds complexity to Kademlia protocol. So, due to all these complex natures of Kademlia, we can confidently consider Kademlia a complex system. In CAS (Complex Adaptive Systems) [4], multiple nonlinear components interact with each other which leads to emergent behavior. So this emergent behavior requires modeling Kademlia as CAS, or fully as CACOONS (Complex Adaptive COmmunicatiOn Networks and environmentS) [5].

To thoroughly understand the complex system, system modeling is must [6]. Modeling CAS through traditional solutions is impossible because CAS are highly robust [4] and have many variables. The emergent behavior for each entity which is involved in a system can be better understood by complex system modeling. A system

[1]COSMOSE Research Group Computer Science Department, COMSATS University Islamabad, Pakistan
*Both authors contributed equally.

can be logically represented by a "Model" with complexity having different levels (normally real systems are more complex).

In lieu of complex systems, hybrid modeling approach has been proposed by Batool and Niazi [7] for the Internet of Things (IoT). They used the combination of complex-network based and agent-based approaches for modeling of complex scenarios in IoT. A normal way of modeling a P2P system is to include its different components like seeders or leechers (peers) as nodes and communication between them can be described as edges or links. After completing complex network model, different tools can be used to compute centrality measures and other metrics on P2P networks. Then by using the analysis from above, global behavior of components can be better understood.

In the past, a number of works have been done to implement Kademlia protocol in Object Oriented domain and other domains. But to the best of our knowledge, nobody has developed the ABM (agent-based modeling), DREAM (DescRiptivE Agent-based Modeling) and ODD (overview, design concepts and details) specification for Kademlia protocol. In this chapter, we will be implementing Kademlia protocol in NetLogo. And then, we will be presenting ODD and DREAM specification for Kademlia.

The objective of writing this chapter is twofold. First, ABM model for the Kademlia is being proposed using complex networks-based and agent-based approaches combination. While developing the model of Kademlia, we will use the already developed complex networks like Line-walking Turtles, Communication-T-T network and Random Network. For validation purpose, we will present a hybrid routing algorithm based on centrality. By this, we can say that a communication has taken place end to end between users. Second, ODD [8] and DREAM [3] are two techniques that we have used for ABM specification. Then, we will also give a comparative analysis of ODD and DREAM.

Some of our main contributions are as follows:

- To model and simulate Kademlia, an agent-based and complex network modeling approach is proposed.
- ODD specification technique for Kademlia's ABM model.
- DREAM specification technique for Kademlia's ABM model.
- An algorithm based on hybrid centrality measure.
- A comparative analysis of ODD and DREAM specification techniques.

Rest of the chapter is structured as follows: Section 10.2 presents background and literature review; in Section 10.3, the model design is given; Section 10.4 describes results and discussion, and Section 10.5 describes the conclusion and future work.

10.2   Background and literature review

In this section, we will be presenting CAS overview, basic concept of cognitive agent-based computing, architecture and the basic working principle of Kademlia.

10.2.1   Complex adaptive systems

CAS consist of agents interacting with each other which are inhomogeneous, can adapt any environment and evolve with that. The important point to note here is that there is no such defined line between systems and environment. Instead, the main concept is, the system is linked with other making an echo system. Based on complex behaviors, artificial systems, natural (biological, ecological and societal) [2] systems and other complex systems can be characterized as "CAS." CAS analysis is done by combining theoretical, applied and experimental methods. In experimental methods, computer simulations and mathematical modeling are done.

10.2.2   Cognitive agent-based computing

Back in 2011, a unified cognitive agent-based computation framework was presented by Niazi and Hussain [9] for complex network and ABM. The proposed framework follows different levels of modeling like

● the exploratory ABM
● the DREAM
● model validation
● complex-network model

The exploratory approach involves the use of agents for exploring the complex systems, identifying the best suited agent-based model for the required problem, then developing concept's proof and data required for verifying and validating the model. The descriptive agent–based model comprises pseudocode, the complex network of model and social network analysis. There are some other levels like verification and/or validation of model, etc.

10.2.3   Complex network modeling

Complex network modeling is the initial level of framework used for complex systems modeling [10]. It is useful when data required for communication between nodes is available. So, along with classification of network, this framework level helps in building the complex models of networks. Also, this framework level helps in getting precious network information by determining local and global quantitative measures. All other models (mathematical and statistical) have no such ability for providing details about complex network patterns and emergent behavior but we can achieve this by using complex network modeling.

10.2.4   Architecture of the "Kademlia" protocol

10.2.4.1   Introduction

Each participating node has a node Id which is opaque 160 bits long random number. The node Id is not only used to identify the node by using some algorithms but also to locate the resources. Hash function is used to compute the Key from value which usually is a 160-bit wide number. Using this, any node can retrieve the value which is associated with the given key. Node IDs which are close to key are used to store the particular <Key, Value> pair. If algorithm has to search for the particular value, the algorithm has to know the key for that value and iterate the network in multiple steps. Every step will be used to find the nodes which are closer to key unless nodes return the needed value or no more nodes closer to keys are found. Near any target, key servers can also be located using the route ID–based algorithm. Kademlia uses XOR metric to calculate the distance—a reason for many benefits which we get by using Kademlia. Due to symmetry property of XOR metric members of Kademlia get queries from same distribution of nodes containing in the routing tables.

Systems like Chord do not gain information like routing from the queries received because they do not use the XOR metric property [1]. This asymmetry leads to inflexible routing tables. Within an interval, Kademlia can send query or asynchronous parallel queries to get any node routes which are based on latency. In Kademlia nodes that are near to some particular ID are located using one routing algorithm from end to end. Whereas old systems use separate algorithms to go near the particular ID and for last hops another algorithm is used. Of all the existing systems, Kademlia resembles the most with Pastry's first phase. But Pastry in the second phase switch toward the difference between IDs in numeric terms. Pastry nodes that are closer to second metric are fairly distant from the first, which creates discontinuity at node IDs and creating negative effect on performance.

10.2.4.2 System description

Approach explained in this section is general which all other DHTs use. As explained earlier, node IDs are 160 bit wide random numbers [1]. In the context of this chapter, it is only assumed that node Ids are random numbers. Binary tree leaf is equivalent to node ID in Kademlia, every node id's position can be found out by exclusive ID prefix which is shortest.

In Figure 10.1, in example tree node position with prefix "0011" is shown. For each node, we have divided the main binary tree into series of subtrees which have no nodes. So the tree at the highest position contains binary tree's half portion which does not contain any node. And the next successive subtree contains the remaining subtree which does not have any node and that process goes on, on the other subtrees. So the example which is explained below for the node 0011, all the subtrees are shown in circles and have nodes that have prefixes 1, 01, 000 and 0010 in this order.

In Kademlia protocol, it is ensured that each node should know of at least one node from every subtree but the condition here is that subtree should have at least one node. Now it is explained that any node can be located by some else node using node ID. In Figure 10.2, we have shown the example of node 0011 trying to locate the 1110 node by querying fairest node from its connections for finding the contacts from the lower and other lower subtrees; due to all these efforts, it finds the node which was needed.

In the remaining section, we will be explaining and presenting the techniques for making the lookup algorithm good. In the first section, we described the ID and its closeness notion to the key, then we spoke for looking and storing <Key, Value> pair on closest k nodes. Then, a lookup protocol was presented which works even

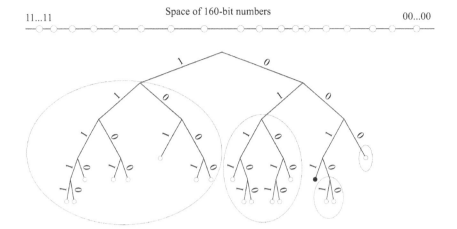

Figure 10.1   Kademlia Binary Tree. 0011 is represented by black dot. All subtrees
where 0011 have contact are shown by Gray Dots. Adapted, with
permission, from Reference [1]

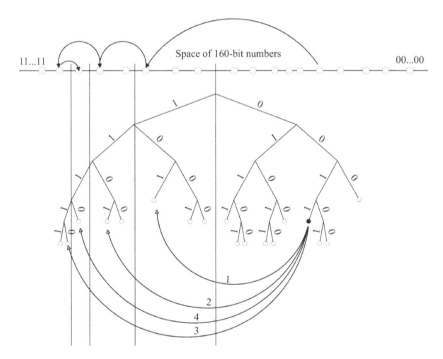

Figure 10.2   Node locating by ID. 0011 prefix node finds 1110 by learning and
querying. Line space on top shows 160-bit Ids. RPC messages made
by 1110 are shown below. Say first RPC is at 101 known to 1110
already. Next RPCs are to nodes returned by RPCs. Adapted, with
permission, from Reference [1]

when there is no node which shares a unique/particular prefix with the key. Also, in situation when some of the given node's subtrees are empty.

10.2.4.3 Distance calculation

Kademlia's particularity was using XOR metric to calculate the points of distance in the keys space [1]. In this section, we will be explaining how pairs of <Key, Value> are assigned to distinct nodes. Suppose, we have two identifiers x and y, each 160 bit wide. In terms of Kademlia, the distance between these two identifiers is calculated as XOR $\oplus$ between them.

$$d(x, y) = x \oplus y \tag{10.1}$$

Here the point to note is the reason why XOR is chosen [11]. It is because of its following common properties with geometric distance formula:

- Distance will be zero between node and itself. $d(x, x) = 0, d(x, y) > 0$ if $x \neq y$
- Distance from x and y will be the same as distance from y and x. $d(x, y) = d(y, x)$
- XOR also obliges the triangle property. Let's suppose three points are given then distance between z and x will be $\geq$ to sum of distance from x to y to z. $d(x, y) + d(y, z) \geq d(x, z)$

In binary tree's system's sketch XOR has distance implicit notion. In completely filled binary tree having 160-bit wide IDs, the distance from one ID to other will be smallest subtree height which has both of those IDs. In case when tree is not completely filled, then the closest leaf to an ID x will share the largest prefix which is common from x. There may exist more than one leaf with common longest prefix if the tree has empty branches. This is a case where leaf closest to "x" is also the nearest leaf to "x" (ID) which is generated by bits (similar to tree' s null branches) flipping in x.

XOR is unidirectional like Chord's clockwise circle metric. For a point x and the distance $\Delta > 0$, there exists one point y where $d(x, y) = \Delta$. Unidirectionality ensures that all the lookups being performed should coincide on same path irrespective of node from where they originated. Hot spots are being alleviated by caching the <Key, Value> pairs. So dissimilar to Chord but same to Pastry, XOR is balanced

$$d(x, y) = d(y, x) \quad \text{for all } x \text{ and } y \tag{10.2}$$

10.2.4.4 Node

For routing query messages or contacting other nodes, nodes store information about every other node. For each $0 \leq i < 160$, every node have <IP address, UDP port, Node Id> triplets for each node which is in between distance of $2i$ and $2i + 1$. Here these set of lists is called K-Buckets. k-buckets always have sorted output. Buckets are sorted with respect to time seen. Tail has the most recently seen node, whereas head has the least recently seen node. At time when node receives any message, it updates particular K-bucket for sender's Node ID. Also for small value of i, kBuckets are generally empty. List grows up to k size for large i values.

Assume a case when a node which is sending exists in a receiver's bucket, then the receiver changes its position to list's tail. Assume a case when a sending node does not exist in particular K-bucket and there are also fewer than k-entries in bucket, then new sender is just inserted at list's tail. If a particular K-bucket is full, then the recipient just pings the least recently seen node and if it fails to pong, then the node is evicted and the new node is inserted at the tail of the list, and in contrast, if least recently seen that node responds in time, then that node is shifted to the end of the list and received node is abandoned. All these nodes which are live are never removed from the list.

10.2.4.5   Protocol

Four (RPCs), namely, PING, STORE, FIND_NODE, FIND_VALUE constitute Kademlia protocol.

- **PING**
  In this RPC, a node sends a message to another and if gets reply from that, both the nodes have to update the particular k-buckets. Basically, this RPC is used to check is node is live.
- **STORE**
  Recipient node have to store a <Key, Value> pair received from sender. So that it can be used later for retrieval.
- **FIND_NODE**
  In this RPC, sender node sends a 160-bit key and recipient has to return K-triples that are closest to the received key. In the best case, it has to return k triples but it can send less if it has knowledge of few triples.
- **FIND_VALUE**
  This RPC is equivalent to FIND_NODE if the corresponding value is not present. If it is present then only the value is returned.

10.2.4.6   Node Look up

To locate node, algorithm which Kademlia protocol performs is called "Node Look up." Members of Kademlia perform a recursive operation to locate k closest nodes. Look up starts from picking non-empty k-bucket's α (global concurrency parameter) nodes. Asynchronous parallel requests are sent to chosen nodes. In recursive step, requests are again sent to nodes which are found out live by previous RPCs and nodes who fail to reply are dropped. After every parallel search, the closest node is updated. This process continues until returned node is no longer than the closest already found or node initiating has accumulated k probed nodes. If FIND_NODE process fails, then it sends the request to closer nodes which are already not queried.

10.2.4.7   Routing table

The structure of basic routing table given the protocol is fairly straight forward, but for unbalanced trees, a minor refinement is needed. Routing table can be represented by a tree which is binary, here k-buckets are represented by leaves. The position of kBucket in a binary tree is the prefix, and the nodes that k-bucket contains have common prefix in their IDs. So, in this way, every k-bucket at least covers a part of the 160-bit ID space. All k-buckets with no overlap cover the whole range of 160-bit space [1].

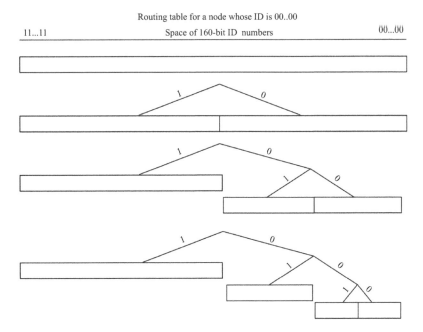

Figure 10.3   Routing table evolution. Initially, it was single k-bucket for node. Bucket whose range cover Node Id splits as the k-bucket fills. Adapted, with permission, from Reference [1]

Allocation of nodes in routing tree are dynamic. The process is shown in Figure 10.3. For example, we take a node u, at start routing table of u will contain only one node that is covering whole 160-bit ID space. When a new node contacts with u or u itself find out a new node, then u tries to add new node into a suitable k-bucket. Assume k-bucket has space; then for adding brand-new node, a simple insert operation is performed else-if k-bucket range is big enough to include ID of u then we divide the bucket into two brand-new sub buckets and also the previous content is divided between them. After division, the insert operation is performed to add new node. In the scenario where different range of k-bucket is full, new coming node is simply discarded.

10.2.5   Literature review

Due to security problems [13] and privacy [14], we cannot use P2P systems for commercial applications. Bottleneck problem occurs if few CAs have to distribute CRL to network users (Figure 10.4). The proposed solution, as discussed in [15], is to distribute segmented CRL using overlay (just another Peer), without compromising the peer's storage. The results achieved show that in segmented CRL, Peak Request rate is not reduced, but file size is less which nodes store. I think proposed system's execution time will be higher due to extra step of CRL's segmentation.

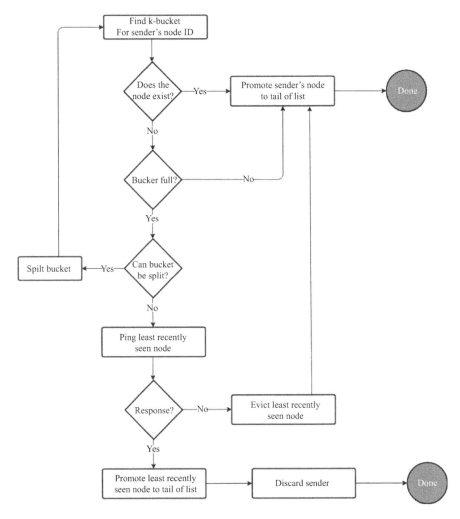

Figure 10.4   Adding contacts with bucket splitting. Adapted, with permission, from Reference [12]

P2P concept has been introduced in mobile networks and has a lot of applications. But due to highly robust, random network topology and low transmission range information retrieval problem arises. In the current paper [16], the authors proposed economical and automatic information retrieval approach based on cache, which is updated by seeing relevant factors. NS2 simulation experiments show that this approach has better results than old ones. Average response time from query and network messages are reduced. One thing missing was they could have extended their method to content based search in these networks (Figure 10.5).

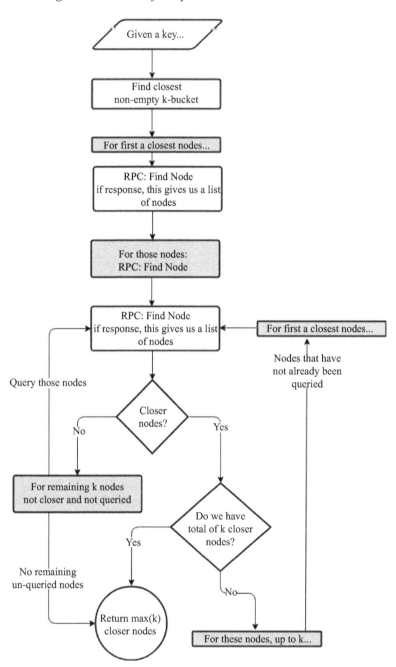

Figure 10.5   Node lookup algorithm. Adapted, with permission, from Reference [12]

Botnets became popular in recent years and these have many applications, but Botnets have become apparent as mostly extreme cyberattacks in near years. To solve this backdrop, as discussed in [17], the authors have proposed new botnet called AntBot in which C&C information is spread across all bots. aMule-based distributed simulator is used for implementation. Doing enough simulations, it is proved that against pollution-based migrants AntBot operated resiliently. They could have used few other defense mechanism also which they did not.

The paper [18] is about management of content retrieval in KAD. Content search is implemented in KAD by using Kademlia DHTs. In P2P systems, information loss happens if node churn occurs. KAD already deals with this by publishing multiple redundant copies of information. Here [18] they have only tweaked some parameters and showed performance. Their results show that lookup performance can be increased by coupling lookup and content retrieval but latency will be same. What more they could have done is to use more design parameters i.e. requests in parallel, timeout, round-trip delay, etc.

P2P networks can be used in applications like file sharing. They have to share and gather huge chunk of computing resources due to which their average energy usage is higher. There are other [19] proxy [20] based approaches used to solve this. The paper [21] addresses energy problem by proposing a two-layer model [22]. The lower layer is composed of files for sharing, whereas the upper is composed of DHTs, the work of which is to index peers and files and show availability of each. Simulation result shows, in proposed system 50% less energy is used as compared to others [23,24], and without any other delay, more than 80% of files start downloading. They could have shortlisted peers (which have the highest availability ratios) among other peers.

Efficient and fast search operations are possible in P2P networks [25]. For querying or disseminating valuable information, their topology is useful. While broadcasting when a network fails to link or bad node joins network problem occurs. To solve this problem, the paper [26] presents a replication broadcast algorithm for Kademlia [27]. This algorithm only uses entries in the routing tables of Kademlia. The results show that replication can be used for enhancing speed and reliability. The problem is, replication also increases networking traffic. Trade-off between time for broadcast and network traffic cost could have been a solution for this.

In structures P2P systems, each peer may join or leave at any time. This is referred to as churn. In distributed systems and in P2P systems due to high dynamicity routing failures, saved data loss or random peer view occurs so minimizing churn is necessary. To minimize churn, the paper [28] proposed a zone replication technique which is implemented on DHTs top and consists of three steps: publishing, maintenance protocols and searching. PeerfactSim.Kom was used for simulation; the result shows that the proposed solution produces smaller routes and good resistance to churn as compared to others. They divided the proposed solution into three parts which I think will increase the execution time as compared to old systems.

Clustering has become a hot topic for P2P systems in recent years. Plain DHTs have many applications but still they lack for some kind of applications, i.e., multimedia, etc. Also, they addressed whether clustering improves querying performance or not? For supporting wide applications range in the paper [29], they presented Echo. Echo (based on the Cayley graph model [30]) is a framework which improves query efficiency by combining DHTs functionality [31], homogeneity of load with Clustering. Also, they examined the effect of prefix clustering on congestion by using "Congestion-freeness" [32] notation. The results show that as this model is Cayley dependent, other DHTs can also be implemented using this. Where they lost was they could have maintained peers statistics to make network strong and firm.

Kademlia and chord [33] provide an effective way for finding other nodes in P2P networks. But lookup queries for finding resources along the path can be disrupt easily by malicious nodes. In [34], a Reputation for Directory Services (ReDS) framework is described that first track the lookup requests presented by other nodes and then based on tracking it enhances lookups in redundant DHTs. The author also explores how shared reputation can work with ReDS in the context of free-rider prevention [35]. By using ReDS, simulations showed that over an extensive range of situations lookups, success rates enhanced by 80% or more for Kad and Halo.

For distributed multimedia services in 2008, IETF P2PSIP was designing a protocol to combine the SIP functionality of media session and for P2P resources localization and decentralized distribution. At that time, infrastructure that was considered for P2PSIP scenarios was single domain for inside connectivity. The paper [36] proposes a peer called "super peer," the architecture is hierarchical based among multi P2PSIP domains for interconnection. Every domain will have one minimum "super peer" which will be stable peer among all other peers, and together, all super peers from different domain make an upper overlay layer. To validate the routing state and routing performance of analytical model PeerFactSim.Kom simulator [37] was used. The results show that peers routing entries significantly (approximately 50%) decrease as associated routing states and we increase in number of domains. A realistic scenario is simulated by setting up the churn as explained in [38]. The selection mechanism of super-peers can increase the performance.

In the last few years, P2P network architecture has gain very much popularity in variety of applications and services, for example, VoIP streaming and collaborative computing applications. The essential requirement of deploying such networks are security and integrity. Withstanding various malicious peers in the network also called Sybil attack can poison the routing tables and may make the retrieval, storage and routing process time consuming and extremely difficult. The paper [39] proposes a new trust and reputation-based technique which makes more secure the retrieval, storing and routing the resources in Kademlia network. For simulation, a discrete event simulator DEUS is used. In results, it is noticed that when the trust-based algorithm is applied on Kademlia's RPCs (PING, STORE, FIND), the networks perform 20% better than pure RPCs. In simulation 200 Sybil nodes vs 1000 true nodes are used. In future, it will be interesting to find a mathematical close form for finding optimal value of balancing factor.

The P2P networks play an important role in communication-oriented systems. But the performance of P2P networks is significantly affected by the churn phenomenon particularly in mobile environment. Until now, no proper analysis is presented for structured P2P networks. In [40], the authors conduct an evaluation for the Kademlia-based P2P system in the communication-oriented environment in the presence of churn. For simulation, Nethawk EAST simulator is used and the simulation is conducted on mobile platform. It is noticed that for robust system under different levels of churn 3 degree is enough for both lookup parallelism and resource replication, and in terms of CPU load, 200 bytes or less were used for optimal energy consumption. In future, it can be applied on some larger settings to confirm the conclusion draw in this paper.

In the literature, many DHTs have been vigorously studied, and for peer organization in DHTs, many diverse suggestions have been made, but in real systems, very few DHTs implemented on huge scale. Previously, developed crawler–based p2p system has duration limitation of crawls to few days at best. The system developed, as discussed in [41], is able to operate at the rate of per crawl every 5 min and peer behavior is measured in terms of up-time distribution and churn rate. Their findings conclude that in high resolution document sharing peers (of kad) leave and join the networks with a binomial distribution (negative), while session time of peer will be same as Weibull distribution. In this paper, ID repetition has not been taken into consideration.

Kademlia and chord are two relevant DHTs that are used in different P2P applications to provide decentralized services. In the literature, quite a lot piece of work has appeared that evaluate the performance of both DHTs, but the results are neither concluding nor consistent because a different churn model is used and neglects the key point that churn occurs since the beginning of the DHTs lifetimes. In [42], a realistic and fair framework is integrated by the following:

1. Executing the churn model at the time of peer creation,
2. Performance evaluation methodology considers different DHTs parameters are not equal, and
3. A churn metric that keep track of rate of change of P2P population.

Successful lookup ratio metric is used for performance evaluation, and it is noticed that under the similar scenario, Kademlia shows better performance than chord. The only problem here is that simulations were conducted on small scale. For real world, it was not tested.

In any P2P network, user enters or leaves network continuously. This behavior is called churn. It has become important to understand the resilience properties when the churn is changing on high rate. In paper [43], the authors in first step find the dynamic churn model that is lifetime based for P2P system and have reached stationarity that is reducible to a uniform node failure model. In the next step, a reachable component method is developed, and then using this method under different level of churn rates, routing performance of P2P networks [44] is evaluated. The results show that de Bruijn graph based [45] routing networks show outstanding resilience under tremendously high rates of nodes turnovers. The routing networks that were tested include randomized-Chord, Chord, CAN and Kademlia.

Structured overlay networks have gained much popularity in our daily applications. The main problem of such networks is vulnerable to attack that aimed to damage the functionality and structure of networks. In past, many proposals [46] for secure architecture design were presented, but comprehensive and broadly acknowledged solution is lacking. In [47], the authors present a new solution called Layered Identity-based Kademlia-Like Infrastructure (Likir) which aims to secure implementation of P2P network based on DHT. For performance evaluation, both Likir and Kademlia overlay nets were run on PlanetLab networks, and lookup operations in Likir are greater than Kademlia because of cryptographic efforts spent during a node session, but Likir is more secure than Kademlia. In evaluation, small nets were run on PlanetLab, but in real, there are millions of peers and it will further increase the lookup time for Likir.

Kademlia, based on keys, is one of the most effective routing protocols. In Kademlia's routing phase it needs to contact log (N) nodes. Due to this bottleneck problem occurs. To reduce the quantity of nodes that participates in lookup process, the paper [48] proposed an algorithm named "Shades." In the proposed algorithm, nodes have their own caches that will reduce time in cache hit case. Other algorithms may store a set but they cannot count the number of times items are inserted [49]. Simulation results show that Shades have reduced median quantity of nodes contributing to each lookup by 22%–36% compared to others when tested and 30%–40% reduced when compared to Kademlia. In real time, it may corrupt the data. They have taken data corruption problem into consideration.

In the paper [50], another implementation of Kademlia DHT protocol is discussed which have over 1 million nodes. All nodes are concurrent and use eDonkey file sharing protocol. There are many design weaknesses in Kad that attackers can exploit to fail the find and search mechanism. They measured that cost and other parameters of those attacks against 16,000 nodes which are kad connected. The attack they have described in their paper has two phases. One is "Preparation Phase" and the other one is "Execution Phase." DVN, simulation which was quite large in magnitude, scaled up to 200k nodes. And after all, they found out their attacks are more cost effective.

Based on DHTs, Kad network is the most popular P2P network and due to network's scalability and reliability, the user base of this network will continue to grow in future. In the literature, many decentralized P2P architectures have been proposed but still most existing system running on centralized architecture. In article [51], two popular network types are examined one is server-based eDonkey and other decentralized DHT Kad. In comparison to these networks, it is discovered that eDonkey receive all 10% server request that conclude that eDonkey is more popular, but in other results, it is noticed that Kad is more robust as compared to eDonkey. These comparisons are not done on very large scale. So if it was done on large then results may vary.

10.3   Model design

In this section, we will give a detail description of ODD and DREAM models for our Kademlia protocol. We will also present the Unified Modeling Language (UML) diagrams of our protocol.

10.3.1 ODD model of "Kademlia"

ODD was first developed by Grimm *et al.* [8], and later, it was updated in [52]. Basically, ODD is a textually representation technique for the specification of ABM system. In ODD, there is a checklist which describes significant features of ABM system. ODD divides the ABM system into three main sections and then these sections are further divided into subsections.

10.3.2 Overview

- **Purpose**
  The main purpose of this model is to understand and learn that how agent-based modeling techniques can be applied to simulate Kademlia protocol on huge scale. Further to compare Kademlia performance with already existing non-ABM models.
- **Entities**
  There are three types of entities involved in this model. The first one is the entities that are in the network, the second that want to join the network and the third that are leaving the networks. All three types of entities are represented by nodes. Each node owns some variables like every node has a node id of 160 bits and a list of kBuckets to store the contact and key, value pair to store data.
- **Process overview**
  At first, all three types of nodes are generated and placed randomly, and then using protocol's RPCs, they start contacting each other and start forming a network. When one node contacts to another to join the network, the recipient node checks their buckets. If it is not full, the recipient node simply adds the new node at the end of one of the buckets, which depends on the XOR distance between their ids, but if the bucket is full, then the recipient will ping the node on top of bucket, and if gets response from the pinged node, then it will discard new node, otherwise it will add the new node by removing the pinged node. kBuckets are sorted by time, the last seen nodes will be on top and the recent seen will be in the bottom of buckets.

10.3.3 Design concept

- **Basic principles**
  The main hypothesis of our model is that Kademlia ABM performance is better for large-scale system. We used already developed complex network models like Line-walking Turtles, Communication-T-T network and Random Network as a starting point and extended it to build our Kademlia P2P protocol.
- **Emergence**
  In this type of features, we tell that what type of outputs of ABM is modeled. In simple words, we can say that what results we are expecting from the model. So, from Kademlia model, the key outputs that we expect are how messages behave as we increase the size of network that means increase the number of nodes

and second output that we want to analyze how latency changes as we increase network size.

- **Adaptation**
  It is basically how agents adapt environment changes. So, we can say it is a test of decision-making capabilities of agents when environment changes. Constraints are already well defined and accordingly the environment changes. We test our model in two scenarios—one is in the presence churn and the other is in the absence of churn.

- **Objectives**
  In adaptive environment as environment changes, the individual node or agent also gets the effect or reward from the change for their adaptive behavior to accomplish their own tasks. So, the core objective of our model is to analyze how the messages and latency behave as we change network size when churn is present and when churn is absent.

- **Sensing**
  There are some properties related to each node that help in decision making when nodes are communicating each other. These properties also increase the performance of system. In our case to find the data, the node does not contact every node in the network, but it only contacts to the nodes that are in their buckets. This behavior reduces the number of messages and also the latency to find the required data.

- **Interaction**
  The nodes that are in buckets can contact each other to find or store a value.

- **Stochastically**
  The model is developed using the protocol specification from the Markov paper [1].

- **Observation**
  In the simulation on every step, the following information are collected:
  1. Network size or the number of nodes
  2. Number of nodes joining the network
  3. Number of nodes leaving the network
  4. Number of messages

10.3.4   Details

In this section, we will describe how the model is initialized, what are the initial states of model, we will also tell about the input data if any used, and at the end, we will also tell about the model's parameters and their values.

- **Initialization**
  NetLogo tool is used for model implementation. NetLogo is a tool used for ABM. It is free and open source. The initialization of model is done by calling the "setup" procedure, and in this function, randomly nodes are created and a random id is assigned to them and all nodes are placed randomly. Some walkers are generated also position on some of the nodes. kBuckets of every node are also initialize with nobody.

Table 10.1   Evaluation metrics

Parameter	Value
Body of the text	61×61
No. of Nodes	100, 500, 1,000
Hash Degree	2^{20}
Initial Seeds	10
Alpha	3
K	20

- **Input data**
  No input data is feed to model.
- **Submodels**
  The parameters that are used in the model and their values are given in Table 10.1.

10.3.5   Activity diagrams of "Kademlia"

In Figure 10.6, the activity diagram of lookup process is presented. The lookup operation is same as described above in the Kademlia architecture section (10.2.4).

In Figure 10.7, the activity diagram of add contact process is shown. In this process, a node receives any message (request or reply) from other node. More information can be found in the Kademlia architecture section (10.2.4).

10.3.6   DREAM model of "Kademlia"

In this particular section, we will be giving a detailed overview of DREAM (Figure 10.8). DREAM [3] is a cognitive ABM technique. It is an ABM specification technique in which we have to develop a complex network model of the ABM, pseudo-code based description of the model and also we have to do social network analysis of the network model. It also offers a visual-based analysis.

10.3.7   Network model

As already been mentioned in the chapter, Network's Paper model can be made by extending baseline network model to initiate a DREAM. Network's submodel can be used/imported in network modeling tool. The results are shown in Figure 10.9.

10.3.8   Pseudo-code description

- **Agents and breeds design**
  In the simulation model of this design, we have used two types of agents, i.e., Nodes and walkers. The overall complex network consists of nodes which are joined by links. For routing purpose, agents of type walkers are used. Walker type

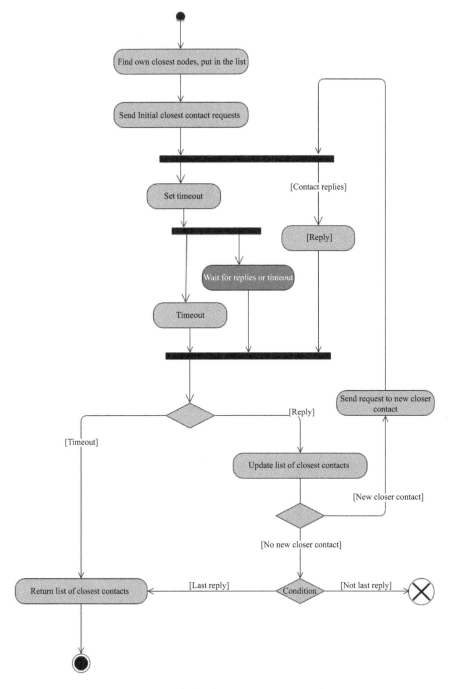

Figure 10.6   Node lookup process activity diagram

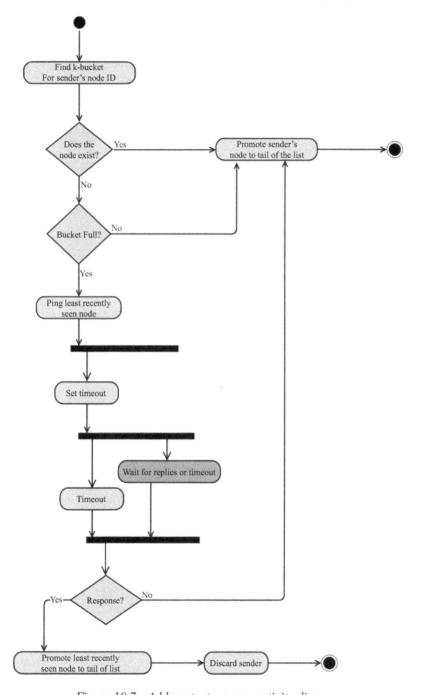

Figure 10.7   Add contact process activity diagram

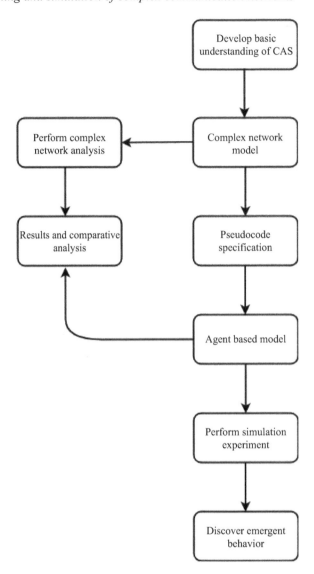

Figure 10.8   DREAM methodology for ABM. Adapted, with permission, from Reference [3]

agents can move across the network. Here first, we will be defining the node type agent (Table 10.2).

We have implemented this model in NetLogo (Toolkit for ABM). For one node agent, there are four internal variables, hid that is node id and an opaque 160-bit number, which we have used as a label for a particular node. kBuckets are for node's all bits

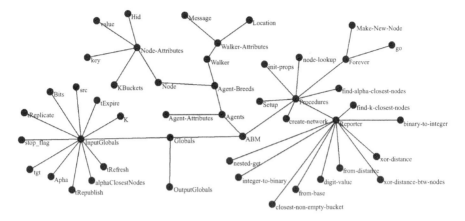

Figure 10.9   Network view after manipulating algorithms

Table 10.2   Breed nodes details

Breed **Node**: This breed represents nodes present in Kademlia network

---

Internal Variables: <hid, key, value, kBuckets>
   hid: Node Id of 160 bits. Also used as node label
   key: Node key
   Value: Node value
   KBuckets: Each node from $0 <= i < 160$ has a list of <IP address, UDP port, Node ID> triples for nodes of distance between 2^i and $2^i + 1$ from itself.

Table 10.3   Breed walker details

Breed **Walker**: This agent is used for walker that can move across the network.

Internal Variables: <location, message>
   location: This variable is used to represent the current location of walker
   message: Walker message

---

because every node for each $0 <= i < 160$ keeps a list of <IP address, UDP port, Node ID> triplet for those nodes which have distance between 2^i and $2^i + 1$ from itself. Next, we will be defining the walker type agent.

Here Walker breed represents walker type agents (Table 10.3). These walkers are used for the purpose of routing to move in a network. Next, we presented the internal variables used in this type of network. The location variable is used to store the current location of a walker. The message is being sent to Walker and does action on the basis of that. Now as we have developed a specification model of the agent

Table 10.4   Input Globals details

Input Globals: <Population, Hash_Degree, initial-seed>

Sliders
    Population: Used for specifying initial number of nodes
    Hash_Degree: Used to specify the size of hash
    initial-seed: Used to specify initial seed size in network

Table 10.5   Setup procedure details

Procedure **Setup:** Creating environment for simulation.

Input: User Interface Global parameters are used
Output: Initial setup of simulation environment
Execution: Called in the start
Context: Observer
begin
    1. Clear screen
    2. Clear all outputs
    3. Reset ticks
    4. Set value for global constants
    5. Create nodes, set labels for all nodes and also set "cur" variable value
    6. Create walker
    7. Create network and set value for "src" and "tgt" node

all breeds, next we will be developing model for global variables used in the code. In simulation, different configurations are dependent on Global variables. So these global variables are important.

- **Globals**

  In this section, we will be describing the global variables in the simulation model. For simulation setup, there are eight key input variables. In the specification model below, key global variables are described.

Here we have three variables which are used as input, and we can clearly note that all these variables are sliders (GUI element in NetLogo) (Table 10.4). The population is the input provided by the user to specify the initial number of nodes. Hash_Degree is the input provided by the user to specify the size of hash, that is, each node will be given a random hash id between 0 and $2^{(Hash\_Degree)-1}$. Initial_seed is the input which will be used to specify initial seeds size in a network.

- **Procedures**

  After describing breeds and global variables in this section, we will be describing procedures that are part of our model. There can be different types of procedures. The first procedure is setup procedure (Table 10.5).

Table 10.6   "init-props" procedure details

Procedure **init-props:** Setting up nodes properties

Input: User Interface's Global parameters and global constants
Output: Changing nodes properties
Execution: Called by setup
begin
    1. Set shape as circle
    2. Setting x-coordinate of node
    3. Setting y-coordinate of node
    4. Setting hid of node
    5. Setting kBuckets
end

For every ABM, a setup procedure is the most important one. Models can have different names for setup procedure, but more or less similar kind of procedure is needed to set up simulation scenarios. Here point to note is that remaining quantity of old simulation should not be disturbing future simulations. That's why old results are cleared before the next simulation. By doing this, we can make sure that our simulation is running in a clean fashion. Then, some global constant variables, which are used in the simulation, are assigned constant values. Number of nodes depend on the "Population" slider value. Then, each node is traversed and each node's label is set. Node ID is used as node label. Also for each node, another procedure "init-props" is called that will be explained later. In the next step, for all nodes cur is set as "self." After this a walker is created, walker's color is changed to one-of base colors. Next walker's location is set to one of the nodes and the initial location is moved to walker's location.

"init-props" is the procedure which will set each nodes shape, label, store x and y-coordinate, and id (Table 10.6). Also at the end kBuckets for each node is set as nobody. To arrange nodes in some order and shape, there is button present on screen called "Make Circle" that can be used. Next, we will be describing procedure named "go."

"go" is the procedure which is being called repeatedly for each simulation (Table 10.7). This function gets called when "go (forever)" button is clicked or "go (single step)" is clicked. First, in this procedure, ticks counter value is reported and then another procedure named "node-lookup" is called.

"integer-to-binary" is the procedure used to convert an integer to a binary number (Table 10.8). First, we start the loop which runs till number is not equal to zero. First, set "rem" as number % 2. In the next step, set bitList, now add rem in the start of bitList. Next, set "num" as number % 2. At last, "number" is set as a floor of "num" to return.

Table 10.7   "go" procedure details

Procedure **go:** Creating network and setting ticks

Input: No specific input parameters used
Output: For each simulation agent this equates to single step
Execution: Repeatedly called simulation execution
begin
    1. Report current value of ticks counter
    2. node-lookup
end

Table 10.8   "integer-to-binary" procedure details

Procedure **integer-to-binary:** Converts integer or decimal number to binary

---

Input: Internal variable containing decimal number
Output: Return binary number equivalent of input
Execution: Called from xor-distance
begin
    1. Define local variables
    2. Do this while number is not equal to 0 [
    rem = number % 2
    In this step set bitList now add rem in the start of bitList
    Then we will set num = number / 2
    At last number = floor of num   ]
    3. After converting integer to binary return that number
end

Table 10.9   "nested-get" procedure details

Procedure **nested-get:** Internal function to get value from list

---

Input: Internal variables
Output: Return value from list
Execution: Called from create-network
begin
    1. Get list and index
    2. Return value from second list
end

"nested-get" is the procedure which is used to get value from the list on the basis of index (Table 10.9). It takes list and index as input and return value from the list. It is called from "create-network" for each node.

"node-lookup" is the procedure which is used to do node lookup procedure (Table 10.10). Lookup is started by first picking alpha nodes from closest non-empty

Table 10.10 "node-lookup" procedure details

Procedure **node-lookup:** Internal function to do node lookup

---

Input: User Interface's Global parameters and global constants
Output: Doing node lookup
Execution: Called from go
Context: Observer
begin
    1. Find alpha closest nodes
    2. For each alpha closest node make a link with src. Advance tick by 1
    3. Ping each node and find K closest nodes
    4. Check if nodes from K closest nodes reply or not
    5. For each k closest node make a link with src from alpha nodes
    6. Find xor-distance btw node from "K" and "tgt"
    7. If distance is 1 "stop" go procedure & advance tick by 1
end

Table 10.11 "find-alpha-closest-nodes" procedure details

Procedure **find-alpha-closest-nodes:** Internal function to get alpha closest nodes

---

Input: No specific input parameters used
Output: No output
Execution: Called from node-lookup
begin
    1. Call procedure named "closest-non-empty-bucket"
    2. Set value of global variable named "alphaCLosestNodes"
end

---

kBucket. If closest non-empty kBucket has fewer entries, it just picks alpha nodes it has knowledge of. The source then sends to find the node RPCs to alpha nodes, where alpha is typically 3. Then each alpha sends request to further K nodes and check if they reply or not. If alpha = 1, then Kademlia resembles chord.

"find-alpha-closest-nodes" is the procedure which is used to get alpha closest nodes from bucket (Table 10.11). It calls "closest-non-empty-bucket" procedure to get the closest non-empty bucket, and then from the output of that procedure, we get alpha nodes.

"find-k-closest-nodes" is used to get k closest nodes from bucket (Table 10.12). It calls "closest-non-empty-bucket" procedure to get the closest non-empty bucket, and then from the output of that procedure, we get k nodes to return.

"xor-distance-btw-nodes" is a procedure which is used to find xor distance between two nodes (Table 10.13). It calls "xor-distance" procedure to find xor distance between src and tgt ids.

Table 10.12   "find-k-closest-nodes" procedure details

Procedure **find-k-closest-nodes:** Internal function to get k closest nodes

---

Input: Internal variables
Output: Return value from list
Execution: Called from node-lookup
begin
    1. Call procedure named "closest-non-empty-bucket"
    2. Return k nodes from output of "closest-non-empty-bucket"
end

Table 10.13   "xor-distance-btw-nodes" procedure details

Procedure **xor-distance-btw-nodes:** Internal function to find XOR distance between nodes

---

Input: Internal variables
Output: Return value from list
Execution: Called from node-lookup
begin
    1. Declare and define local variables
    2. For src & tgt inputs set respective ids
    3. For src & tgt id's calculate XOR distance
    4. Return the dist
end

"create-network" is the procedure in which in start some local variables are declared and assigned values (Table 10.14). Next for each node, set "cur" variable as self and nodeId as hId. "Self" is NetLogo keyword, which means "me." Then change walker properties like set the location as "cur," color and move to new location. Then for this particular node ask all the other nodes. Set "newNode" as self. After this for that particular node, calculate XOR distance using procedure named "xor-distance" which we will be explaining later. Populate bucket and then set "is-in-bucket" flag based on which there is next if condition. In "bucket" at index 0 replace-item with "cur." Then for each walker define new variable "dest" as newNode. Move the face to "dest" and check if distance "dest" is less than "speed" and do some operations to change walker properties there. In the next step, Kademlia protocols have been implemented. When some node sends a message to some other node, receiver node updates the particular kBucket of node ID of sender node. The protocol has some conditions. If the sending node is already present in recipient's kBucket, then recipient moves it to the end of that list. If node is not present in the particular kBucket and that bucket has less than "k (Global constant variable)" entries, then recipient only inserts new sender at the end/tail of list. If the particular kBucket is already full, then recipient pings the kBucket's last-recently seen node just to decide the future of new

Table 10.14   "create-network" procedure details

Procedure **create-network:** Creating network

Input: User Interface's Global parameters and global constants
Output: Creating network
Execution: Called by setup
Context: Observer
begin
1.   Local variables declaration. All next steps will be done for each node
2.   For each node set "cur" variable as self, set node Id as hid & change
walker properties
3.   Ask all other nodes for this node & set newNode as self
4.   Then for that "newNode"
       Find XOR, fill "bucket" with "nested-get" procedure output
       Check if "cur" is "bucket" member then set "is-in-bucket" true
       In "bucket" at index 0 replace-item with "cur"
       Then change each walker properties
5.   Here check if "is-in-bucket" is true or not
    If yes then do this [
       "currPos" &"lstPos" as position of "cur" &"nobody" in bucket
       Check if "lstPos" is false?
       If yes then do this [
         From bucket remove item at "currPos" & put "cur" at tail
         To walkers send message "ok"]
       else do this [
         From bucket replace item at the "lstPos" with "cur"
         From bucket remove item at "currPos" & put "nobody" at tail
         To walkers send message "ok"]
    ] else do this [
       Define local variable "lstPos" as position of nobody in bucket
       Check if "lstPos" is false? If yes then [
         To walkers send message "fail"]
       else do this [
         From bucket replace item at the "lstPos" with "cur"
         To walkers send message "ok"]
       Then for each walker [
         Change walker properties
       Define local variable named "flag" and set its value as false
       At last for each walker end message the value in flag
       check if "flag" is "ok" or "fail"
  For "newNode" from kBuckets replace at "dist" with "bucket"

---

node and least recently node. If there is no reply from least recently seen node, then recipient just removes it from the particular kBucket and at the tail new sender is inserted. Otherwise, if there is a response from last recently seen node, it is shifted to tail of list, and new sender is discarded.

"closest-non-empty-bucket" is used to get the closest non-empty bucket from buckets list (Table 10.15). It calls "xor-distance" procedure to find xor distance between src and tgt ids. Then from buckets list get the closest non-empty bucket and return it.

Table 10.15   "closest-non-empty-bucket" procedure details

Procedure **closest-non-empty-bucket:** Internal function to get closest non-empty bucket

---

Input: Internal variables
Output: Return value from list
Execution: Called from find "k" & "alpha" closest nodes
begin
    1. Declare and define local variables
    2. For src & tgt inputs set respective ids
    3. For src & tgt ids calculate XOR distance
    4. Return bucket got from "nested-get" procedure
end

Table 10.16   "xor-distance" procedure details

Procedure **xor-distance:** Calculates XOR distance between two numbers

---

Input: User Interface's Global parameters and global constants
Output: Return XOR distance between two numbers
Execution: Called from create-network
begin
    1. Define local variables
    2. if bin1Len = bin2Len then [
        if bin1Len > bin2Len then [
          zeroPadding = bin1Len - bin2Len
          Then make a list of "0" and pad it to "zero"
          Make a list containing zeros and bin2. Put that in bin2
        ] else this [
          zeroPadding = bin2Len - bin1Len
          Then make a list of "0" and pad it to "zero"
          Make a list containing zeros and bin1. Put that in bin1
        ]
    ]
    3. for loop on each of the arrays bin1 and bin2
    In loop's each iteration
        If bin1 each bit is not equal to bin2 then increment dis by 1
end

"xor-distance" is the procedure which is used to calculate the distance between two numbers (Table 10.16). It takes two decimal numbers as input and first convert them to binary numbers. Then we find the length of both numbers and using "length" function of NetLogo. First, we check if both numbers are of equal length or not. If not of equal length, then again we check if first number is greater than second or not if yes then we will set "zeroPadding" as subtraction of bin1Len and bin2Len. Then, make a list of zeros and pad it to list named "zero" and, at last, make a list containing zeros and bin2 and put that in bin2. In else, we will set "zeroPadding" as subtraction

Table 10.17   "binary-to-integer" procedure details

Procedure **binary-to-integer:** Converts Binary number to integer

---

Input: Internal variable containing bits
Output: Return Integer number equivalent of input
Execution: Called from create-network
begin
   1. Bits as input
   2. After converting bits to integer return that number
end

---

of bin2Len and bin1Len. Then make a list of zeros and pad it to list named "zero" and, at last, make a list containing zeros and bin1 and put that in bin1. After main if condition we have declared a local variable "dist as 0" and in last run for loop on each of the arrays bin1 and bin2. In loop's each iteration, we compare each bit of bin1 and bin2, if bits are not equal, then increment "dis" by one. And return dist.

"binary-to-integer" is the procedure which is used to convert binary number to decimal number (Table 10.17). It takes bits as input and convert them to a decimal number.

10.4   Results and discussion

10.4.1   Evaluation metrics

To evaluate the performance of Kademlia, we used a number of messages as our metric. It is defined as the number of messages exchanged between nodes for node lookup operation. We simulate the messages behavior as we increase the network size. Then, on the basis of this behavior, the comparison is performed between simulations of PeerSim and NetLogo. The parameters that are used to set simulation environment are shown in Table 10.1.

10.4.2   Power law plots of centrality measures

Initially, we started out with performing manipulation of the raw network. Degree centrality measure is analyzed first. After the degree measure is done for each node, it is added to corresponding node as an attribute. Centrality measure is used to resize and colorize every node. Resultant network is shown in Figure 10.10. We can get compelling overview of network by degree centrality, there are some other quantitative measures as well shown in Table 10.18.

For further details, Table 10.18 can be used. Plots for Degree Centrality (%) (Figure 10.11), closeness centrality (Figure 10.12) and betweenness centrality (Figure 10.13) are shown below. Also, for each type of centrality measure, we have also plotted poweRlaw plots using "R" language. Corresponding poweRlaw plots are

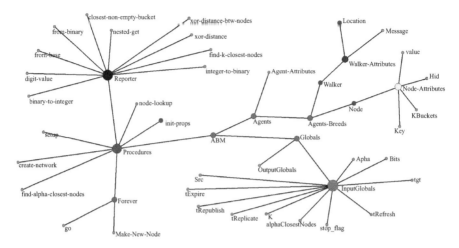

Figure 10.10   Kademlia Complex Network Model, resized and colorized according to degree centrality

shown below in the figures: Figure 10.14 shows poweRlaw plot of Degree, Figure 10.15 shows poweRlaw plot of Closeness centrality and Figure 10.16 shows poweRlaw plot of Betweenness centrality (Figures 10.17 and 10.18).

10.4.3   PeerSim simulation using existing code in PeerSim

In PeerSim, Kademlia module is implemented using the event driven model approach. The implementation is little bit different from the actual Kademlia protocol. The implementation of store and search values are skipped and only the finding the node is implemented. Messages exchange and few timeouts events are the focal events that run the protocol. Timeouts events are mainly used for the nodes failure in the network. In the next section, details of how protocol runs and messages will be explained.

1. **Messages**
    In Kademlia protocol, the messages exchanged are the events linked to PeerSim. For every message, we create an instance of Messages class that extends PeerSim class of SimpleEvent.
    i.   **MSG_FINDNODE:** Find node action is stared using this message.
    ii. **MSG_ROUTE:** This message is used to query about the target node from the neighbors.
    iii. **MSG_RESPONSE:** In response to MSG_ROUTE, this msg is sent that contains the information of k nodes that are nearest to the target node.
2. **Code comments**
    i.   **Bootstrap process**
         In the core of PeerSim, the WireKOut class contained. Two classes StateBuilder and CustomDistribution are implemented to facilitate network's bootstrap and initialization process.

Table 10.18 Table of Degree, Betweenness Closeness and Eigen centrality

Label	Degree	Closeness	Betweenness	Eigen
ABM	3	0.34	0.66	0.41
Globals	3	0.29	0.41	0.42
OutputGlobals	1	0.23	0	0.12
InputGlobals	12	0.25	0.41	0.96
Apha	1	0.20	0	0.27
Bits	1	0.20	0	0.27
K	1	0.20	0	0.27
tExpire	1	0.20	0	0.27
tRefresh	1	0.20	0	0.27
tReplicate	1	0.20	0	0.27
tRepublish	1	0.20	0	0.27
Agents	3	0.29	0.39	0.19
Agent-Attributes	1	0.22	0	0.06
Agent-Breeds	3	0.25	0.36	0.14
Node	2	0.21	0.19	0.11
Node-Attributes	5	0.18	0.16	0.14
Hid	1	0.15	0	0.05
key	1	0.15	0	0.05
value	1	0.15	0	0.05
KBuckets	1	0.15	0	0.05
Walker	2	0.20	0.12	0.08
Walker-Attributes	4	0.17	0.08	0.06
Location	2	0.14	0	0.03
Message	1	0.14	0	0.03
Procedures	9	0.32	0.59	0.80
Forever	3	0.25	0.08	0.28
go	1	0.20	0	0.08
Make-New-Node	1	0.20	0	0.08
Setup	1	0.24	0	0.22
create-network	1	0.24	0	0.22
init-props	2	0.24	0	0.22
nested-get	1	0.21	0	0.27
binary-to-integer	1	0.21	0	0.27
integer-to-binary	1	0.21	0	0.27
xor-distance	1	0.21	0	0.27
from-binary	1	0.21	0	0.27
from-base	1	0.21	0	0.27
digit-value	1	0.21	0	0.27
Reporter	11	0.27	0.38	1
alphaCLosestNodes	1	0.20	0	0.27
src	1	0.20	0	0.27
tgt	1	0.20	0	0.27
stop_flag	1	0.20	0	0.27
node-lookup	1	0.24	0	0.22
find-alpha-closest-nodes	1	0.24	0	0.22
find-k-closest-nodes	1	0.21	0	0.27
closest-non-empty-bucket	1	0.21	0	0.27
xor-distance-btw-nodes	1	0.21	0	0.27

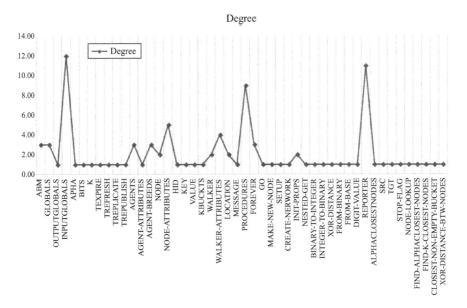

Figure 10.11   Plot showing the Degree centrality

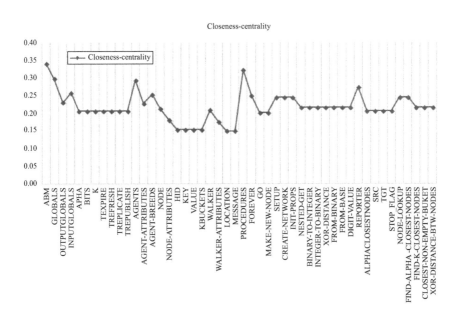

Figure 10.12   Plot showing the poweRlaw of Degree centrality

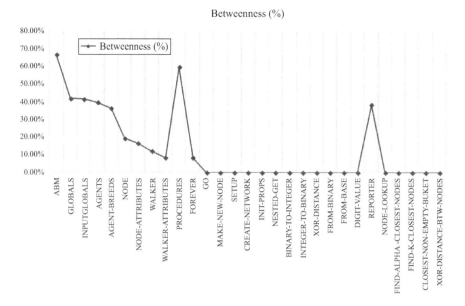

Figure 10.13   Plot showing the closeness centrality

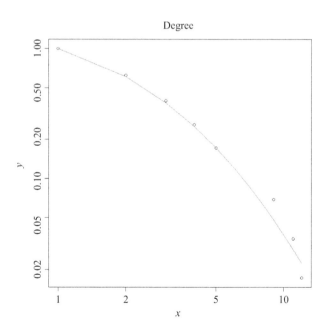

Figure 10.14   Plot showing the poweRlaw of Closeness centrality

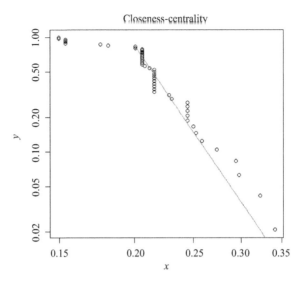

Figure 10.15   Plot showing the Betweenness centrality

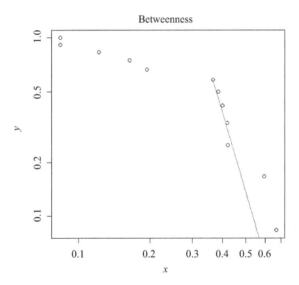

Figure 10.16   Plot showing the poweRlaw of Betweenness centrality

When the process starts, the WireKOut class randomly creates the links between nodes. It creates a virtually overlay network. After this, CustomDistribution class is used to initialize the network, for example, for every node, unique ids are assigned in a range between $0 \ldots 2^{BITS}$. Here BITS is coming from the Kademlia protocol and usually the value

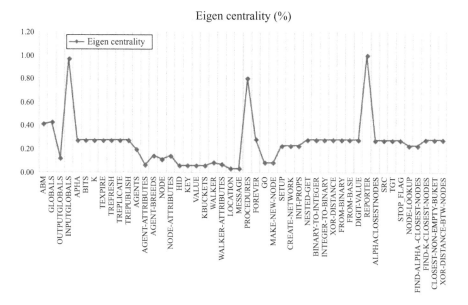

Figure 10.17   Plot showing the Eigen centrality

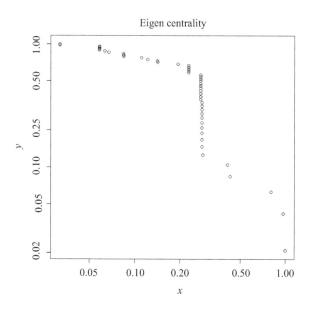

Figure 10.18   Plot showing the poweRlaw of Eigen centrality

is 160. After network's initialization, bootstrap process is performed using the StateBuilder class. The main objective of this class is to fill up the node's k-buckets.

ii. **Find a node**

The class TrafficGenerated generates a message MSG_FINDNODE that initiates the process of find the node. The first step of this process is to create an object of FindOperation class that has all the data related to specific process. In particular, this object has the information of closest neighbor set and information of queried nodes and about requests that are outstanding. In this class, there are two variables nrHops and timestamp that keep track of duration and hops operation count respectively. Rest is the same process that are already explained in the overview section.

iii. **Joining the network**

This process is also the same as described in the overview of this section. When a new node is created using PeerSim, a new id is assigned to it. Initially, this node has an empty bucket and then a random node is selected and added to the newly created node's k-bucket. Then this node sends an MSG_ROUTE to random node and this way it starts filling their k-bucket.

3. **Simulations**

There is a configuration file that is used to customize the Kademlia module. It is a simple txt file that is easy to edit, and it defines the whole protocol by setting the different controls and parameters. In the next section, a brief explanation is given of the module different components.

i. **The Protocol**

There are system-wide three main parameters that can be used to customize the protocol.

BITS: It is used to define the node id length. The default value is 160.

K: It is used to define a single k-bucket length. The default value is 20.

ALPHA: It is used in lookup operation. It defines the simultaneous lookup process. The default value is 3.

ii. **The controls**

Two different types of control classes are provided; one is Turbolence and the other is TrafficGenerator. The TrafficGenerator select random source to generate random find node messages to search. The Turbolence class used the given probability to add or remove nodes from the network. The following parameters are used to configure probabilities:

p_idle: The default value is 0, it means the current execution neither removes the node nor adds them in the network.

p_add: The default value is 0.5. This is probability of addition of new node in current execution.

p_rem: Default values is 0.5. This is the probability of failure of a node that exists in the network.

iii. **The Observer**

In this module, there is a class called KademliaOberver. The main purpose of this class is to gather statistics of protocol behavior. It keeps track hops

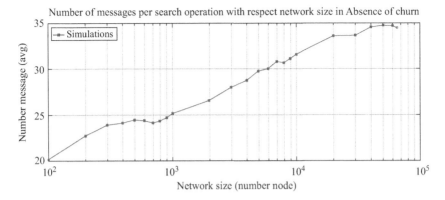

Figure 10.19   Number of messages with respect to network size per search operation in churn absence

count, delivers messages count and finds operation count. The STEP is the only allowed parameter that used the stderr to define the output frequency measure.

4. **Results**

The performance of Kademlia protocol in search operations is analyzed in both scenarios, that is, in the presence of churn and in the absence of churn by varying the network size. The result of average messages exchanged during the search operation is presented.

i. **Simulation Parameters**

In order to simulate Transmission Control Protocol (TCP)-like connection-oriented protocol, a reliable channel was selected. In simulation, the network size was varied from 128 to 65,536 nodes. In each round, the number of nodes was doubled and total simulation time was around 1 h. The observer step was 100,000 and traffic step was calculated from the following equation:

$$\text{TrafficStep} = \frac{\text{Simulation Time}}{\text{Network Size}} \tag{10.3}$$

The turbulence step was calculated from the blew equation

$$Turbulence\ Step = (Simulation\ Time * 20)/Network\ Size \tag{10.4}$$

5. **Plots**

The results from the java simulation were written in text file, and then using MATLAB®, the results were plotted.

In Figure 10.19, system's search operation performance is evaluated in the stable environment, which means in the absence of churn. In this plot, the number of messages required per search operation is compared with network size. You can see that as network size is increasing, the number of messages is also increasing. The turbulence is not used in this plot.

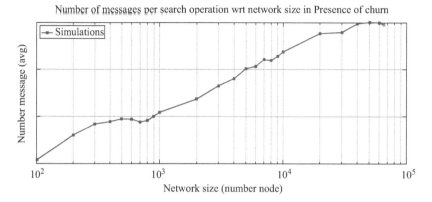

Number of messages per search operation wrt network size in Presence of churn

Figure 10.20   *Number of messages with respect to network size per search
operation in churn presence*

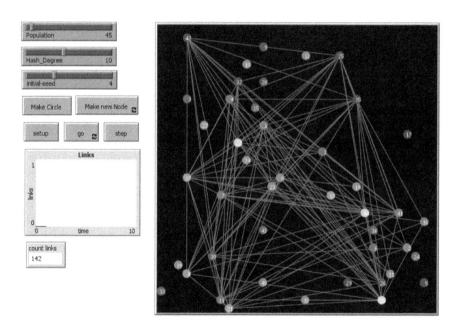

Figure 10.21   *NetLogo user interface of Kademlia random shape. It has sliders,
buttons, plot, etc.*

Figure 10.20 is the same graph as the previous one. The only difference is, now
the churn is present. You can see that there is very little difference in the graph as
compared with the graph from absence of churn. In this graph, the turbulence control
is activated. The probability of nodes for leaving or joining the network is same. It is
also an evidence to show that Kademlia handles churn very well (Figure 10.21).

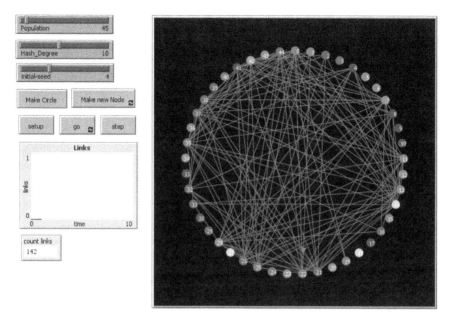

Figure 10.22   NetLogo user interface of Kademlia circle shape. It has sliders, buttons, plot, etc.

10.4.4   ABM simulation

To simulate Kademlia P2P network, a model is developed in NetLogo using ABM approach. NetLogo setup is shown in Figure 10.22. NetLogo is a unified modeling environment and programming language for ABM. It is an open source and can be downloaded freely from NetLogo website. The implementation is little bit different from the actual Kademlia protocol. The implementation of store and search values are skipped and only the finding the node is implemented. Messages exchange is the focal event that run the protocol.

There is a method "setup" that is called before the simulation start. It is the core function that is used to initialize and bootstrap the process. In this function, first all the global variables are initialized and then peers or nodes are created and placed randomly. After that, random ids are generated in the range of $0 \ldots 2^{20}$ and assigned to each node. There is button "make-circle" that can be used to arrange the randomly placed nodes in a circle just for better visualization. At the end of setup, the method "create-network" is called. This method used the routing algorithm and filled each node's k-buckets.

Once the environment is setup then the "go" button is used to run the simulation. In "go" method, "node-lookup" method is called that run the lookup algorithm to find the target node. Source node and target node are randomly set in the "setup" function. The implemented lookup algorithm is the same as explained in the Kademlia

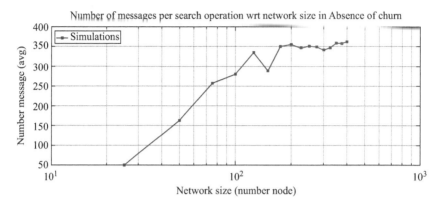

Figure 10.23   Number of messages with respect to network size per search operation in churn absence

architecture section. We keep track of links that are made between the nodes in the process of lookup and take them as messages.

10.4.4.1   Configuration

Like PeerSim in ABM, there are system-wide three main parameters that can be used to customized the protocol. The first one is the BITS that defines the node id length and the default value is 20. In PeerSim simulation, the default value of BITS is 160, but in NetLogo, it is set to 20 because of limiting the simulation scale and to avoid overflow errors in hashes operation. The second one is K that is used to define the single k-bucket length default value is 20. The third one is ALPHA that is used in lookup operation, and it defines the simultaneous lookup process. The default value is 3.

10.4.4.2   Results

As described earlier, the performance of Kademlia protocol in search operation is analyzed. The performance is analyzed in both scenarios that are in the presence of churn and in the absence of churn by varying the network size. The result of average messages exchanged during the search operation is presented. We used BehaviorSpace tool of NetLogo for our experiment. The experiment consists of 200 simulations, 100 in the presence of churn and 100 in the absence of churn. The network size is changed from 25 to 400 with a difference of 25. The results from the NetLogo simulation was written in csv file, and then by using MATLAB, the results were plotted.

Figure 10.23 shows the results of lookup operation in the absence of churn, which means stable environment. In this plot, the number of messages required per search operation is compared with network size. You can see, as the network size increases, the number of messages required is also increasing.

Figure 10.24 is the same graph as the previous one. The only difference is, now the churn is present. You can see that there is very little difference in the graph as

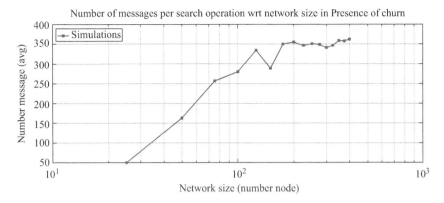

Figure 10.24   Number of messages with respect to network size per search
operation in churn presence

compared with the graph from absence of churn. The probability of nodes for leaving or joining the network is same. It is also an evidence to show that Kademlia handles churn very well.

10.4.5   Comparison of PeerSim and ABM results

In this section, we will do comparison of both NetLogo and PeerSim simulations. The environment setup in both simulations was not exactly the same. For example, the BITS length that is used to generate node id was different in both environments. In PeerSim, it is 2^{160} and in NetLogo it is 2^{20}. The second main difference was the network size. In PeerSim, the network size was starting from 128 and ending at 65,000. But in NetLogo due to resources constraint, it was not possible to set network this much big as it was used in PeerSim. In NetLogo, the network size was starting from 0 to 400 with the interval of 25.

In Figure 10.25, the comparison of lookup operations is shown in the stable environment, which means the absence of churn. In the plots, you can see that the behavior of both simulations is comparable. As we are increasing the network size, the number of messages required for lookup operation are also increasing.

In Figure 10.26, the comparison of lookup operation in the presence of churn is shown for both simulations. As we have discussed earlier, the behavior of Kademlia does not change much in the presence of churn. So, in result, the plots are almost same as the above one. You can also see that the behavior of both plots are comparable.

10.4.6   Discussion

10.4.6.1   Comparison of ODD and DREAM

In this part of the chapter, we will present comparison of ODD [8] and DREAM [3] from all perspectives.

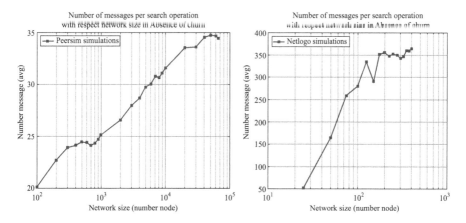

*Figure 10.25   The left side is PeerSim simulation and in the right NetLogo
simulation*

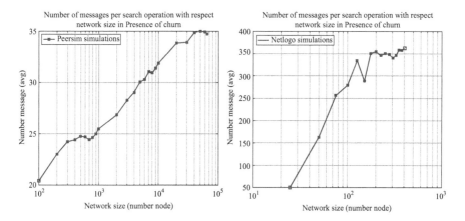

*Figure 10.26   The left side is PeerSim simulation and in the right NetLogo
simulation*

ODD only provides a textual description of ABM for the purpose of making the model more readable and ODD also promotes rigorous model formulations. To describe ABM, it provides checklist covering key features. ODD has some limitations that are described next. According to a survey conducted by Grimm [52], for the publications which have used ODD only, 75% have used it correctly and the remaining 25% have some flaws, even some parts of the protocol were compromised. So, in [52], the author concludes that using ODD protocol, it is not an easy task to write down ABM specification.

An issue with ODD is, some of the ODD specifications are overlapping in different sections. For example, introduction and purpose sections have overlaps. The same thing that is asked in the purpose section is also included in the introduction section. The submodel section also has some similarities with the design concept section. The submodel section is also described again in the scheduling and process section.

ODD is not suitable for comparison between different ABMs and to describe large ABM. Some large ABMs have a lot of features and ODD has less description of them, which is not enough to cover the whole ABM. Comparison between different ABMs is not possible in ODD due to the lack of quantitative assessment. The only way to do is to prepare an ODD checklist for both ABMs and put them is a table and find the differences and similarities between them. Therefore, using ODD comparison of two ABMs is not an easy task and also reviewing of many ABMs is a hectic the task.

ODD specification has redundancy, replication and less information. Sometimes, the same ABM with a different version is published in different publications. But in these publications, ODD specifications are almost same just with little change in process section and entities. Another issue with ODD is, using it, someone cannot replicate ABM because it provides a very specific description of the model. So, to replicate ABM with such specific, redundant and ambiguous information is quite difficult.

As compared to ODD, DREAM provides a detailed description of ABM. Using DREAM, we can develop a complex network model of any ABM, describe through pseudo-code specification and also can make steps for network analysis. So, by using DREAM, we can easily get detailed description and pseudo-code based design of ABM which will be quite helpful for us to compare ABMs and replicate ABMs of any domain.

DREAM allows to understand and analyze visually the complex network of any ABM without going into much code details. It provides quantitative measurement after performing network analysis, which lacks in ODD. Therefore, by using these measurements, one can easily understand, replicate and compare ABM from different domains.

DREAM is independent of scientific domains and applicable to any ABM research domain. DREAM allows comparing different models that are developed in different scientific domains. For example, to compare two ABMs of different domains, first we have to develop the complex networks of both models and then we can analyze and compare them in the same manner. So, we can say that by DREAM, we cannot only get detailed description but we can also compare ABMs from different domains.

DREAM further provides pseudo-code specification of ABM. Using this specification, anyone can understand ABM regardless of discipline. This specification translates to code and then ABM development. Hence, using DREAM, we can confidently say that it is an easy task to understand and to replicate any ABM.

In [53], the authors did a very good empirical analysis of both protocol methodologies. They used 13 features to evaluate the methodologies and calculate the rank of each methodology. The averaging result of DREAM was 1.76 and ODD was 0.69.

Hence, the results show that DREAM provides very detailed specification, and it is the best suitable choice for ABM understanding, comparison and replication.

10.4.6.2   Kademlia relation with theory of computation

The theory of computation has three portions, i.e., computability, complexity theory and automata theory. In this section, we will try and relate Kademlia with "Complexity." Now we will try and find the time complexity for Kademlia FIND_NODE operation. Let's say we are looking for an ID that is furthest away from our own node ID. For this, we first query alpha closest nodes than from the result of those alpha nodes we further query K more nodes and this process continues until there are no closer nodes present (Node Look up operation is defined already in detail in Node Look up section). Since our each hop sweeps off minimum distance of 1 bit at one time, therefore $O(\log(n))$ hop count is needed (here "n" is network size).

From the above paragraph, the point to note is, we will need less steps to find the required node if the node id to find is closer to our own node id. As we already know, "k" is a constant variable (nodes present per bucket), therefore $\log(k)$ is also a constant. If nodes in a bucket are doubled, then we will have double the resolution of given keyspace area and therefore will (probabilistically) get a node that is at least 1 bit closer to target than a bucket of size $k/2$. Another point here is, there are some optimizations which are needed, like for handling unbalanced trees, each node splits kBuckets as per requirement to satisfy that they have knowledge of trees having k nodes at minimum. Therefore, at last, we can conclude that Kademlia being a P2P system is complex and $O(\log(n))$ hop count is needed for each operation to complete.

10.5   Conclusion and future work

In the past a number of works have been done to implement P2P protocols in Object Oriented domain and other domains. In this chapter, we have implemented Kademlia, a P2P protocol in NetLogo. We have presented ABM and Complex-network based model for Kademlia. In the results section, we have compared our results with the results of Kademlia simulation in PeerSim. As we have already written in the results section, our results have matched with the results of PeerSim by some error margin.

In future what can be done to improve model is due to the resource constraint, we have to only run the simulation for network sizes of up to 500. More network sizes can be added in simulation to get further insight of model and do better comparison with already present models. Another addition can be, we implemented only the FIND_NODE operation from the original Kademlia protocol. Other operations like STORE value and SEARCH value can also be implemented.

References

[1]   Maymounkov P, Mazieres D. Kademlia: A peer-to-peer information system based on the XOR metric. In: International Workshop on Peer-to-Peer Systems. Springer; 2002. p. 53–65.

[2]   Niazi MA. Complex adaptive systems modeling: A multidisciplinary roadmap. Complex Adaptive Systems Modeling. 2013 Mar;1(1):1. Available from: https://doi.org/10.1186/2194-3206-1-1.

[3]   Niazi MA. Towards a Novel Unified Framework for Developing Formal, Network and Validated Agent-Based Simulation Models of Complex Adaptive Systems. arXiv preprint arXiv:1708.02357. 2011.

[4]   Niazi MA, Hussain A. Complex Adaptive Communication Networks and Environments: Part 2. SAGE Publications Sage UK: London, England; 2013.

[5]   Niazi MA, Hussain A. Complex Adaptive Communication Networks and Environments: Part 1. SAGE Publications Sage UK: London, England; 2013.

[6]   Epstein JM. Why model? Journal of Artificial Societies and Social Simulation. 2008;11(4):12.

[7]   Batool K, Niazi MA. Modeling the internet of things: a hybrid modeling approach using complex networks and agent-based models. Complex Adaptive Systems Modeling. 2017 Mar;5(1):4.

[8]   Grimm V, Berger U, Bastiansen F, *et al.* A standard protocol for describing individual-based and agent-based models. Ecological Modelling. 2006;198 (1–2):115–126.

[9]   Niazi MA, Hussain A. Cognitive Agent-Based Computing-I: A Unified Framework for Modeling Complex Adaptive Systems Using Agent-Based & Complex Network-Based Methods. Springer Science & Business Media; 2012.

[10] Riaz F, Niazi MA. Validation of Enhanced Emotion Enabled Cognitive Agent Using Virtual Overlay Multi-Agent System Approach. arXiv preprint arXiv:170801628. 2017.

[11] Bonani M, Furlan D. A Kademlia module for PeerSim. Università di Trento, (2009/2010) Web site: http://peersim sourceforge net [seen: 01/08/2014]. 2010.

[12] Clifton M. The Kademlia Protocol Succinctly; 2018. Available from: https://www.syncfusion.com/ebooks/kademlia_protocol_succinctly

[13] Aikebaier A, Enokido T, Takizawa M. Trustworthy group making algorithm in distributed systems. Human-Centric Computing and Information Sciences. 2011;1(1):6.

[14] Teraoka T. Organization and exploration of heterogeneous personal data collected in daily life. Human-Centric Computing and Information Sciences. 2012;2(1):1.

[15] Caubet J, Gañán C, Esparza O, *et al.* Certificate revocation list distribution system for the KAD network. The Computer Journal. 2014;57(2):273–280.

[16] Xu Q, Shen H, Chen Z, *et al.* Hybrid information retrieval policies based on cooperative cache in mobile P2P networks. Frontiers of Computer Science in China. 2009;3(3):381–395.

[17] Yan G, Ha DT, Eidenbenz S. AntBot: Anti-pollution peer-to-peer botnets. Computer Networks. 2011;55(8):1941–1956.

[18] Steiner M, Carra D, Biersack EW. Evaluating and improving the content access in KAD. Peer-to-Peer Networking and Applications. 2010;3(2):115–128.

[19] Jin H, Chen H. SemreX: Efficient search in a semantic overlay for literature retrieval. Future Generation Computer Systems. 2008;24(6):475–488.

[20]   Anastasi G, Giannetti I, Passarella A. A BitTorrent proxy for Green Internet file sharing: Design and experimental evaluation. Computer Communications. 2010;33(7):794–802.

[21]   Trunfio P. A two-layer model for improving the energy efficiency of file sharing peer-to-peer networks. Concurrency and Computation: Practice and Experience. 2015;27(13):3166–3183.

[22]   Zhao BY, Huang L, Stribling J, *et al.* Tapestry: A resilient global-scale overlay for service deployment. IEEE Journal on Selected Areas in Communications. 2004;22(1):41–53.

[23]   Enokido T, Aikebaier A, Takizawa M. A model for reducing power consumption in peer-to-peer systems. IEEE Systems Journal. 2010;4(2):221–229.

[24]   Park K, Valduriez P. Energy efficient data access in mobile p2p networks. IEEE Transactions on Knowledge and Data Engineering. 2011;23(11):1619–1634.

[25]   Androutsellis-Theotokis S, Spinellis D. A survey of peer-to-peer content distribution technologies. ACM Computing Surveys (CSUR). 2004;36(4):335–371.

[26]   Czirkos Z, Hosszú G. Solution for the broadcasting in the Kademlia peer-to-peer overlay. Computer Networks. 2013;57(8):1853–1862.

[27]   Urdaneta G, Pierre G, Steen MV. A survey of DHT security techniques. ACM Computing Surveys (CSUR). 2011;43(2):8.

[28]   Trifa Z, Khemakhem M. A novel replication technique to attenuate churn effects. Peer-to-Peer Networking and Applications. 2016;9(2):344–355.

[29]   Sánchez-Artigas M, López PG. Echo: A peer-to-peer clustering framework for improving communication in DHTs. Journal of Parallel and Distributed Computing. 2010;70(2):126–143.

[30]   Akers SB, Krishnamurthy B. A group-theoretic model for symmetric interconnection networks. IEEE Transactions on Computers. 1989;38(4):555–566.

[31]   Chung F, Coffman E, Reiman M, *et al.* The forwarding index of communication networks. IEEE Transactions on Information Theory. 1987;33(2):224–232.

[32]   Xu J, Kumar A, Yu X. On the fundamental tradeoffs between routing table size and network diameter in peer-to-peer networks. IEEE Journal on Selected Areas in Communications. 2004;22(1):151–163.

[33]   Stoica I, Morris R, Liben-Nowell D, *et al.* Chord: a scalable peer-to-peer lookup protocol for internet applications. IEEE/ACM Transactions on Networking (TON). 2003;11(1):17–32.

[34]   Akavipat R, Al-Ameen MN, Kapadia A, *et al.* ReDS: A framework for reputation-enhanced DHTs. IEEE Transactions on Parallel and Distributed Systems. 2014;25(2):321–331.

[35]   Hoffman K, Zage D, Nita-Rotaru C. A survey of attack and defense techniques for reputation systems. ACM Computing Surveys (CSUR). 2009;42(1):1.

[36]   Martinez-Yelmo I, Bikfalvi A, Cuevas R, *et al.* H-P2PSIP: Interconnection of P2PSIP domains for global multimedia services based on a hierarchical DHT overlay network. Computer Networks. 2009;53(4):556–568.

[37]   Darlagiannis V, Mauthe A, Liebau N, *et al.* An adaptable, role-based simulator for P2P networks. In: MSV/AMCS; 2004. p. 52–59.

[38] Steiner M, En-Najjary T, Biersack EW. A global view of KAD. In: Proceedings of the 7th ACM SIGCOMM Conference on Internet Measurement. ACM; 2007. p. 117–122.

[39] Pecori R. S-Kademlia: A trust and reputation method to mitigate a Sybil attack in Kademlia. Computer Networks. 2016;94:205–218.

[40] Ou Z, Harjula E, Kassinen O, *et al.* Performance evaluation of a Kademlia-based communication-oriented P2P system under churn. Computer Networks. 2010;54(5):689–705.

[41] Steiner M, En-Najjary T, Biersack EW. Long term study of peer behavior in the KAD DHT. IEEE/ACM Transactions on Networking (ToN). 2009;17(5):1371–1384.

[42] Medrano-Chávez AG, Pérez-Cortés E, Lopez-Guerrero M. A performance comparison of Chord and Kademlia DHTs in high churn scenarios. Peer-to-Peer Networking and Applications. 2015;8(5):807–821.

[43] Kong JS, Bridgewater JS, Roychowdhury VP. Resilience of structured P2P systems under churn: The reachable component method. Computer Communications. 2008;31(10):2109–2123.

[44] Sen S, Wang J. Analyzing peer-to-peer traffic across large networks. In: Proceedings of the 2nd ACM SIGCOMM Workshop on Internet Measurement. ACM; 2002. p. 137–150.

[45] Fraigniaud P, Gauron P. D2B: A de Bruijn based content-addressable network. Theoretical Computer Science. 2006;355(1):65–79.

[46] Boneh D, Franklin M. Identity-based encryption from the Weil pairing. SIAM Journal on Computing. 2003;32(3):586–615.

[47] Aiello LM, Milanesio M, Ruffo G, *et al.* An identity-based approach to secure P2P applications with Likir. Peer-to-Peer Networking and Applications. 2011;4(4):420–438.

[48] Einziger G, Friedman R, Kantor Y. Shades: Expediting Kademlia's lookup process. Computer Networks. 2016;99:37–50.

[49] Fan L, Cao P, Almeida J, *et al.* Summary cache: A scalable wide-area web cache sharing protocol. IEEE/ACM Transactions on Networking. 2000;8(3): 281–293.

[50] Wang P, Tyra J, Chan-Tin E, *et al.* Attacking the KAD network—Real world evaluation and high fidelity simulation using DVN. Security and Communication Networks. 2013;6(12):1556–1575.

[51] Locher T, Schmid S, Wattenhofer R. eDonkey & eMule's Kad: Measurements & Attacks. Fundamenta Informaticae. 2011;109(4):383–403.

[52] Grimm V, Berger U, DeAngelis DL, *et al.* The ODD protocol: A review and first update. Ecological Modelling. 2010;221(23):2760–2768.

[53] Akram W, Niazi MA, Iantovics LB. Towards Agent-Based Model Specification in Smart Grid: A Cognitive Agent-based Computing Approach. arXiv preprint arXiv:171003189. 2017.

Chapter 11

Descriptive agent-based modeling of the "BitTorrent" P2P protocol

Abdul Saboor[1], Nasir Khan[1], and Mubariz Rehman[1]

11.1 Introduction

BitTorrent designed by Cohen [1] is a peer-to-peer (P2P) file-sharing protocol developed to distribute data in such a manner that the actual distributer would be capable to decrease the bandwidth use, and he will be able to reach the same amount of users as well. In BitTorrent [1], data is divided or broken down into small chunks, and every single user using those chunks would be available to upload to other people in a swarm to save bandwidth where swarm is a group of peers connected to single torrent. It is one of the most efficient protocols for replication and distribution of files over the internet [2]. BitTorrent facilitates users to upload and download files at the same time with each other as it uses principal known as tit-for-tat where BitTorrent client act as a client and server. BitTorrent is the complex networks which improve the performance of the communication systems.

In a network, many clients communicate with each other in a complex P2P network using BitTorrent. In BitTorrent, peers work as seeders or lechers which act both as a client and a server [1]. The main complexity [3] of the BitTorrent model is that the whole model follows the power law distribution mechanism. In this mechanism, when the total number of seeder involved in communication increases with respect to network size, it tends to decrease in the network clustered. High-network clustering values are only obtained when the swarm size of the network increases exponentially.

Agent-based modeling leads to the most efficient paradigm for the complex adaptive systems (CASs) [4] and for the large-scale complex adaptive communication networks and environment. One of the main limitations of the other models is that it does not offer the flexibility in terms of the complex connectivity of P2P network. Agent-based modeling for the complex systems provides many benefits like the overall throughput of the system increases and nodes interaction become easy for communication.

In the modern era where the network sizes increase as well as the complexity and the dynamically increase in the different network communication like wireless

[1] Cosmose Research Group, COMSATS University, Pakistan

networks, mobile ad hoc networks, P2P communication. This all leads to the increase in the complex network-emergent phenomena. These effects cause the network congestion and the overall communication cost increases. To avoid these effects, modeling of the network communication is needed instead of CASs and complex adaptive communication networks and environments (CACOONS) [5]. Agent-based modeling of the complex networks overcomes these issues and increases the flexibility of the systems. Agent based modelling (ABM) is commonly used to model the dynamic behavior in CASs. These techniques are more effective for the modeling of complex networks because the interaction of the networks is handled using the visualization techniques. As the simulation experiment of the complex systems increases the overall complexity of the systems, so there is a need of specific modeling approach for the large-scale networks. ABM allows a clear, concise and the unambiguous representation of the complex network model.

In the past few years, the unpredictable growth with respect to scale and complexity of the network is observed. Networks sizes are increasing rapidly which results in increase in their complexity. In the large complex networks, it is very difficult to test and evaluate each module and its implementation. For the above aforementioned purpose, simulation in terms of modeling plays a vital role in the development and the design of the distributed system networks. For the agent-based modeling, a very interactive and popular tool NetLogo is used. It provides the visual simulations for the complex adaptive networks. One of the main features of the NetLogo models is that it provides user friendly environment.

From the previous studies, is no descriptive agent based model (DREAM) [4] for the BitTorrent P2P networks is developed. To overcome the complexity of distributed networks, the basic Described agent-based model is proposed in [4]. In this chapter, we have modeled DREAM of BitTorent. Overview, design concepts and details (ODD) is used for textual representation of BitTorent. We have shown the comparison between the ODD model [6] and DREAM model. As a result, we conclude that for agent-based modeling, the NetLogo environment is extremely flexible with very less space complexity and the DREAM model provides the basic description of agent-based modeling.

In the BitTorrent P2P network, the main benefits are as follows:

- In the decentralized system, the file which is to be downloaded is not present in central server system. If the source 1 is not active, you can download it from the other available source, but in the centralized system, if the main server downs, the downloading process stops.
- During the downloading process, if the connection interrupts after establishing connection, the downloading process continues from where it stops and not from start.
- Even with the slower network bandwidth torrent provide user friendly environment for downloading the big files.
- BitTorrent provides flexible environment for user whether it is downloading or searching process. Prevents the tempered or corrupted file for sharing.
- Ability to download the larger files (Figure 11.1).

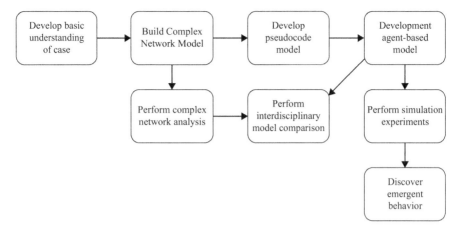

Figure 11.1   Descriptive agent-based modeling level. Adopted, with permission, from Reference [4]

The main contribution of the BitTorrent in ABM is

1. Agent-based modeling by using NetLogo provides the wide range of benefits such as real-time analysis and statistical measurements.
2. Implementation in NetLogo is more efficient and in visualized form. Complex network simulation is easier as compared to Network Simulator 2, etc.
3. In the NetLogo simulator, only the active environment is to be showed, so there will be less space and time complexity.
4. NetLogo provides the user friendly environment in which it is easy to understand how communication between the nodes occurs.

BitTorrent is a complex P2P network, which improves the performance of the communication systems. Designing agent-based model for the complex system provides multiple benefits as well as nodes interaction becomes more easy for communication. Proactive agent design is the famous used approach for the agent-based modeling. These are example of the complex systems [4] in which agents can communicate and take decisions intelligently by itself. In terms of P2P networks, agent-based modeling also supports the mobility of the nodes in the network. However, the overall performance also increases using the mobility and adaption of the peers (Figure 11.2).

11.1.1   Contributions

The main contribution of the BitTorrent in ABM is

- Agent-based modeling by using NetLogo provides the wide range of benefits such as real-time analysis and statistical measurements.
- Implementation in NetLogo is more efficient and in a visualized form. Complex network simulation is easier as compare to Network Simulator 2, etc.

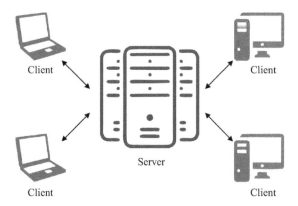

Figure 11.2   Client/Server architecture. Adopted, with permission, from Reference [7]

- In the NetLogo simulator, only the active environment is to be showed, so there will be less space and time complexity.
- NetLogo provides the user friendly environment in which it is easy to understand how communication between the occurs.

11.2   Background and literature review

11.2.1   Complex adaptive systems

In [5], the authors have presented a novel framework known as FABS for agent-based modeling. They introduced FABS for developing wireless sensor network–based simulation and experimental approach, for monitoring complex adaptive environments. Furthermore, the authors explained the advantage of existing model by implementing a boids model. Initial simulation and experiments has proved how describe specification can be efficiently utilized to make simulation models of the CAS phenomena along with the involvement of wireless senor networks.However, wireless network model for complex system takes high computation complexity.

The authors [4] demonstrated multiple advantages of NetLogo, a modeling tool for the agent-based environment as well as describe how to compile a random-walk function in the simulation. Furthermore, the authors defined a systematic technique for the use of agent-based modeling to compile and simulate a model for the complex wireless senor network's applications along with demonstrating the effectiveness and efficiency of agent-based modeling, in both forms as a general and NetLogo, specifically for modeling and designing different types of communication networks.However, the authors conclude that dynamically simulation parameters are not to be supported by proposed NetLogo model.

In [4], the authors presented a broad overview of complex communication networks modeling and analysis using techniques from social network analysis (SNA). The authors also examined that how network dynamics can be characterized using complex networks. Furthermore, the authors have used a complex network approach to analyze these networks in a more effective way to identify important nodes, the nodes creating congestion, bridges, network centrality and the broker nodes. The authors have also presented a complete overview of the common network-analysis techniques.

The authors [6] show the first steps toward the answer of a set of nontrivial problems related with the colossal task of emerging a combined framework applicable to multidisciplinary and interdisciplinary CAS research trainings. Furthermore, the author proposed a grouping of complex network approaches and agent-based modeling, two standards which are predominant in the CAS modeling.

11.2.2   Modeling and simulation of CACOONS

Due to the invention in the field of the networks, the network size and the complexity also increases. In the networks, the invention in not only in term of scale but also the new approach for the communication is explored such as the wireless sensor networks, ad hoc mobile networks and distributed networks such as P2P networks. With the emergence of these phenomena, the complexity of the networks also increases. Mostly, the effects of these technologies are untraceable in the single network.

In this research article [8], the authors proposed a new approach for the real-time intruder detection while using the lightweight software agent–based technique for visual surveillance model. The main objective of the research work is to obtain the maximum throughput with the precision and accuracy while minimizing the complexities such as computation and storage of the system.

The authors in [9] perform the mathematical analysis and the formal framework for determining the inconsistencies in the evaluation and validation of the complex networks simulations in the field of the wireless sensor networks. The outcome of this research article is that the simulation-based system validation is more enhanced and empowered using the efficient and specific frameworks and their statistical and mathematical analysis.

In the paper [10], the main focus of the researchers is on the object-oriented communication framework or model. In the multiple-tiered networks, the individual behavior of the agents is determined by the modeling and simulating environment. For the efficient simulation support, a multiple number of agent behaviors are defined using the various best outcome results algorithms and different dynamic data structures.

11.2.2.1   Agent-based modeling

With the help of the high computation systems, the complex problems in the fields of physics, chemistry, biology and economics are to be solved. The new inventions in terms of computing power, software establishment and the computer visual graphics lead to solve complex problem by using the agent-based modeling. With the help of the agent-based modeling, the development of large and complex system simulations in real time is easier. Different agent-based modeling tools used for the simulation of some of them are swarm, macon, NetLogo, metaabm, etc. Agent-based modeling has direct relation with the CASs.

11.2.2.2   Cognitive agent-based computing

The authors in [4] proposed a versatile multi-agent scheme for the dynamic resources distribution in scalable manufacturing system by using the resource agents. The main work of these intelligent agents is the dynamic distribution of the tasks or called operation, and there is no interaction between the agents. Agents perform the simulations on the real-time environment. The main functionalities of these resource agents are scheduling, detection and error management.

In this research article [11], the authors proposed a new agent-based model for the dynamic resource allocation and distribution in the steel-manufacturing process. In this agent-based modeling, the agents act as various stages in development processes. Different metaheuristic techniques are implemented for the efficient scheduling on the real time-agent based modeling environment.

11.2.2.3   Complex network modeling

In [12], the researchers defined and described a multidisciplinary road map along with the summary of agent-based modeling and complex networks for the network modeling and experimental simulation of CASs. Furthermore, the author introduced CAS modeling, a new multidisciplinary Open/BioMed Central journal with objectives and scope in both of these mentioned areas. CAS modeling reflects a tremendous advantage peer-reviewed location for multidisciplinary scientific researcher in the field of agent-based modeling and the complex networks.

11.3   BitTorrent peer-to-peer protocol

P2P networks have attracted a lot of audience in the past few years due to its idea of connecting many end hosts also known as peers in an ad hoc manner. It is an efficient and scalable way of sharing file which do not need and sophisticated central software for communicating the data across the sender and receiver. P2P is distributed architecture for sharing digitized content like audio music, videos and e-books over the network which adds benefit of no single point to failure. It is performed by the usage of files transfer on per chunk-by-chunk and per host-by-host bases. In recent, it has seen that some advance application have been shifted to P2P architecture such as media streaming, online gaming as well as it is providing secure environment to business users for data sharing and collaboration. One of 2012 survey showed over 5,374 petabytes P2P traffic has been consumed over a month considered to be 20.5% of internet traffic consumed as per Cisco estimation [4]. In short, compared to traditional client/server architecture, P2P is more preferred in many aspects. There are a lot of P2P protocol currently available, and one of the most well-known P2P protocol currently under discussion is BitTorrent protocol. BitTorrent is one of most widely used protocol in P2P networks [13] (Figure 11.3).

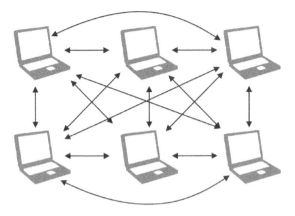

Figure 11.3   Peer-to-peer network. Adopted, with permission, from Reference [14]

BitTorrent was developed by Bram Cohen who an American computer programmer and was born on October 12, 1975, in New York City. Cohen designed BitTorrent to distribute data in a manner that the actual distributer would be capable to decrease the bandwidth use and he will be able to reach same amount of users as well.

The basic idea behind BitTorrent revolves around pieces or seeds in which data is divides or broken down into small chunks and every single user using those chunks would be available to upload to other people in a swarm to save bandwidth where swarm is a group of peers connected to single torrent. It is one of the most efficient protocols for replication and distribution of files over the internet. In traditional sources centralized server was used which allowed users to download files from that server and with the increase in download requests resulted in bandwidth usage which in turn increases traffic cost and also bandwidth exhaustion. While comparison with BitTorrent, it facilitates users to upload and download files at the same time with each other as it uses principal known as tit-for-tat where BitTorrent client act as a client and server.

11.3.1   BitTorrent history overview

Till the late 1990s, after working for several dot com companies, Cohen was inspired from the idea of his last project he worked on named MojoNation; the idea was to break confidential files into small pieces to keep them safe and secure and encrypting and storing them into different locations. This idea took Cohen to build file-sharing system by dividing files into small pieces or chunks which is now known as Bit-Torrent. The first beta version of BitTorrent was released in summer 2001, which was later presented at the conference in 2002. Cohen's goal behind BitTorrent was to provide users a simpler and quicker way of swapping and distributing Linux software online. But later on, BitTorrent potential got recognition, and in 2004, pirate copies of different files started to dominate the BitTorrent traffic which resulted in huge growth. Furthermore, BitTorrents protocol is open source and completely free [1].

11.3.2   Content publishing in BitTorrent

In order to make content available for users in BitTorrent, the publisher makes a .torrent file linked to its content. After making the .torrent file, the publisher of a content uploads it to the portal of a BitTorrent. There are few portals of BitTorrent, and The Pirate Bay is one of them which is indexing millions of torrents and getting millions of day-to-day visits. These portals offer comprehensive information concerning each indexed torrent, and this information differs slightly from portal to portal, but generally it comprises content category, number of associated files, whole content size in the torrent, complete file name, upload date, username of torrent uploader, number of leechers and seeders participating in the swarm.

11.3.3   Joining swarm and peers discovery in BitTorrent

To download a particular content, a BitTorrent user looks for .torrent file associated with that particular torrent and downloads it. To open the downloaded torrent, Bit-Torrent client is used. Upon opening .torrent file, BitTorrent client is connects to the Tracker. When a new peer initially contacts to the Tracker, it uses an announce started request which is replied by the Tracker with the total number of leechers and seeders contributing in the swarm including the IP addresses of peers selected randomly. These peers make the initial neighborhood of the new node. Furthermore, if a peer's neighborhood size is below a particular threshold which is 20 typically, then an announce started request is sends again to the Tracker to get new neighbors. Finally, when a peer wants to leave the swarm, it generates an announce stopped request to the Tracker which removes that particular peer from the participants list in the swarm.

11.3.4   Delivery procedure BitTorrent

In BitTorrent, communication between two peers is carried out by using the peer wire protocol. Initial handshake is used to start every communication. Just after the initial handshaking sequence is finished, BIT FIELD message is used by the peers to exchange the bit fields. These bit field indicates chunks of the file that a peer has downloaded already. Furthermore, each time when a peer obtains a new chunk, it uses a HAVE message to notify its neighbors. Hence, each peer is aware of the chunks that their neighbor has at that particular moment. BitTorrent also uses the Tit-for-Tat strategy for the delivery mechanism in which basically every leecher uploads chunks to those leechers from whom they are downloading more chunks. The choking algorithm is the one for providing this behavior. It is a repeated operation in which a single leecher unchoke M other leechers from its neighborhood for whom it uploads chunks to every 10 s. These M unchoked leechers are those from whom more chunks are downloaded by the peer during the last 20 s. The remaining of the neighbors are choked. In the case of a seeder, it unchokes M leechers to whom it has uploaded more chunks in the last 20 s. Furthermore, to the regular unchoke operation, BitTorrent puts into practice the optimistic unchoke operation. In which leechers and seeders randomly select one choked neighbor every 30 s to upload chunks to. Finally, when a neighbor is a leecher, it tries the Rarest First Policy in order to select which chunk to demand to this neighbor. As leechers have complete knowledge about every chunks availability in its neighborhood, it always demands the oddest one.

11.3.5   *BitTorrent architecture and working*

To understand working and operation of BitTorrent, we must first look into the terminology that bit BitTorrent uses in order to operate; these terminologies are given in the following sections.

11.3.5.1   Peer

A peer is an active client of BitTorrent to whom other peers can connect with for the purpose of sharing content or it can be stated as peer is an instance of a client than is running on any computer.

11.3.5.2   Swarm

Swarm represents the whole networks of users or peers which are connected to a same torrent.

11.3.5.3   Tracker

Tracker is a server whose responsibility is keeping track of all the peers or communication between the peers which are in the swarm by using BitTorrent protocol.

11.3.5.4   Leecher

Leecher is referred to that peer which has downloaded some complete content and which never uploads content or data.

11.3.5.5   Seeder

Seeders refer to those who have done downloading content and leaves connection open for uploading parts of content to leechers.

11.3.5.6   Mechanism and architecture

To share content or group of files, initially a peer creates a file which is known as .torrent file. This .torrent file is a metadata file which describes the properties of content that is going to be shared; these properties are URL (uniform resource locator) of the tracker length of chunks or pieces, name, etc., After all this BitTorrent file is then divided into small pieces of 512 kb by default, for which SHA-1 checksum in .torrent file is specified, where SHA-1 checksum is used to check error detection.

Initial steps in posting a file on BitTorrent are as follows: we first create a .torrent file which is a metadata file containing information of content which is going to be shared. The information includes URL of the tracker, length of chunks or pieces, name, file size and Hashing information, etc. This torrent is needed to download the file, and the torrent file can be distributed or shared by e-mail, http, etc.

Second, other most important thing you need is BitTorrent client in order to download or seed a file. This BitTorrent client application is used to administrate the download process. There are multiple BitTorrent client applications available supporting standard BitTorrent protocol but may differ in few certain features [12]. Now to start downloading, .torrent file is downloaded and opened in BitTorrent client.

When the user requests for the web page, the computer system sends the request to the download server where the web page is present. The computers through which the web page is obtained are the central servers. This scenario describes how much of the traffic is on the web [12]. To prevent these from these traffic issues, P2P protocols are introduced. BitTorrent is classified as a P2P protocol; the computers involved in the process are called BitTorrent swarm (a set of computers which upload and download the same torrent). These computers send and receive the data with the central server involvement.

Mostly, .torrent file is uploaded in the BitTorrent client by joining the BitTorent Swarm. There is tracker embedded in the .torrent file with the help of this BitTorent contacts. The specific server which contains the address of the connected computers is called a tracker. The Tracker has a specific IP address which is to be shared to the swarm for connection establishment.

Once the connection is established between all the systems, the BitTorrent client download the numerous small pieces of files in the torrent. Once the downloading process completes, it can share the data to the other BitTorrent clients in the swarm. In this special procedure, every system can download and upload the data on the same torrent. This process enhance the downloading speed; if there are 10,000 request for a single file to download on the same server by using P2P protocol, it does not cause the traffic load on the server but also enhances the response time of the server. It also speeds up the systems.

When the user requests for downloading from the swarm, it is called peers or leechers. After downloading the complete file from the swarm, the user remains connected to server. Contributing maximum bandwidth so the other user involved can easily download the file. This process is called seeders. The torrent which has complete set of file must join the swarm (group of computers) so that the other systems can start the downloading process. If there is no seeder in the P2P protocol process, then the downloading process will not to be started. It is to prefer the clients who utilize their more upload bandwidth other the transferring the files to others clients which is a very slow process. The users who contribute more toward the upload bandwidth have speed up the swarms download process. The nature of BitTorrent files is a flood like architecture in which many nodes have the file. The output of the systems increases if mores nodes are to be attached to swarm. This will be in result of reducing the computation cost of the system and also bandwidth/resource utilization. Some of the others P2P protocol enhances the redundancy issues, while BitTorrent also get rid of these issues. BitTorrent is used for reducing the distribution cost of BOINC client server system [8].

11.3.5.7   Limitations of BitTorrent

- If the file user wishes to download but have no seeds, the downloading process stops.
- The user is not aware of file which is being downloaded whether it is original or not. It totally depends on the user feedbacks.

- Both the downloading and uploading cause congestion in bandwidth, but it can be avoided if you have fast internet.
- While downloading the file from torrent, everyone can have access of each system IP Address. This will lead to a lack of security. To enhance the security, use the virtual private networks simply.
- The computer performance decreases while using the P2P BitTorrent software.

11.4   BitTorrent literature review

The authors [16] analyzed BitTorrent traffic characteristics of IPv4 and IPv6 using real network traces, and to analyze the difference between IPv4 and IPv6 traffic, authors used obtained traces of 150 mbps link among Japan and USA. Results of aforementioned autocorrelations and power spectral densities indicated more similarities than differences in sense of inter-arrival time of packet and size for IPv4 and IPv6 BitTorrent traffic. However, the authors only considered single P2P protocol for aforementioned work.

In [17], the authors suggested a technique that increases the obtainability of fragments in a BitTorrent P2P system by assuming a neural network and a mathematical model. Furthermore, for remembering the features of how users will behave, wavelet transforming of the time series of peer availability and then by feeding results as input to a nonlinear autoregressive neural network are used. However, the authors proposed model computational cost due to modeling, and prediction is very high to the tracker.

The Piece selection algorithm known as IIPS is presented in [18] to pick uncommon parts which, if transferred, would tolerate data transmission and increase piece mixture among cooperating peers unlike previous last problem which reduces the performance of BitTorrent. Results of proposed algorithm showed that in terms of system robustness, IIPS overtakes the BitTorrent's LRF algorithm furthermore successfully results in 28%–60% lesser existences than LRFs last piece problem. However, the authors neither did nor implemented IIPS practically to analyze performance metrics in an open-source BitTorrent.

In [19], the authors offered an innovative investigation methodology that takes into account specific, sole features of BitTorrent. Furthermore, the authors suggested an investigation method for illegal file-sharing constructed on characteristics of the file distribution process by means of BitTorrent. By following this procedure, an investigator can more successfully conduct an inquiry about illegal file distribution. However, upgrading of the law nearby P2P investigations in respects to the application of the procedure is desired.

The authors [20] stated that the dimension of free riders in the BitTorrent setting is growing and extended up to 16.8% of the total population of peers and analyzed four different types of BitTorrent clients which are low bandwidth strategic and nonstrategic clients and high bandwidth strategic and nonstrategic clients to communicate with other peers in a network. Results show that the Tit-for-Tat accomplishes well in distribution cooperation between peers in BitTorrent. However, optimistic unchoke signifies the single serious limitation of the Tit-for-Tat process, and thus BitTorrent.

In [14], the authors implemented broad trace analysis and modeling to understand the behaviors of such systems and found that the current BitTorrent system delivers poor service availability, unfair services to peers and fluctuating downloading performance. Hence, the authors proposed a new architecture design where the different torrent tracker sites are organized into an overlap to assist inter-torrent association.

BitTorrent plays an essential part in the Internet but has required an up-to-date knowledge of its Ecosystem. The authors [13]provided a broad picture of English-language BitTorrent community Ecosystem and identified over 4.6 million exclusive torrents and 38,996 trackers of maximum widespread torrent-discovery sites in period of 9 months. Furthermore, the authors developed multitracker crawler and found that BitTorrent ecosystem is most successful open application by many measures. However, the fame of BitTorrent content is delicate to its age.

Complex network model is proposed in [21], for BitTorrent-like networks and validated analytical computers using simulations. Furthermore, the authors have shown that their proposed model is consistent by using BitTorrent simulator with BitTorrent protocol. However, for heterogeneous configurations, authors have ignored to generalize the model and extensions like peer exchange.

In [22], the authors have developed scalable methodologies by gathering wide measurements of BitTorrent demand demographics, laying their resultant traffic matrix and showing that a huge fraction of small ISPs do not have plenty means to limit traffic after studying real ISPs. However, area yields win circumstances for average and huge size ISPs which is confined by unlocalizable torrents which have insufficient native neighbors.

A new phenomena-based friendship and trust model for the scheme of P2P file-sharing networks design is presented by the authors [23] and executed the TRI-BLERP2P file-sharing system. Furthermore, the authors described that how to make social overlay and semantics on top of the BitTorrent protocol. The authors have shown that how several TRIBLER mechanisms can produce good performance with respect to current results and addressed major challenges in P2P research. However, the authors ignored to extend reputation system with TRIBLER, application-level multicasting and tag-based navigation. However, proposed system is unable to predict users file concern and public relationship and practical file replication and recommendation.

The authors [24] confirmed by analyzing BitTorrent trace that by clustering nodes with common interest and physically close can polish file searching proficiency in P2P system and proposed SOCNET that integrates multiple components and proved that proposed system SOCNET performed better than other systems in trustworthiness, dynamism resilience, system overhead and file searching efficiency.

In [25], the authors considered BitTorrent network and conducted an agency service in which agents are used for priority evaluation for users requested tasks by prioritizing each task in descending order. Furthermore, the authors have designed new scheme to avoid duplicate files transferring and for activation of free riders. Results showed that proposed idea performed better for more downloads compared to original under limited time and bandwidth.

The authors [15] addressed distribution problem and defined collaborative file distribution system including transmission and possession matrix while gathering theoretical bound required for minimum time distribution. Furthermore, the authors developed several types of algorithm which decides that to whom and which file pieces are to be sent with in a scheduling problem. Results of proposed algorithm which is a weighted maximum flow algorithm show better results than other algorithms by returning optimal solution in many cases.

The authors [26] proposed methodology to motivate peers for contribution of resources in the network in which every peer in an access link shares his upload and download streams. Furthermore, presented allocation scheme is implemented on each peer in a distributed manner. Results of proposed scheme showed improvement in peers performance in heterogeneous system, and with comparison with BitTorrent it exhibits significant advantage in few terms like dynamic capacity allocation and seeds motivation to for contributing and remaining in the system.

In [27], the authors have discussed some motivations of BitTorrent protocol and focused on its two main components which are piece revelation strategy and unchoking algorithm. Furthermore, the authors have used game theoretic approach and show that BitTorrent does not use tit-for-tat and proposed another model known as auction-based mode as well new bootstrap mechanism for ping. Results of proposed model attains objectivity and robustness deprived of any wireline alterations to the BitTorrent protocol. However, the components considered in this work are treated orthogonally and focused on incentives in a swarm instead of between swarm.

For multimedia broadcasting, the authors [7] have proposed strategy based on quality over P2P network based on PBS and showed that those peers who have greater service level provides greater and stable quality of multimedia. Results showed that aforementioned technique is more efficient for multimedia broadcasting of P2P then TFT, which is used by BitTorrent currently. However, the authors ignored to carry simulations on other video-streaming applications. In [28], the authors proposed improved version of BitTorrent for P2P communication in cloud computing.Instead if sharing the complete file, the peers and seeders can share the segments of the data with other clients and they proposed a centralized tracker in order to increase the security. In this way, a peer do not know the identity of the client.Results show that their proposed model show significant results in cloud computing. However, the authors did not compare the results with other P2P protocols to show its significance.

In P2P protocols, BitTorrent is mostly used for file sharing. For congestion control, TCP was introduced in BitTorent which was later on replace by uTP which control the congestion on application level. In [29], the authors studied the completion time of torrents using both TCP and new uTP protocol in order to compare their performances. The results show that uTP performed better in terms of torrent time completion. However, the authors did not consider multiple peers on a single machine.

In [30], the authors performed analysis and measurement study on very famous P2P protocol BitTorrent. The researchers made all the data publicly available so that people can verify it truthfulness. Due to BitTorrent work in global components, it ensures both reliability of content and metadata. They also found at that decentralization makes the metadata more exposed. The authors ignored decentralization issue in this study.

Researchers in [31] performed simulations to study bit torrent protocol. They studied the protocol and evaluated its performance under different workloads. In their study, they considered several performance measuring metrics like file download time, peer-link consumption and distribution of trackers among various peers. The results show that BitTorrent protocol performs optimally in terms of peer-link consumption and file download time except some extreme conditions.

P2P networks are mostly the top-listed networks for sharing the information on the internet. It is a very difficult task to have a proper check and balance for the data shared through P2P networks, whether the information is legitimate and illegitimate. The existing P2P networks follow the simple handshake protocol in which authentication services are not included. The authors [32] proposed an XTRA-P2P framework through which covert channel communication is avoided. XTRA-P2P is robust for the security attacks on P2P networks. Efficient outcomes against eavesdropping attacks and brute force attacks of proposed technique are observed. In this paper, the authors focus on three parameters: handshake message, bit field message and piece message; however, the author did not consider other parameters of P2P protocol.

The BitTorrent protocol is designed for the content with time insensitive. The author [33] enhanced the current approach for the video streaming as it is time sensitive.The piece have higher download priority. In the basic BitTorrent protocol, the pieces are downloaded randomly; however, priority-based approach is introduced by the authors to enhance the system for streaming.However, the time complexity of the proposed system is much higher than the general P2P network system. Effectiveness and the streaming model robustness in real network is still a challenge.

BitTorrent is a scalable P2P distributed system, without overwhelming the capacity of the server, large files can be shared. Resource utilization is one of the challenges in BitTorrent systems. The authors [34] introduced the discrete event simulator for BitTorrent systems that equalized the overall performance of newly joined nodes that have less or more blocks than average nodes. Delay issued for pre-seeded nodes are also mentioned in paper. By combining the bandwidth matching tracker and pairwise block level, the unfairness among nodes in BitTorrent system decreases.

BitTorrent is the most powerful and complex P2P protocol. To evaluate P2P protocol simulation and experimental evaluation is the basic methodologies. In this paper, the author [35] introduce a Torrent Lab, a specific test bed in which live experiments and BitTorrent simulation are performed.It allows the new agents to involve in the simulation; however, in this simulation, the Torrent lab is machine dependent. Hardware Resources and Machine Processing Capacity are the limitation of this work done.

In this paper, the authors [36] proposed a multi-agent based modeling of BitTorrent P2P protocol. The authors proposed a system in JADE platform where each peer act as agent and each agent can exchange the information with other agent as BitTorrent Platform works. The proposed models allow more than 1,000 systems to involve in simulation at the same time. The results were analyzed on multiple parameters: number of peers, abort rate, exit rate, etc. Limitation of this proposed work is that JADE agent run autonomously while real and precise agent-based models run on any real BitTorrent network.

The authors [37] proposed a simple fluid model (P2P model) in which different parameters such as scalability, efficiency and the performance is evaluated.In the proposed work, the model have high efficiency with respect to capacity utilization;

however, the model proposed used global knowledge as a selection of peers, while in real P2P network system, the peers have limited view of other peers.

In this article, the author [38] presents a discrete event simulator using a BitTorrent simulations. The main objective of the proposed work is to achieve multiple peer up to 5,000. The proposed work focused on the upload capability of each node and data size each node serves. However, the limitation of this work is that the size of individual peer is less as compared to usual. Overall behavior of P2P protocol is affected from set size because the selection strategy is based upon set size of peer.

In this paper, the authors [39] simulate P2P protocol at the packet level. Packet-level approach is more complex than the flow-level simulations of P2P protocols. The simulation is done on NS-2 network simulator which allows using multiple algorithms for peer and piecing selection. Constant peer population and flash crowd algorithm is evaluated. However, the complete scenario on packet level approach is not evaluated even when the first seed appears until all the peer in network can get the shared content.

The authors [40] proposed a general P2P simulator for BitTorrent modeling. GPS supports the modeling of the download component. Complex networks, large size of files is complex in model-based approach. However, the proposed simulator can achieve maximum efficiency by modeling the communication at message-level approach. Geographical peer selection and bloom filter usage approach is involved in BitTorrent GPS model. However, the run-time efficiency of the model is much high than the P2P simulator.

BitTorrent P2P system that generates a large amount of ISP traffic through which the cost increases. To overcome this issue, a new approach is implemented to control cost and enhance BitTorrent traffic locality. The authors [41] implemented a technique in which biased neighbor selection is focused. In biased neighbor selection, peer choose the majority but not all neighbors as a peer. By comparing with other approaches bandwidth limiting, gateway peer and caching biased neighbor selection not required dedicated server and can be implemented on big P2P networks. In the future, the proposed technique is integrated with bandwidth limiting and caching to improve the overall performance of the system.

In the research article [42], the deep concepts behind the BitTorrent working is discussed.The main objective is to understand the behavior of the P2P protocol under heavy traffic load. In the research article, authors are concerned about the two parameters: one is the download speed and the other is the availability. The comparison between different P2P protocols is also discussed in paper.

In [43], different P2P protocols are studied which are used both in industry as well as self-used.The main of this paper is to discuss the different technical issues such as network flow control, delivery, etc. P2P technologies used in the future generation computing are also discussed. Traffic measurement is the key focus of the P2P network protocol studied in this article.

In [44], the authors inspected BitTorrent protocol for data diffusion in the environment of Computational Desktop Grid. The authors designed a prototype and finds out that even if Desktop Grid architecture depends on centralized coordination, even then they can simply incorporate this P2P technology deprived of fundamental variations on their model of deployment. Furthermore, the authors have shown experimental performance and evaluation on a LAN cluster that BitTorrent performed well for large data files transfers and scalable with increasing number of nodes but suffers from overhead while transmitting small data files.

Working of BitTorrent is studied by the authors [45] and used several mechanisms to attain optimal performance of the protocol. The authors investigated the influence on download rate by number of peers by using libtorrent client and NS-2 simulator. The authors found that overall Download Time decreases with increase in number of peers because of higher availability of files through multiple peers. However, Download Time increases due to increase in traffic.

Three modification are presented by the authors [46] to improve fairness of Bit-Torrent protocol. According to the authors presented models, all models provide some level of improvements. Furthermore, the authors ranked all presented modification according to improvement to fairness, and this ranking also demonstrates that how each proposal modifies the BitTorrent.

To show the topology of P2P networks, the authors [47] have presented plane graph model by using the BitTorrent protocol. For evaluation, different parameters are calculated such as clustering, betweenness, shortest path, coefficient, etc., for revealing the topological features of BitTorrent. The authors showed that high clustering value is obtained with larger swarm, which permits peers to have additional adjacent neighbors. Furthermore, the authors achieved positive correlation between node strength and betweenness. However, the authors did not consider analysis of the dynamical progress of BitTorrent networks.

The authors [48] have presented BitTorrent-based data sharing for mobile devices. The authors presented work analyzes different solution based on cloud which uses remote server for downloading via BitTorrent and moving it to mobile devices in energy efficient way. Furthermore, the authors presented model is evaluated via measurements carried out on mobile devices, which showed that energy consumption is minimized, traffic is minimized and the content on the cloud server can be accessed in multiple ways such as streaming or HTTP. However, savers in the clouds are not feasible both from business and architecture side.

11.4.1 *PeerSim*

PeerSim is the simulator of the P2P network models which provide the scalability, flexibility and the efficient environment for P2P simulations. In the real environment, the experimental simulation of P2P systems is very costly and their results are not reproductive. To avoid all of these issues, a java-based simulator is introduced called PeerSim. Testing of the specific protocol is also supported by the PeerSim. The main features of the PeerSim are described in the following sections.

11.4.1.1 Scalability

PeerSim provides the scalable real-time environment for the simulations of P2P network model. In PeerSim simulator, the network size is not fixed, it depends upon the P2P network protocol for the network simulations. The large simulation can also performed in very cost-cost effective manner.

11.4.1.2 Modularity

In the simulation environment of the PeerSim, all of the components are freely involved in the simulations and are easy to configure. The components enrollment in the simulation is dynamic and controllable.

11.5 Model design

11.5.1 ODD approach

In the ODD model, there is basic description of entities, variables, attributes and by which these all are characterized. The basic purpose of the proposed model is an agent-based simulation of complex P2P networks. BitTorrent is a complex file-sharing protocol which is involved in the 40%–70% of the total internet traffic. To avoid the complexity of the internet traffic protocol is defined which reduces the network congestion as well as enhanced the overall performance of the whole network. The proposed model describes the real-time simulation of the network communication by using multiple agents. The proposed model described the several major characteristics of proposed model. These characteristic describe the effects of parameters.

11.5.1.1 Entities, state variables and scales

In this proposed agent-based models, the turtles are the entities. Each turtle acts as an individual seeder or leecher in the whole network. These entities describe the overall behavior of the BitTorrent-proposed model. Also, with the help of these entities, the overall runtime behavior of the whole network is observed. The turtle describe the communication behavior of the whole network. Seeds/Leeches are the attribute of the entities. A type of turtles that contains the all file part at the beginning of network simulation and present in the simulation for sharing uploading bandwidth is called seeders. These are to be shown as a green circle in the proposed environment. The number of initial seeders are controlled using the appropriate slider parameter. A type of turtles that initialize with no file parts and then getting different pieces from other clients and lechees are called lechers. Initially, in the environment, the color of these turtles types are red and it turns to blue after getting at least one file piece. When they get all the complete pieces of the file, then they turn into green and change their type to seeders. The total amount of these turtles is controlled by the slider parameter. Time represents the scale of the proposed model, its values carrying from 1 to 10.

Parameter
Selfishness is the parameter which controls the performance of the whole agent-based model. The probability of the death of the green turtle is called selfishness. This is the specific parameter of the BitTorrent that some users are selfish, they do not care that other users are still downloading the file parts. The selfishness of the BitTorrent P2P agent-based model is controlled by the slider from 0% to 100%.

11.5.1.2 Process overview and scheduling

At the beginning of the simulation, the total numbers of the seeds and leeches are appearing. The amount of these leeches is controlled by the slider parameter. The file is divided into total ten chunks, each chunk represents a segment, so there are total ten segments involved in the simulation. Each turtle have their own ten variables each of which corresponds to a possession of a particular file segments. For example, if the segment 5 variable for a given turtle is sent to be 0, it means the current turtle cannot possess this segment. If on the other hand it is to be set to one, then we conclude as the given turtle can possess the segment. Depending on this condition, the seeders begin with all ten segments variable set to one while leechers have zero. The simulation can be finished in two ways: whether all turtles turn green which means they have downloaded the complete file entirely or the second way is that seeds drop out from the simulation and remove one or more segments from the network. In this case, it would be not possible for anyone to finish the simulation properly by downloading completely.

11.5.1.3 Design concepts

There are different designs concepts, each of which are describing as follows:

Basic principles
The basic principle of the BitTorrent model is to design a P2P agent-based model which can simulate the performance with less computation complexity and increase the run time of the simulations. The comparisons with other simulation platforms like NS2 have more time complexity as well as space complexity. To overcome this entire limitation, agent-based modeling is to be adopted. Multiple numbers of clients can contribute and download the data from the network. One of the main advantages of the proposed model is that it does not contain congestion in the network model.

Emergence
The main idea behind the proposed network model is to control the environment dynamically. The turtles involved in the agent-based model and other parameters like selfishness are to be controlled using the slider parameter. The outcome of the model is unpredictable by changing the behavior space and complexity. The results are mostly dependent on the rules defined inside model than on the individual or environment.

Adaptation
In the proposed network environment, there are some specific rules that are defined. There must be at least a seeder in the simulation. If the probability for the simulation drop out is more, then there will be more chances of simulation failure. To avoid the failure outcome, these rules are to be adopted by the network model. When lecher gets all the segment of the file then they turn its behavior to the seeder for supporting the environment. If there will be more seeders involved in the simulation, then the run time of the simulation is minimized. In the other way, we called seeder value as a fitness factor of the network.

Objectives
There is a clear negative correlation between the number of initial seeds and the total time it takes for all turtles to completely download the file. Furthermore, torrents with few initial seeds also exhibit greater volatility in torrent activity (roughly analogous to

average download speed). Both these behaviors are consistent with real-world observations. This simulation also beautifully reproduces typical torrent speed behavior
whereby downloads begin slowly (with few seeders) then rapidly speed up as more
file segments get distributed. This is closely related to the observed pattern whereby
some segments are distributed much more widely than others initially (when there
are only 1 or 2 seeds), a behavior that is also reproduced by this simulation.

Learning

In the proposed model, the agents cannot change their behavior with respect to time or
previous history. The behaviors depend upon the downloading segments; if there will
at least one segment, then the color of the turtle changes to blue. While downloading
all the complete segments its color changes to green. These are the main learning
concepts of the proposed agent-based model.

Interaction

There is the direct interaction between the seeder and leeches. According to the
main concept of the P2P network, there is one-to-one connection between each node.
Depending on these interactions, the segment values are changed and seeders upload
the segment to other leeches. The interaction communication between each nodes are
represented as line in the graphical user interface.

Initialization

At the initial stage of the model, there are two entities initialized: seeders and leeches.
The values of the seeders and leeches can be changed by using the slider parameter.
It is not necessary at each simulation the values of seeders and leeches are same.
Its change depends upon the required network specifications. The initial conditions
have much effect on the model and simulation. If the value increases, the run time
complexity also increases.

Input data

The proposed model does not contain the input data to represent process like time-
varying, etc. Although the input is not in terms of initial value of state variable or
parameter values, in the dynamic system model, the input is to be obtained by the user
by the state variables. Depending upon these state variable values, the simulation runs
but the internal variables are not affected by these input data. Success/failure rate of
the network model depends upon the state-variable values. The simulation is said to
have succeeded when all the agents acquire all file segments (i.e., all turtles end up
green). The simulation fails when users are unable to obtain some file segments. This
will occur when seeds drop out (due to selfishness) before certain file segments are
in distribution amongst leeches.

11.5.2   Overview of the proposed model

This model simulates the dynamics of a P2P BitTorrent network. BitTorrent is a
popular protocol for distributing large files amongst multiple users over the internet,
and studies have shown it to account for approximately 70% of total IP traffic. It
works by breaking a file into many segments (usually 100 or more depending on

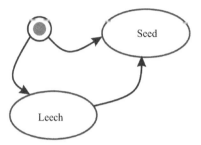

Figure 11.4   States of node agents

the size of the file) and allows individual users who are downloading the file to simultaneously upload completed segments to other users. This means that users are downloading from each other as well as from the server or the initial distributors. Compared to traditional web hosting where everyone downloads directly from the server, BitTorrent significantly reduces bandwidth usage and hardware requirements for the initial distributor, thereby resulting in substantial cost savings. In some cases (as this model will demonstrate), it is possible for the initial distributors (seeds) to drop out completely without undermining the network, provided enough file segments are in distribution.

11.5.2.1   Problem statement

This simulation aims to capture several major characteristics of BitTorrent behavior outlined below and should enable the measurement of the effects of certain parameter changes in terms of their impact on the time it takes for users to download the file, and on the probability that the system will fail because there are not enough file segments in distribution. In particular, this simulation seeks to understand the effect of the number of initial seeds, the type of seed algorithm and the selfishness of users on the time it takes for all users to complete downloading the file.

11.5.2.2   Node agents

In this study, for ABM we require the placement of network node agents. The agents are randomly deployed. The network nodes can communicate with each other. In ABM for BitTorent, these node agents communicate with each other over the network for the purpose of file sharing. For communication, either all nodes should be present in a single-hop network or the nodes should be accessible in the case of multi-hop network. The goal of these nodes is to share a file over a network in different segments.

11.5.2.3   States of node agents

Node agents of BitTorent have several states as shown in node agents Figure 11.4. First, there are nodes agent in the network that begin with all parts of the file. We call these agents seeds which exist for the sole purpose of uploading content to leeches.

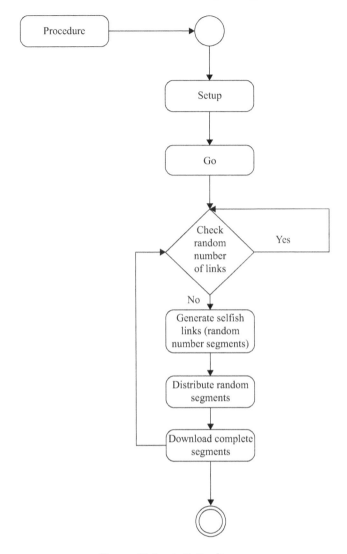

Figure 11.5   Activity diagram

Leeches are node agents that begin with no parts of the file then gradually obtain file segments from seeds and from other leeches.

11.5.2.4   Activity diagrams

In this section, we have shown the overall activities of BitTorrent Network implemented in NetLogo. In Figure 11.5, we have shown how BitTorrent Network perform its operation by doing different activities.

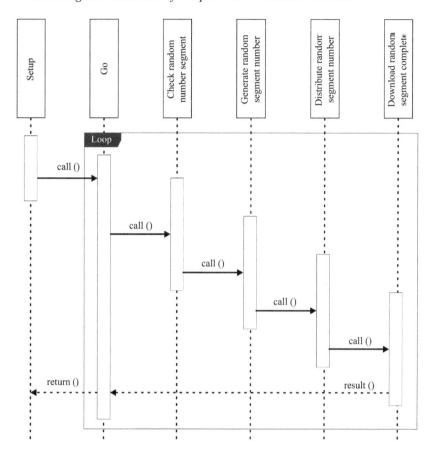

Figure 11.6   Sequence diagram

11.5.2.5   Sequence diagrams

In this section, we have shown the sequence in which BitTorrent is implemented in NetLogo. The sequence diagram of BitTorrent Network implemented in NetLogo is shown in Figure 11.6.

11.5.3   DREAM model

DREAM can be initiated by developing a paper model of network designed by extending network model. We draw the network model using network modeling tool called (Gephi). The result can be seen in Figure 11.7. In the initial view of the network, it is difficult to comprehend. For this, we need to manipulate the network. We manipulate the network using various algorithms, the resulting network can be viewed in Figure 11.8.

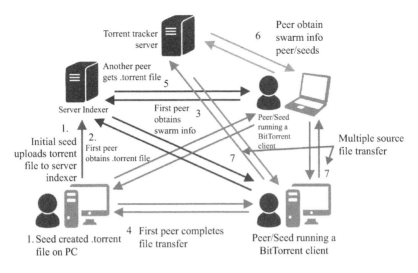

Figure 11.7   BitTorrent architecture. Adopted, with permission, from Reference [15]

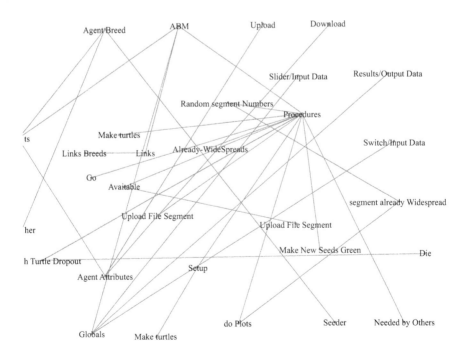

Figure 11.8   Network model of ABM

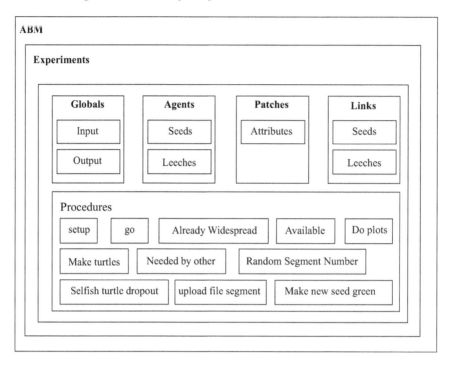

*Figure 11.9   An overview of ABM constructs in relation with the specification
model constructs*

This network model will be described in detail and a CNA will be performed
next. The result will be discussed in the results section of the chapter (Figure 11.9).

11.5.4   Pseudocode-based specification

In this section, we describe detailed translated specification model expanded from
this network sub-modeled as a means of describing the model in detail. Initially, we
will describe the agents and breeds.

11.5.4.1   Agents and breeds

The model has two breeds and their definitions are given next in the form of a
specification model as below:

Breed **Seeds:** This breed represent the seeds in the network.

Internal Variables: Random segment number, segments
random segment number: This number is randomly generated at each
iteration and determines which file segment a seed will attempt to share
segments: Data chunks of a complete file and depends on the size of the file

---

Now, the "Seeds" breed here is first described in the specification model. Afterwards range we note the internal variables that will be used in the simulation. There are two specific internal variables here. One is the "Random_segment_number" variable. This variable is used to generate a randomly generated number at each iteration and determines which file segment a seed will attempt to share.

The other variable is "Segments" which represents data chunks of a complete file and depends on the size of the file. The seeds will have all the segments of a file.

Breed **Leeche:** This breed represent the leeches in the network.

Internal Variables: Random segment number, segments
random segment number: This number is randomly generated at each iteration and determines which file segment a leech will try to get or share
segments: Data chunks of a complete file and depends on the size of the file

---

Unlike seeds, the leeches are the agents which try to get segments from the seeds or leeches. After developing a specification model of the agent breeds, we next develop a model for various globals variable. These variables play an important role for the configuration of the simulation model.

11.5.5   Globals

In this section, we describe the global variables for the simulation model. Now, in this model, the key input global variables here are four in number. These are described next in the specification model given below.

Input Globals: $<p, s, l, smart_{s}eeding>$

Sliders:
p: The probability that, each time interval, seeds will drop out of the system.
s: The number of seeds that the simulation begins with.
l: The number of leeches the simulation begins with.
Switch:
Smart Seeding: Whether seeds operate intelligently, or randomly.

In this specification, we can clearly see that three of the variables, viz., "p," "s" and "l," represent inputs which are provided by the user via a "slider" UI element in the NetLogo simulation model. Whereas "Smart_seeding" is a switch or represents a Boolean input variable. "p" is used for the probability (%) that, each time interval, seeds will drop out of the system. "s" is the number of seeds that the simulation begins with. "l" is the number of leeches the simulation begins with and the "Smart_seeding" is whether seeds operate intelligently, or randomly. The slider variable "s" and "l" are used to initialize the number of seeds and leeches for the particular simulations. Greater the value of "s," the network will converge faster, and greater the number of "l," greater will be the execution time of a simulation.

These globals are randomly placed when initial simulation screen is setup having "s," "l" and "p" as a sliders used for gradual change in an input configuration required in different experiments. It also assists in minimizing the effects of our configuration. So by changing the number of agents and repeating the number of simulations several times will not skew the eventual results.

The switch "smart_seeding " is used for seeding intelligently or randomly. When this is switched off, seeds randomly select a segment to upload. When smart seeding is switched on, seeds will be significantly more likely to upload a segment that is not already in distribution amongst leeches.

11.5.6   Procedures

The procedure makes seeds "smart" in which they are highly unlikely to seed parts that are already in distribution amongst leeches. If more than one leech has the segment corresponding the random number that was initially generated by a seed, random numbers will continue to be generated up to three times until a segment is found that is not yet distributed. From personal experience, real-world torrents exhibit both random and "smart" behavior depending on tracker parameters.

Procedure **Check-if-segment-is-already-widespread:**

Input: Uses global parameters from the User Interface
Output: All agents and patches are setup for simulation
Execution: Called at the start
Context: Observer
Begin
    1. ask seeds with [random-segment-number].
    2. check-if-segment-is-available
End

11.5.6.1   Check-if-segment-is-available

This checks to see if the random segment number that has been selected corresponds to a file segment that the turtle does not possess. If this is the case, the turtle is instructed to keep generating numbers (up to 100 times) until it finds one that corresponds to a file segment that it DOES possess. In other words, it ensures leeches are uploading segments that they actually have.

Procedure **Check-if-segment-is-available:**

Input: Uses global parameters from the User Interface
Output: All agents and patches are setup for simulation
Execution: Called at the start
Context: Observer
Begin
    1. ask leeches with [random-segment-number].
    2. check-if-segment-is-needed-by-others
End

11.5.6.2 Check-if-segment-is-needed-by-others

This procedure checks to see if the random segment number that has been selected corresponds to a file segment that every other turtle already possesses. If this is the case, the turtle is instructed to keep generating numbers (up to 100 times) until it finds one that corresponds to a file segment that another turtle needs. For example, if everyone else already has segment 2, a turtle would not upload segment 2.

Procedure **Check-if-segment-is-needed-by-others:**

Input: Uses global parameters from the User Interface
Output: All agents and patches are setup for simulation
Execution: Called at the start
Context: Observer
Begin
    1. ask turtle with [random-segment-number].
    2. repeat 1 hundred times
    3. upload file segment
End

11.5.6.3 Do-plots

Procedure **Do-plots:**

Input: Uses global parameters from the User Interface
Output: All agents and patches are setup for simulation
Execution: Called at the start
Context: Observer
Begin
    1. set-current-plot "Links"
    2. set-current-plot-pen "number-of-links"
    3. plot count links
    4. set-current-plot "Segments"
    5. set-current-plot-pen "segment"
    6. export all plot
    7. export all interface
End

This procedure creates the histogram showing the number of links during each "tick" and a plot showing the number of segments completed with time. This is an accurate representation of total torrent activity; however, it may not directly correspond to an individual turtle's activity.

11.5.6.4 Generate-random-segment-number

"ask leeches with [random-segment-number]" is a "workaround" to overcome a problem whereby leeches who generated a segment number (say 8) who did not possess that segment would immediately upload it (during that tick) if they happened to receive it from another turtle. Their random segment number is set to 88888, so they would not try to upload anything until the next tick (i.e., after they have generated another random number).

Procedure **Generate-random-segment-number:**

Input: Uses global parameters from the User Interface
Output: All agents and patches are setup for simulation
Execution: Called at the start
Context: Observer
Begin
    1. ask turtle with color
    2. set random segment number
    3. check-if-segment-is-already-widespread
End

11.5.6.5 Go

Procedure **Go:**

Input: Uses global parameters from the User Interface
Output: All agents and patches are setup for simulation
Execution: Called at the start
Context: Observer
Begin
    1. if success = 1 [stop]
    2. generate-random-segment-number
    3. makes-new-seeds-greenwidespread
End

In this procedure, the simulation starts and P2P communication begins. If the complete file is downloaded by at least one, it generates a random segment number and the seed turn to green.

11.5.6.6 Make-turtles

Procedure **Make-turtles:**

Input: Uses global parameters from the User Interface
Output: All agents and patches are setup for simulation
Execution: Called at the start
Context: Observer
Begin
    1. create-leeches initial-leeches [set color red]
    2. create-seeds initial-seeds
    3. repeat max-pxcor − 3 [layout-spring turtles links 0.2 5 1]
End

In this procedure, initially all leeches are to be defined and the color of these leeches at beginning is set as a red. Then at the Step 2, the seeds are also settled. The leeches and seeds are settled in a circle.

11.5.6.7 Makes-new-seeds-green

Procedure **Makes-new-seeds-green:**

Input: Uses global parameters from the User Interface
Output: All agents and patches are setup for simulation
Execution: Called at the start
Context: Observer
Begin
    1. Ask turtles with [segment − 0 = 1]
    2. Set color green
    3. End
End

This makes turtles that have finished downloading all ten segments turn green. It does not technically change their breed, but this does not really matter in terms of the performance and accuracy of this simulation.

11.5.6.8 Selfish-green-turtles-dropout

Procedure **Selfish-green-turtles-dropout:**

Input: Uses global parameters from the User Interface
Output: All agents and patches are setup for simulation
Execution: Called at the start
Context: Observer
Begin
    1. ask turtles with [color = green]
    2. if random 99 < green-turtle-selfishness
    3. [die]
End

In this procedure, if the seeds those download and participate in other leeches downloading process, these are called selfish turtles. If the selfishness is more than the leeches, then the simulations stops.

11.5.6.9 Setup

For this model to work with NetLogo's new plotting features, clear-all-and-reset-ticks should be replaced with clear-all at the beginning of your setup procedure and reset-ticks at the end of the procedure.

Procedure **Setup:**

Input: Uses global parameters from the User Interface
Output: All agents and patches are setup for simulation
Execution: Called at the start
Context: Observer

Begin
    1. clear-all-and-reset-ticks
    2. set-default-shape turtles "circle"
    3. make-turtles
End

11.5.6.10 Upload-file-segment

Assuming turtles actually have the segment they are trying to upload (highly likely given the "check-if-segment-is-available" above), they will seek out a leech without that segment and share it with that turtle. The visible link is partly for esthetic purposes and simply to show where the upload activity is during each "tick." This process is repeated for every possible random number (0–9) and therefore every possible file segment.

Procedure **Upload-file-segment:**

Input: Uses global parameters from the User Interface
Output: All agents and patches are setup for simulation
Execution: Called at the start
Context: Observer
Begin
    1. ask turtle with random segment number
    2. ask one of leeches with segment 0
    3. set random random segment number 99999
    4. set color Blue
    5. create-link from myself
End

All the procedures and their communication links are shown in Tables 11.1 and 11.2.

11.5.7 Experiments

As expected, there is a clear negative correlation between the number of initial seeds and the total time it takes for all turtles to completely download the file. Furthermore, torrents with few initial seeds also exhibit greater volatility in torrent activity (roughly analogous to average download speed). Both these behaviors are consistent with real-world observations. This simulation also beautifully reproduces typical torrent speed behavior whereby downloads begin slowly (with few seeders) then rapidly speed up as more file segments get distributed. This is closely related to the observed pattern whereby some segments are distributed much more widely than others initially (when there are only 1 or 2 seeds), a behavior that is also reproduced by this simulation. As selfishness increases, the probability of failure also increases, although the random nature of this agent behavior creates extreme variance in this relationship (i.e., total

Table 11.1   ABM network linkage table

Node from	Node to
ABM	Globals
ABM	Procedure
ABM	Links
ABM	Agents
Global	Slider
Global	Switch
Global	Results
Procedure	Already Widespread
Procedure	Available
Procedure	Needed by other
Procedure	Do plots
Procedure	Random Segment Number
Procedure	Go
Procedure	Make turtles
Procedure	Make new seed green
Procedure	Selfish turtle dropout
Procedure	setup
Procedure	upload file segment
Links	Link Breeds
Agents	Agents Breed
Agents	Agents Attribute
Attribute	Uploading
Attribute	Downloading

failure is an extreme occurrence). This variance highlights the important conclusion that torrents will likely fail when people behave selfishly (which might happen when they have to pay for their upload quota). Selfishness also increases the volatility of total torrent activity and hence reproduces this aspect of real-world torrent behavior.

Experiment vary-nodes: Experiment with the effects of number of seeds and leeches

Input: s, p, l, smart seeding?
Setup procedures: Setup
Go procedures: Go
Context: Observer
Parameters:
  p: 0.6
  s: 30
  l: 400
  smart seeing: false
End *Stop condition:* Ticks $= 28$
Final commands: None

Table 11.2   Network statistics

Label	Degree	Eigencentrality	Eccentricity	Betweenness
ABM	4	44.8982	3	258
Globals	4	19.9087	4	81
Procedures	12	100	4	309.5
Agents	3	20.5066	4	147
Links	2	14.6366	4	28
Slider/Input Data	1	6.4495	5	0
Switch/Input Data	1	6.4495	5	0
Results/Output Data	1	6.4495	5	0
Links Breeds	1	4.5308	5	0
Already-WideSpreads	1	27.3111	5	0
Available	2	29.8494	5	28
Upload File Segment	1	8.2762	6	0
Needed by Others	1	27.3111	5	0
do Plots	2	32.3933	5	13.5
Random Segment Numbers	2	32.3933	5	13.5
segment already Widespread	2	17.8165	6	0.5
Go	1	27.3111	5	0
Make turtles	1	27.3111	5	0
Make New Seeds Green	1	27.3111	5	0
Selfish Turtle Dropout	2	29.8494	5	28
Die	1	8.2762	6	0
Setup	2	29.8494	5	28
Make turtles	1	8.2762	6	0
Upload File Segment	1	27.3111	5	0
Agent Breed	3	9.9394	5	55
Seeder	1	3.7863	6	0
Leecher	1	3.7863	6	0
Agent Attributes	3	9.9394	5	55
Upload	1	3.7863	6	0
Download	1	3.7863	6	0

Experiment vary-nodes: Experiment with the effects of number of seeds and leeches

Input: s, p, l, smart seeding?
Setup procedures: Setup
Go procedures: Go
Context: Observer
Parameters:
  p: 0.3
  s: 25
  l: 300
  smart seeing: true
End *Stop condition:* Ticks = 28
Final commands: None

At lower levels of selfishness, a greater number of initial seeds can to some extent compensate for selfish behavior. Once selfishness reaches around 75%, then the system will fail unless the number of seeds is high (relative to the number of file segments). These findings have significant implications for the design of seeding algorithms. Clearly, everything should be done to ensure that users behave in a way that is not "selfish." In other words, incentives should be provided for users to keep sharing files after they have completed downloading all segments. Many trackers and BitTorrent clients already do this through the use of minimum upload/download ratios and by withholding the last segment from users in order to keep them active for longer (described further below). Future additions to this simulation could be made to analyze the effect of such incentives on the probability of failure and the total download time. In a typical torrent, some file segments are much more widely distributed than others initially. This simulation reproduces this behavior when there are few seeds to begin with, as the first segments to be seeded become widespread much more rapidly.

In this simulation, adding leeches does not seem to significantly alter download speed (time to completion). This is contrary to real-world observations. One likely explanation for this is that in the real world, many leeches contribute nothing or very little in terms of upload capacity, whereas in this model, the upload capacity of leeches is just as great as that of the initial seeders. If leech upload capacity were to be restricted (e.g., they could only upload once every four ticks), then adding leeches would, in theory, significantly increase the total time to completion.

The effect of smart seeding highlights the importance of seeding algorithms in the overall performance of the torrent. It took as much as 30% longer to complete file distribution with random seeding compared with "smart seeding." In the real-world different trackers and BitTorrent, clients create a range of interesting behaviors. For example, some seeds will not upload to leeches with only one incomplete segment remaining to be downloaded. This effectively prevents them from dropping out (being selfish) and ensures that they will continue to contribute to the network, thereby increasing its viability. The effects of optimizing BitTorrent performance using peer selection algorithms are examined by Huang, Wang, Zhang and Liu. Their study found that a dynamic upload quota allocation scheme based on the principle of investment return successfully reduced file download time during simulations. Further improvements to the seeding algorithm, particularly of the kind that rewards users with high upload ratios, are discussed below among other suggested additions to this experiment.

11.5.8   Results and discussions

11.5.8.1   Metrics table and description

In this section, we are discussing NetLogo Based BitTorrent Protocol simulated results by implementation and comparing our results with the results of PeerSim. For our comparison with PeerSim, we have implemented BitTorrent in NetLogo as shown in figure. Our BitTorrent Interface Design consists of a graphical window, switch, monitors, graphs and a multiple sliders. On left side we have three sliders, first

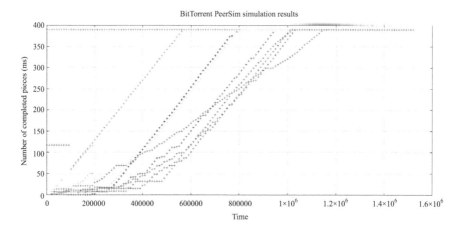

Figure 11.10   PeerSim simulation result

one represents the selfishness, second slider is used to set number of initial seeds and third slider is used to set a number of initial leechers. Furthermore on left side we also have a switch which is used for the purpose of smart seeding and a graph showing links. On the right side, we have multiple monitors showing segments and a segment graph. To appropriately notice performance of BitTorrent in NetLogo we have simulated a similar network setup and parameters as explained in PeerSim. Furthermore, the results of both NetLogo and PeerSim are described in subsequent sections.

11.5.9   PeerSim results

In Figure 11.10, performance of BitTorrent protocol with single shared torrent in a fixed size network is shown in which each colored lines within a graph shows down-loaded pieces for every node in a network. In starting phase distribution of pieces among nodes for simulations is chosen randomly: few nodes begin with greater num-ber of downloaded pieces and remaining have fewer completed pieces. As simulation begins, torrent is shared by the nodes among their neighbors downloading parts. Node will only start to share when the simulations start: some nodes will be waiting for their selection (Pink nodes) probably because these nodes are chocked due to start with more or less 120 segments at time 0.

11.5.10   ABM results

In this section, we have discussed the results of ABM; for this, we have performed a simulation. The parameters of the simulations are shown in Table 11.3.

Initially, we setup the simulation after which leeches and seeds appear whose quantity along with other parameters can be increased or decreased by using sliders

Table 11.3   Simulation parameters

Parameters	Values
Initial leeches	400
Green turtles selfishness	0
Initial seeders	50
Smart seeding	On

parameters. The file which is to be distributed has 10 segments and each agent owns 10 variables and each variable corresponds to particular segment in which 1 indicates possession of particular segment and 0 indicates that turtle does not possess that particular segment. In this model, all leeches are set to 0. In this experiment, we have set initial seeds to 50, leeches to 400 and set selfishness probability of dropping out of each green turtle to 0 with smart seeding set to ON. We start our experiment by calling go procedure. During each and every tick, each turtle who owns a file segment (initially seeds) produces a random number from 0 to 9 which matches to the file segment which is going to be uploaded. They will only generate numbers that are required by others. With smart seeding switched ON, seeds will generate random numbers up to three times until they find one that is possessed by 1 or fewer leeches to prevent the probability of uploading the same seeds multiple times. The upload method starts by requesting a turtle to select that leech which does not own file segment to be uploaded and then a particular turtle is asked to set variable to 1 which shows that they now own that segment. Maximum of one segment is uploading per tick and no limit for downloading. Furthermore, in real-world scenario, download capacity hugely beats upload capacity for maximum users. Once a leech owns single segment, it goes blue, and finally when a leech owns all segments, they go green which can be seen in figures that all the leeches turns green after getting all the segments. After leeches go green, there is a likelihood that these green turtles will drop out which signifies selfish manners, and the drop out probability for individual green turtle can be set by means of slider parameter. In this scenario, the simulation can finish in either two ways. First, it can end when all turtles go green. Second, if seeds drop out in a way that completely eliminates one or more segments from circulation (Figures 11.11–11.13).

11.5.11   Comparison of both

In Figures 11.10 and 11.14, the results of PeerSim and NetLogo are shown. In PeerSim-simulated BitTorrent network, simple study has been prepared to understand the performance of the protocol which can be seen in the figure. In the above graph, behavior of the BitTorrent protocol in fixed sized network of 30 nodes with 100 Mb of single shared torrent is shown. On x-axis we have we have time, which is required to complete each piece, and y-axis represents number of completed pieces in a particular time. Furthermore, colored line with in a graph shows the number of downloaded

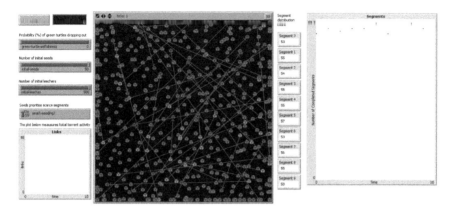

Figure 11.11   NetLogo Interface

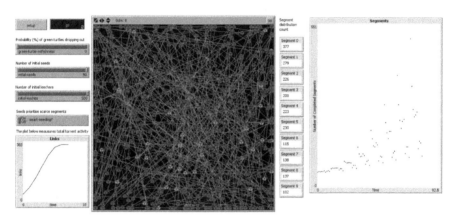

Figure 11.12   NetLogo interface

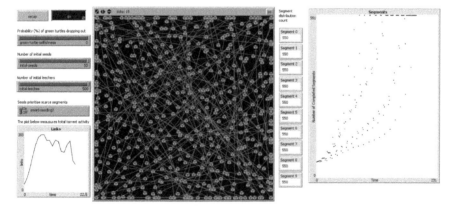

Figure 11.13   NetLogo interface

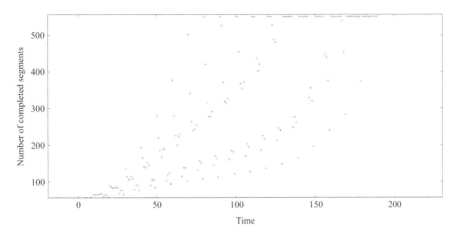

Figure 11.14   NetLogo simulation result

pieces for each node in the network. In PeerSim distribution of the pieces among the nodes is randomly chosen in a beginning. When the simulation begins, few nodes start to distribute torrent with their neighbor nodes, some but not all of the nodes, and when all nodes own file graph shows constant value.

While in our designed agent-based model we have followed the same procedure discussed above. On *x*-axis we have time in which each agent gets all of the segments and on *y*-axis we have number of completed segments which can be seen in Figure 11.4. Furthermore we are distributing segments among agents instead of pieces. When leeches own all the segments, they turn green, which states either selfish behavior or end of simulation which can be seen in the graph when all segments that are distributed among agents graph go constant. Furthermore, detailed results of both ABM and PeerSim are already discussed above in detail.

11.5.12   DREAM network models

In the initial view of the network as shown in Figure 11.7, it is difficult to comprehend. For this, we need to manipulate the network. We manipulate the network using various algorithms, the resulting network can be viewed in Figure 11.8.

11.5.12.1   Plots of centralities

The degree centrality analysis of network is plotted in Figure 11.15. The plot shows that the node containing highest peak of betweenness centrality is "Procedures." Node "Globals" have the second highest betweenness centrality, which represent the inputs, which are entered from the NetLogo user interface model. In Figure 11.16, we have plotted the betweenness centrality in R [49] using power law. The plot show a decaying behavior because there are few nodes in the network which have more connections and more nodes in the network which have few connections (Figure 11.17).

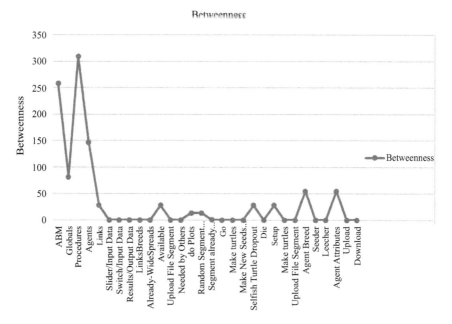

Figure 11.15   Betweenness of NetLogo simulation

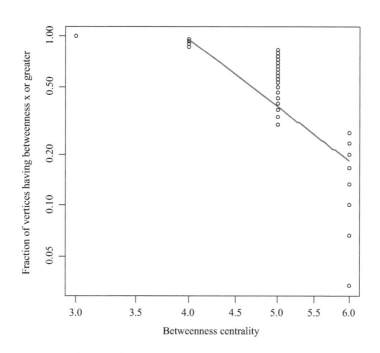

Figure 11.16   Betweenness using power law

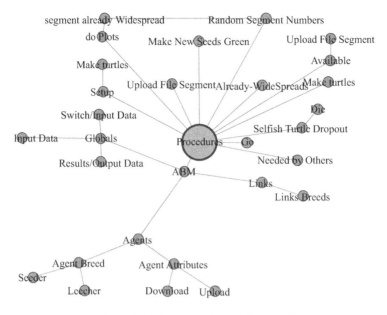

Figure 11.17   Network model of ABM

The plot in Figure 11.18 shows the eccentricity centrality. In the case of eccentricity centrality, following "Procedures" which has the highest Eccentricity centrality value, we find ABM to have the second highest value followed by "Agents," "Agents Breed" and "Agent Attribute." In Figure 11.19, we have plotted the eccentricity centrality in R [49] using power law. The plot show a decaying behavior because there are few nodes in the network which have more connections and more nodes in the network which have few connections.

The plot in Figure 11.20 shows the analysis of eigen centrality of network. The plot shows that the node containing highest peak of eigen centrality is "Procedures" followed by "ABM." However, the nodes containing the lowest eccentricity centrality is "Download" and "Uploads." In Figure 11.21, we have plotted the eigen centrality in R [49] using power law. The plot show a decaying behavior because there are few nodes in the network which have more connections and more nodes in the network which have few connections.

The degree centrality analysis of network is plotted in Figure 11.22. The plot shows that the node containing highest peak of degree centrality is "Procedures." Node "Globals" have the second highest degree centrality, which represent the inputs, which are entered from the NetLogo user interface model. In Figure 11.23, we have plotted the degree centrality in R [49] using power law. The plot shows a decaying behavior because there are few nodes in the network which have more connections and more nodes in the network which have few connections.

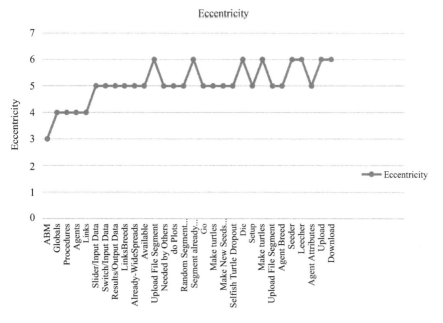

Figure 11.18   Eccentricity of NetLogo simulation

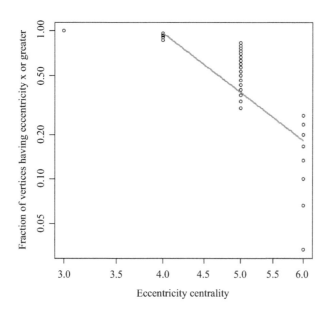

Figure 11.19   Eccentricity using power law

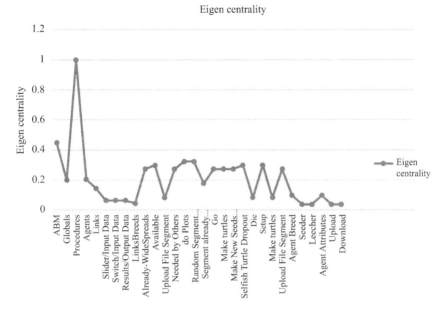

Figure 11.20   Eigencentrality of NetLogo simulation

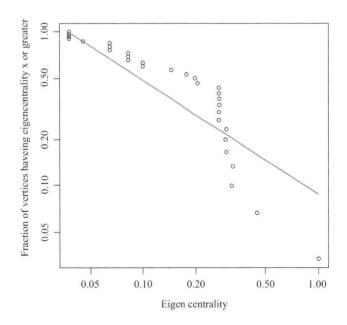

Figure 11.21   Eigencentrality using power law

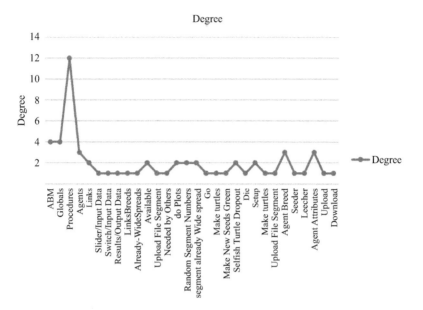

Figure 11.22   Degree of NetLogo simulation

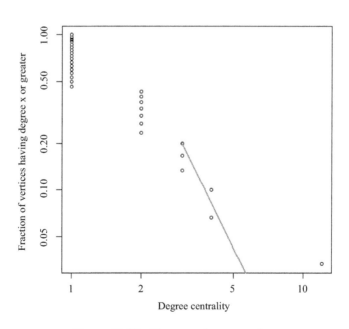

Figure 11.23   Degree using power law

11.6   Discussion (ODD vs DREAM)

Agent-based modeling is the versatile approach for the feasibility studies. The main theme of the agent-based modeling is to make the efficient communication and the comparison between the different communication model. The proposed Descripted agent-based modeling (DREAM) models allow the P2P network simulation and their modeling. DREAM model is the combination of different components such as complex network simulation model, statistical analysis of the P2P communication network and the detail description of the pseudocode-based specific schemes. DREAM models in the agent-based modeling allows the fidelity. One of the main advantages of the DREAM model is that it provides the complete analysis of complex adaptive networks. Another feature of the DREAM model is that it provides the visualized result comparison of P2P networks as well as quantitative results. DREAM model supports the heterogeneity of the agent-based modeling. According to our previous literature knowledge, there is no specific and well-defined Descripted model for BitTorrent P2P network. The proposed Descripted agent-based models for P2P network models provides both the visualized and the qualitative results and their comparison. ODD is the textual techniques of the modeling. ODD is the combination of the Overview, Design concepts and their details. The ODD model is presented by the Grimm *et al.* By the comparison of ODD and DREAM model, there are some limitations of ODD model. These are following:

- ODD model is the textual model for the BitTorrent P2P network. It does not allow the visualized network model in the BitTorrent simulations.
- ODD model does not support the quantitative comparisons, and the statistical measures are not involved in ODD model.
- Agent-based modeling is the visualization modeling of the BitTorrent network. For the comparison between different agent-based models, the visualization modeling is necessary, although ODD does not support visualization modeling.
- In the DREAM model, the pseudocode description is present for the BitTorrent model, while ODD does not allow the pseudocode description of the agent-based model.
- ODD models does not allow the technical description of the algorithms.
- DREAM model supports the activity diagrams, sequence diagrams of the network model while ODD does not include this type of modeling support.

At the end, we conclude that ODD models gives the basic ideas of the modeling, and DREAM model allows the Descripted detail of the agent-based modeling of the BitTorrent. Theory of computation is the basic study of complexity and the automata. Complex theory provides the theoretical study of the complex systems. The agent-based modelers mostly prefer to address a specific theoretical problem by using the agent-based modeling. The main relation between the agent-based modeling and computation theory is that it is mutually more beneficial for the CASs. To solve the theoretical concepts, the complexity provides the fuzzy concepts to the agent-based modeling. As the agent-based modeling enhanced the understanding, concepts of complexity varied from field to field. Theory of computation belongs to the field of the theoretical computer science while agent-based modeling is the visualization interface which provides the solutions for the complex systems or complex networks. Agent-based modeling is the paradigm is still under development process. When the new techniques emerge with ABM, a sufficient time is needed to find all the basic details of the application, their capabilities and limitations.

As the theory of computation provides the proof study of the theoretical problem, agent-based modeling handle these problems with the help of the visualization structure. The most common example for the solution of the complex network system is the agent-based model of Turing machine. A Turing machine have infinite memory and their simulations are more complex. To provide the solution for these complex systems, agent-based modeling is provided. ABM enhances the overall performance of the system and reduces the complexity load of the system.

11.7 Conclusion

Among many P2P networks, BitTorent is very famous for sharing files over a network. In this chapter, we proposed modeling and simulation of the BitTorrent protocol by using a combination of agent-based and complex network-based approaches. The simulation results demonstrate that our proposed ABM-based BitTorrent model performed better. Furthermore, for ABM specification, we followed two approaches, first is ODD and the second one is DREAM methodology. We presented qualitative as well as a quantitative comparison of both ODD and DREAM specification techniques. The comparative study of ODD and DREAM proved that DREAM methodology is the more useful approach for documenting an ABM not only in terms of modeling but also for replication of the models, specifically for P2P networks.

References

[1] Cohen B. "Incentives build robustness in BitTorrent." In Workshop on Economics of Peer-to-Peer systems, vol. 6, pp. 68–72. 2003.
[2] Johnsen JA, Karlsen LE, and Birkeland SS. "Peer-to-peer networking with BitTorrent." Department of Telematics, Norwegian University of Science and Technology (NTNU), Norway. (2005).
[3] Scanlon M and Shen H. "An analysis of BitTorrent cross-swarm peer participation and geolocational distribution." In 2014 23rd International Conference on Computer Communication and Networks (ICCCN), Shanghai, pp. 1–6. 2014.
[4] Niazi MA and Hussain A. Complex adaptive communication networks and environments: part 1. Simulation Transactions of the Society for Modeling and Simulation International 2013;89:559–561.
[5] Niazi MA. Complex adaptive systems modeling: a multidisciplinary roadmap. Complex Adaptive Systems Modeling 2013;1:1.
[6] Grimm V, Berger U, DeAngelis DL, Polhill JG, Giske J, and Railsback SF. The ODD protocol: a review and first update. Ecological Modelling 2010;221(23):2760–2768.

[7]   Niazi MA and Hussain, A. A novel agent-based simulation framework for sensing in complex adaptive environments. IEEE Sensors Journal 2011;11(2): 404–412.

[8]   Batool K, Niazi MA, Sadik S, and Shakil ARR. "Towards modeling complex wireless sensor networks using agents and networks: a systematic approach." In TENCON 2014-2014 IEEE Region 10 Conference, pp. 1–6. IEEE, 2014.

[9]   Khan BS and Niazi MA. Modeling and analysis of network dynamics in complex communication networks using social network methods. arXiv preprint arXiv:1708.00186 (2017).

[10]   Niazi MA. Towards a novel unified framework for developing formal, network and validated agent-based simulation models of complex adaptive systems. arXiv preprint arXiv:1708.02357 (2017).

[11]   Xia Y and Hill DJ. Dynamic Braess's paradox in complex communication networks. IEEE Transactions on Circuits and Systems—II: Express Briefs 2013;60:172–176.

[12]   Pertet S and Narasimhan, P. Causes of failure in web applications (CMU-PDL-05-109). Parallel Data Laboratory. 2005;48:1–19.

[13]   North MJ, Collier NT, and Ozik, J. Complex adaptive systems modeling with Repast Simphony. Complex Adaptive Systems Modeling 2013;1:3.

[14]   Hanna L and Cagan J. Evolutionary multi-agent systems: an adaptive and dynamic approach to optimization. Journal of Mechanical Design 2009;131(1):1–8.

[15]   Cowling PI, Ouelhadj D, and Petrovic S. Dynamic scheduling of steel casting and milling using multi-agents. Journal of Production Planning and Control, 2004;15:1–11.

[16]   Çiflikli C, Gezer A, Özşahin AT, and Özkasap Ö. BitTorrent packet traffic features over IPv6 and IPv4. Simulation Modelling Practice and Theory 2010;18(9):1214–1224, ISSN 1569-190X.

[17]   Napoli C, Pappalardo G, and Tramontana E. A mathematical model for file fragment diffusion and a neural predictor to manage priority queues over BitTorrent. International Journal of Applied Mathematics and Computer Science 2016;26(1):147–160.

[18]   Chiang J-L, Tseng Y-Y, and Chen W-T. Interest-intended piece selection in BitTorrent-like peer-to-peer file sharing systems. Journal of Parallel and Distributed Computing 2011;71(6):879–888. DOI=http://dx.doi.org/10.1016/j.jpdc.2010.12.011.

[19]   Park S, Chung H, and Lee C, *et al.* Methodology and implementation for tracking the file sharers using BitTorrent. Multimedia Tools and Applications 2015;74(1):271–286, SN - 1573-7721.S.

[20]   Zghaibeh M and Harmantzis FC. Revisiting free riding and the Tit-for-Tat in BitTorrent: a measurement study. Peer-to-Peer Networking and Applications 2008;1:162–173.

[21]   Lei Guo, Songqing Chen, Zhen Xiao, Enhua Tan, Xiaoning Ding, and Xiaodong Zhang. 2007. A performance study of BitTorrent-like peer-to-peer

systems, IEEE Journal on Selected Areas in Communications 2017;25(1): 155–169.

[22]   Zhang C, Dhungel P, Wu D, and Ross KW. Unraveling the BitTorrent ecosystem. IEEE Transactions on Parallel and Distributed Systems 2011;22(7),1164–1177.

[23]   Farzad A and Rabiee HR. Modeling topological characteristics of BitTorrent-like peer-to-peer networks. IEEE Communications Letters 2011;15: 896–898.

[24]   Rumín, Cuevas R, Laoutaris N, Yang X, Siganos G, and Rodriguez P. BitTorrent locality and transit TrafficReduction: when, why, and at what cost?" IEEE Transactions on Parallel and Distributed Systems 2014;25:1177–1189.

[25]   Pouwelse JA, Garbacki P, Wang J, *et al.* TRIBLER: a social-based peer-to-peer system: research Articles. Concurrency and Computation: Practice and Experience 2008;20(2):127–138.s.

[26]   Liu G, Shen H, Ward L. An efficient and trustworthy P2P and social network integrated file sharing system. IEEE Transactions on Computers 2015; 64:54–70.

[27]   Teng W-G and Cheng W-H. Exploiting scheduling and free-riding for offline downloading in BitTorrent networks. International Journal of Communication Systems 2013;26:1365–1374.

[28]   Chan JSk, Li VOk and Lui K-s. Performance comparison of scheduling algorithms for peer-to-peer collaborative file distribution. IEEE Journal on Selected Areas in Communications 2007;25(1):146–154.

[29]   Satsiou A and Tassiulas L. Reputation-based resource allocation in P2P systems of rational users. IEEE Transactions on Parallel and Distributed Systems 2010;21(4):466–479.

[30]   Levin D, LaCurts K, Spring N, and Bhattacharjee B. "Bittorrent is an auction: analyzing and improving bittorrent's incentives. In Proceedings of the ACM SIGCOMM 2008 Conference on Data Communication (SIGCOMM'08). ACM, New York, NY, USA, pp. 243–254. 2008.

[31]   Park H and van der Schaar M. "Coalition-based resource reciprocation strategies for P2P multimedia broadcasting. IEEE Transactions on Broadcasting 2008;54(3):557–567.

[32]   Soumya V and Basu A. "Modified BitTorrent protocol and its application in cloud computing environment. International Journal of Computer Science, Engineering and Applications (IJCSEA) 2012;2(5):23–31.

[33]   Testa C., Rossi D., Rao A., Legout A. (2012) "Experimental Assessment of BitTorrent Completion Time in Heterogeneous TCP/uTP Swarms". In: Pescapè A., Salgarelli L., Dimitropoulos X. (eds) Traffic Monitoring and Analysis. TMA 2012. Lecture Notes in Computer Science, vol 7189. Springer, Berlin, Heidelberg.

[34]   Pouwelse J., Garbacki P., Epema D., Sips H. (2005) "The Bittorrent P2P File-Sharing System: Measurements and Analysis". In: Castro M., van Renesse R. (eds) Peer-to-Peer Systems IV. IPTPS 2005. Lecture Notes in Computer Science, vol 3640. Springer, Berlin, Heidelberg.

[35] Bharambe AR, Herley C, and Padmanabhan VN. "Analyzing and Improving a BitTorrent Networks Performance Mechanisms." In INFOCOM 2006. 25th IEEE International Conference on Computer Communications. Proceedings 2006.

[36] Srinivasan A and Aldharrab H. XTRA—eXtended bit-Torrent pRotocol for Authenticated covert peer communication. Peer-to-Peer Networking and Applications 2018;11:1–15.

[37] Vlavianos A, Iliofotou M, and Faloutsos M. "BiToS: Enhancing BitTorrent for supporting streaming applications." In INFOCOM 2006. 25th IEEE International Conference on Computer Communications. Proceedings, pp. 1–6. IEEE, 2006.

[38] Bindal R, Cao P, Chan W, *et al.* "Improving traffic locality in BitTorrent via biased neighbor selection." In Distributed Computing Systems, 2006. ICDCS 2006, pp. 1–9. IEEE, 2006.

[39] Bharambe, AR., Herley C., and Padmanabhan VN. "Analyzing and improving bittorrent performance." Microsoft Research, Microsoft Corporation One Microsoft Way Redmond, WA 98052 (2005): 2005-03.

[40] Barcellos MP, Mansilha RB, and Brasileiro FV. "Torrentlab: Investigating bittorrent through simulation and live experiments." In Computers and Communications, 2008. ISCC 2008. IEEE Symposium on, pp. 507–512. IEEE, 2008.

[41] Costa-Montenegro E, Burguillo-Rial JC, Gil-Castiñeira F, and González-Castaño FJ. Implementation and analysis of the BitTorrent protocol with a multi-agent model. Journal of Network and Computer Applications 2011;34(1):368–383.

[42] Qiu D and Srikant R. "Modeling and performance analysis of BitTorrent-like peer-to-peer networks." In ACM SIGCOMM Computer Communication Review, vol. 34, no. 4, pp. 367–378. ACM, 2004.

[43] CHOE, YungRyn. "Analyzing and improving a bittorrent network's performance mechanisms." In: ACM MM'07. 2007.

[44] Eger K, Hoßfeld T, Binzenhöfer A, and Kunzmann G. "Efficient simulation of large-scale p2p networks: packet-level vs. flow-level simulations." In Proceedings of the Second Workshop on Use of P2P, GRID and Agents for the Development of Content Networks, pp. 9–16. ACM, 2007.

[45] Yang W and Abu-Ghazaleh N. "GPS: a general peer-to-peer simulator and its use for modeling BitTorrent." In Modeling, Analysis, and Simulation of Computer and Telecommunication Systems, 2005. 13th IEEE International Symposium on, pp. 425–432. IEEE, 2005.

[46] R Development Core Team. R: A Language and Environment for Statistical Computing. Vienna, Austria: the R Foundation for Statistical Computing. 2013. Available online at http://www.R-project.org/.

[47] Pouwelse JA, Garbacki P, Epema DHJ, and Sips HJ. "An introduction to the bittorrent peer-to-peer file-sharing system." In 19th IEEE Annual Computer Communications Workshop, IEEE Technical Committee on Computer Communications. 2004.

[48]   Sunaga H, Hoshiai T, Kamei S, and Kimura S. Technical trends in P2P-based communications. IEICE Transactions on Communications 2004;87(10): 2831–2846.

[49]   Wei B, Fedak G, Cappello F. "Collaborative data distribution with BitTorrent for computational desktop grids." IEEE Computer Society. International Symposium on Parallel and Distributed Computing (ISPDC'2005), Jul 2005, Lille/France, 2005.

Chapter 12
Social networks—a scientometric visual survey
Bisma S. Khan[1] and Muaz A. Niazi[1]

12.1 Introduction

"Social networks" is a significant area of research. It has garnered considerable attention from the research community. The literature in the domain of social networks is growing incredibly fast; a corpus of literature is already accumulated. In spite of its importance, one of the key difficulty beginners, academicians, and researchers face in locating information about social networks is the diverse spread of related articles across multiple domains—primarily due to the multidisciplinary nature of the "social networks." Therefore, it is quite difficult to get conversant with the fundamental concepts, future directions, and advent in the field.

It, therefore, seems appropriate to identify emerging trends, temporal patterns, and underlying dynamics in the bibliographic literature of social networks in such a way that is accessible to the beginners, academicians, and researchers. The traditional domain analysis studies consider only a fraction of publications in the domain, which are inadequate to capture diverging trends and transient patterns of the field. To the best of our knowledge, there is no recent review of current literature which uses a scientometric analysis of networks formed from highly cited and key-reviewed articles from the Web of Science (WoS) to explore the general growth of the domain. We retrieved papers from 2001 to 2018 and applied scientometric analysis on the data using CiteSpace. The topic search and cited references in our method provide coverage of the literature at a broader level.

The key contribution of this article is the identification of important nodes occupying significant positions in the scientometric networks, such as highly cited and central articles, articles with citation burst and sigma, clusters of the articles network based on most representative terms, turning points and pivot points, popular keywords, central keywords, keywords with citation burst, most productive, central, and active institutions and countries, and important subject categories in the domain.

The key findings of our research are as follows. First, we have explored key articles in terms of citation frequency, betweenness centrality, citation burst, and sigma through cited-reference co-citation analysis. Then, we have identified core authors based on a number of publications, centrality, burstiness, and sigma

[1]Department of Computer Science, COMSATS University Islamabad, Pakistan

through collaborative analysis of author network. Next, we have revealed core subject categories of the domain in terms of centrality, the frequency of occurrence, bursts, and sigma. Next, we seek productive institutions in terms of betweenness centrality, burstiness, sigma, and frequency of publications. Then, we explored popular keywords in terms of centrality, the frequency of occurrence, co-occurrence burst, and sigma. Subsequently, we identified core countries in terms of sigma, centrality, publication frequency, and bursts. Additionally, we explored clusters of the cited references.

The rest of the paper is organized as follows: Section 12.2 background, Section 12.3 presents the methodology, Section 12.4 demonstrates results, Section 12.5 summarizes the results, and Section 12.6 contains conclusions.

12.2 Background

In this section, we present the background for better understanding of the social networks and visual analysis.

12.2.1 Social networks—an overview

Networks/graphs are about studying interaction or interaction patterns on objects. A network in its simplest form is made up of nodes/vertices and loaded edges/links [1]. A social network is a network in which the nodes are individuals of the population, such as people, group of people, friends, authors, or organizations, and edges are some form of social interaction, such as liking, friendship, belonging, exchange of money, professional relationships, communication patterns, or collaborative interactions. Nodes in the social networks are referred to as actors, and links between actors are referred to as ties (directed or undirected).

Social networks are mostly attributed to online social network services, such as Twitter, Facebook, etc. These sites are designed to help us work together on common interests or activities. Social networks are growing rapidly, at an unprecedented as well as unpredictable scale, because of the exponential increase in the usage of social media. Social networks are dynamic in nature, they evolve over time. In general, they are emergent, self-organizing, and complex. The growth of social networks follows a power law.

Over the past decades, the notion of "social networks" and "social network analysis" has gained considerable interest from multidisciplinary domains, such as social sciences, behavioral sciences, marketing, physics, computer science, and economics. Much of the focus of this interest is the study of relationships between individuals to understand the structure and patterns of the interaction.

Sociologists have introduced social network analysis as a means of analyzing relationships between individuals by studying the exchange of information and resources. Social network analysts rarely draw samples from social networks because social networks are based on relations among actors, therefore actors cannot be sampled.

Social networks exhibit "small world" property also known as "six-degrees of separation" which states that every actor in the world can be reached from every other actor at six hops distance on average. Some social networks are scale free, that is, some actors have many connections with other actors, whereas others have a few connections. These are heterogeneous networks following the power law degree distribution [1].

Some nodes occupy a central position in the network. Information diffuse in the network through central nodes. Diffusion demonstrates the dissemination of events (information, disease, rumor, or virus, etc.) all over a network. Social networks are an active source of information dissemination. On the one hand, useful information can be promoted and propagated efficiently and effectively via social networks. On the other hand, malicious information such as virus and rumor can also propagate uncontrollably in social networks.

12.2.2   Citation networks

Citation networks are the type of information networks. Nodes in the citation network are articles. There will be a directed link from paper A to paper B if paper A cites paper B in its references. Papers are mostly cited to give credit to the prior work of the cited author. The citing article is known as citer and the cited paper is known as citee. Citation of a paper also indicates that citer and citee have relevance [2]. One of the earliest citation networks study was conducted by Price in 1965 [3]. Web of Science, Scopus, CiteSeerX, and Google Scholar, etc., are the well-known examples of citation index databases [2].

The vertices of the citation networks hold information in the form of images and text. Links are formed from "citing article" to "cited article." These networks are acyclic. "An acyclic network is one in which there are no closed loops of directed edges" [2]. Citation networks are acyclic because one paper can only cite another which is already written. These networks demonstrate the information flow between documents. They are constrained in time, so form directed acyclic graphs which own causal structure [4].

12.2.3   Co-citation networks

Two documents are called co-cited if they both are cited by the same another document. Co-citation indicates that co-cited documents are related. In co-citation network, nodes are the cited articles and edges are the co-citation links between node pairs. Unlike citation networks, co-citation networks are undirected symmetric graphs [2]. Other examples of co-citation networks include journal co-citation networks and author co-citation networks. In journal co-citation networks, journals are the nodes, and in author co-citation networks, cited authors are the nodes [2,5].

12.2.4   Bibliographic coupling

Two documents are considered bibliographically coupled if they both cite the same another document. Like co-citation, it also indicates that the documents are related. These are also undirected graphs [2].

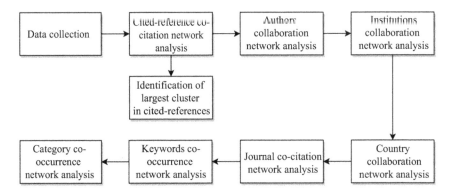

Figure 12.1   Research methodology for the visualization of "social network" to identify trends and emergent patterns in the bibliographic data of the domain

12.2.5   Coauthorship networks

Coauthorship of a document is formed by the collaboration between two or more authors. The coauthors' collaboration forms coauthorship networks. In this network, authors are the nodes, and collaboration between them forms edges. Collaboration could also be between institutions and countries. In that case, institutions or countries will be the nodes and collaboration link between them will be the edges [6].

12.2.6   Co-occurrence networks

These networks are formed by the paired presence of terms within a particular unit of text. Two terms A and B are said to be co-occurred if they both appear in the same document. In co-occurrence networks, the terms are the nodes and co-occurrence links between them form edges. For example, category–category co-occurrence network, keyword–keyword co-occurrence network [5].

After fundamentals of background, we present materials and methods.

12.3   Materials and methods

Here we present data-collection process and identify emerging trends and patterns through data visualization and analysis. We also explore central and highly cited documents, influential collaborative authors, key institutions, core countries, important keywords, and leading research categories. Figure 12.1 demonstrates step-by-step methodology of the scientometric study and visualization of the bibliographic literature in the domain of "social networks."

First, we have collected data from Clarivate Analytics' Web of Science (WoS) using a search query. Then we have analyzed the co-citation network of cited references and we also identify largest connected cluster in the cited reference co-citation

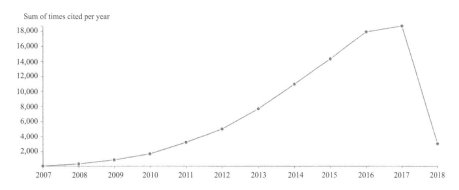

Figure 12.2   Citations history of the scientometric data in the domain of "social networks" over latest 11 years (2007–18)

network. After that, we have performed coauthors' network visualization. Next, we visualize collaborative institution and collaborative country networks. Subsequently, we have analyzed the co-occurrence network of keywords and research categories.

12.3.1   Data collection

To perform substantial co-citation and collaboration analysis of scientific literature, input data is retrieved from the core collection of WoS. An extensive topic search was made and refined by topic = ("social network*"); timespan = 2007–18 years; language = English; document type: articles; and indexes: SCI-EXPANDED, SSCI, A&HCI, ESCI. The date was retrieved is on March 16, 2018. Total 36,469 unique bibliographic records are downloaded by selecting a full record and cited references as record contents.

We first analyze the citation report of WoS. Figure 12.2 illustrates that beginning from few citations in 2007, the domain of "social networks" has risen to almost 18, 000 citations only in the year 2017. It clearly indicates that the domain has garnered considerable attention from the research community in recent years.

Table 12.1 contains published items per year. It is quite interesting to note that the articles published in the domain of "social network*" have surged from approximately 314 (3.001%) in 2001 and reached to almost 5, 710 publications alone in the year 2017, which is 15.570% of total 36,616 records.

12.3.2   CiteSpace—a science mapping tool

In this chapter, CiteSpace—a visual analytic tool is used for visualization of scientometric analysis [5]. This section provides an overview and comparison of CiteSpace with four other science mapping tools.

CiteSpace is a freely available, Java-based bibliometric tool used for visualization of emerging trends and patterns in bibliographic literature. It runs on the Windows, Linux, or Mac-operating systems. It is used for educational and research purposes.

Table 12.1   Publication years by record count

Publication years	Record count	% of 36,616
2017	5,701	15.570
2016	5,335	14.570
2015	4,943	13.500
2014	3,703	10.113
2013	3,230	8.821
2012	2,683	7.327
2011	2,224	6.074
2010	1,774	4.845
2009	1,367	3.733
2008	1,099	3.001

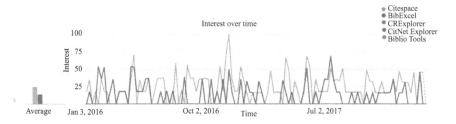

Figure 12.3   Google Trends for the comparison between science mapping tools. Color figure can be viewed at: https://www.researchgate.net

It allows dynamic, spatial, temporal, and interactive visualization. It has a simple and interactive interface.

It could directly import data from WoS, Scopus, and PubMed. CiteSpace takes scientometric information and performs the following bibliometric analysis: author, document, and journal co-citation analysis; coauthors, institutions, and country collaboration analysis; co-terms and keywords co-occurrence analysis. The key features of CiteSpace include timeline view, geographic mapping, and dual-map overlays. It also offers built-in database and could be connected to MySQL on the local host. It can export cited references to Endnote and RIS.

It can also generate a summary report containing key information obtained from the analyzed literature. Its strong documentation is also available, including books, articles, tutorials, demos, manual, and videos. It also has a Facebook page on which near real-time help is available.

We have compared CiteSpace with four other science mapping tools: "BibExcel," "CRExplorer," "CitNetExplorer," and "BiblioTools." Apparently, CiteSpace in Google Trends shown in Figure 12.3 is on top with an average popularity score of 27.

Table 12.2   Comparison between different science mapping tools

Scientometric tool	CiteSpace [8]	BibExcel [9]	CRExplorer [10]	CitNetExplorer [11]	BiblioTools [12]
Availability	Free	Free	Free	Free	Free
Platform	Java	–	Java	Java	Python
Operating system	Windows, Linux or Mac, Java Runtime (JRE)	Windows, Linux	Java Run Time	Windows, any Java supporting OS	Unix, Mac, Windows
Data import	WoS, PubMed, arXiv, ADS, Scopus, NSF Award Abstracts	WoS, Scopus	WoS, Scopus	WoS	WoS
Resolves disambiguation	No	No	Yes	No	No
Bibliometric analysis	Author, institution, countries collaboration networks, document, journal, and author co-citation networks, co-occurrence networks of terms, categories, and keywords, geographic map, overlays; interactive, temporal, and spatial visualization	Bibliometry, bibliometric, citation analysis, bibliographic coupling, shared references, co-citation, cluster analysis	Cited references	Citation networks, interactive visualization	Data parsing, filtering, detecting bibliographic coupling communities, co-occurrence maps (co-words, coauthors, co-citations)
Export	Pajek, Excel, SPSS	Pajek, NetDraw, Excel, SPSS	Pajek	Pajek	Gephi, BiblioMaps
Documentation	Strong	Weak	Weak	Weak	Weak

To compare salient features of CiteSpace with other science-mapping tools, comparative analysis is given in Table 12.2. For a detailed overview and comparison, readers may refer to [7].

After providing materials and methods, we present research results and discussion.

12.4   Results and discussion

In this section, we present results obtained from visual analysis of the scientific literature obtained from WoS. The color figures are available at https://www.researchgate.net/project/Visual-Analysis-of-Scientometric-Networks.

12.4.1   *Cited-references co-citation network analysis*

The goal of our first analysis is the visualization of the document co-citation network to identify key documents in the scientific literature of "social networks." Top 30 nodes per time slice of length 2 years are selected in the timespan of 2001–18. The panoramic view of the network shown in Figure 12.4 comprises 142 unique documents and 453 citation links. Unique colors are used to represent each time slice. Link colors indicate the time period when this link was first established. The blue color corresponds to the earliest years, green color corresponds to the middle years, and the red color indicates the current years. The lighter and darker shades of the same color indicate the earlier and later time slices, respectively. The nodes in the network are resized according to the corresponding degree in the network.

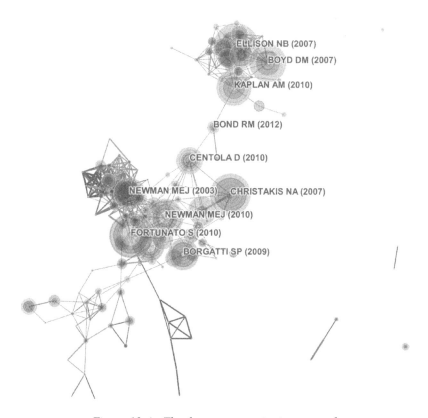

Figure 12.4   The document co-citation network

Table 12.3   Top articles sorted in terms of co-citation frequency

Cited references	Citation counts	GS citation	Cluster #
Ellison NB, 2007, J Comput-Mediat Comm, V12, P1143–1168	558	9,521	2
Fortunato S, 2010, Phys rep, V486, P75–174	542	6,725	0
Christakis NA, 2007, New Engl J Med, V357, P370	501	4,583	0
Boyd DM, 2007, J Comput-Mediat Comm, V13, P210	455	14,702	2
Newman MEJ, 2010, Networks Intro	417	8,381	0
Kaplan AM, 2010, Bus Horizons, V53, P59	387	12,726	2
Borgatti SP, 2009, Science, V323, P892	385	2,493	3
Newman MEJ, 2003, Siam Rev, V45, P167	381	17,398	1
Christakis NA, 2008, New Engl J Med, V358, P2249	375	1,840	0
Snijders TAB, 2010, Soc Networks, V32, P44	359	1,320	3

The largest radius of the node Ellison NB (2007) indicates that it is a landmark node having highest citation frequency. The thickness of purple trims around Kossinets G (2006) shows that it is the most influential article in the domain. The pink trims around the nodes are indicating that their centrality score is ≥0.1.

Further details are given below in tabular form.

Table 12.3 presents top articles ranked in terms of citation counts. The top article is Ellison NB (2007) in Cluster #2 with citation frequency of 558. It has 9,521 citations on GS. The second one is Fortunato S (2010) in Cluster #0 with citation frequency of 542. It obtained 6,725 citations on GS. The third is Christakis NA (2007) in Cluster #0 with citation frequency of 501. It has 4,583 citations on GS. The fourth is Boyd DM (2007) in Cluster #2 with citation frequency of 455. It has 12,726 citations on GS. The fifth is Newman MEJ (2010) in Cluster #0 with citation frequency of 417. The sixth is Kaplan AM (2010) in Cluster #2 with citation frequency of 387. It has 2,493 citations on GS. The seventh is Borgatti SP (2009) in Cluster #3 with citation frequency of 385. It has 17,398 citations on GS. The eighth is Newman MEJ (2003) in Cluster #1 with citation frequency of 381. It has 1,840 citations on GS. The ninth is Christakis NA (2008) in Cluster #0 with citation frequency of 375. The tenth is Snijders TAB (2010) in Cluster #3 with citation frequency of 359. It has 1,320 citations on GS.

Table 12.4 presents top articles listed in terms of citation burst. "The burstiness of the frequency of an entity over time indicates a specific duration in which an abrupt change of the frequency takes place" [13]. The top article ranked by citation burst is Newman MEJ (2003) in Cluster #1 with burstiness of 140.95. The second one is Boyd DM (2007) in Cluster #2 with bursts of 126.59. The third is Albert R (2002) in Cluster #1 with bursts of 120.88. The fourth is Ellison NB (2007) in Cluster #2 with bursts of 111.91. The fifth is Watts DJ (1998) in Cluster #1 with bursts of 100.56. The sixth is Barabasi AL (1999) in Cluster #1 with bursts of 95.75. The seventh is McPherson M (2001) in Cluster #5 with bursts of 83.29. The eighth is Borgatti SP (2002) in Cluster #3 with bursts of 80.51. The ninth is Christakis NA (2007) in Cluster #0 with bursts of 79.06. The tenth is Putnam R D (2000) in Cluster #5 with bursts of 75.83.

Table 12.4 Top articles sorted in terms of citation burst

Cited references	Bursts	Cluster #
Newman MEJ, 2003, Siam Rev, V45, P167	140.95	1
Boyd DM, 2007, J Comput-Mediat Comm, V13, P210	126.59	2
Albert R, 2002, Rev Mod Phys, V74, P47	120.88	1
Ellison NB, 2007, J Comput-Mediat Comm, V12, P1143	111.91	2
Watts DJ, 1998, Nature, V393, P440	100.56	1
Barabasi AL, 1999, Science, V286, P509	95.75	1
Mcpherson M, 2001, Annu Rev Sociol, V27, P415	83.29	5
Borgatti SP, 2002, Ucinet Windows Softw, V, P	80.51	3
Christakis NA, 2007, New Engl J Med, V357, P370	79.06	0
Putnam R D, 2000, Bowling Alone Collap, V, P	75.83	5

Table 12.5 Top articles sorted in terms of betweenness centrality

Cited references	Centrality	Cluster #
Centola D, 2010, Science, V329, P1194	0.58	0
Kaplan AM, 2010, Bus Horizons, V53, P59	0.50	2
Bond RM, 2012, Nature, V489, P295	0.50	0
Onnela JP, 2007, P Natl Acad Sci USA, V104, P7332	0.28	0
Kossinets G, 2006, Science, V311, P88	0.26	0
Brass DJ, 2004, Acad Manage J, V47, P795	0.25	3
Newman MEJ, 2010, Networks Intro, V, P	0.24	0
Newman MEJ, 2003, Siam Rev, V45, P167	0.19	1
Amaral LAN, 2000, P Natl Acad Sci USA, V97, P11149	0.19	1
Snijders TAB, 2010, Soc Networks, V32, P44	0.19	3

Table 12.5 presents top-ranked articles in terms of betweenness centrality. "The betweenness centrality of a node in a network measures the extent to which the node is part of paths that connect an arbitrary pair of nodes in the network" [14]. The top-graded article in terms of betweenness centrality is Centola D (2010) in Cluster #0 with the centrality of 0.58. The second one is Kaplan AM (2010) in Cluster #2 with the centrality of 0.50. The third is Bond RM (2012) in Cluster #0 with the centrality of 0.50. The fourth is Onnela JP (2007) in Cluster #0 with the centrality of 0.28. The fifth is Kossinets G (2006) in Cluster #0 with the centrality of 0.26. The sixth is Brass DJ (2004) in Cluster #3 with the centrality of 0.25. The seventh is Newman MEJ (2010) in Cluster #0 with the centrality of 0.24. The eighth is Newman MEJ (2003) in Cluster #1 with the centrality of 0.19. The ninth is Amaral LAN (2000) in Cluster #1 with the centrality of 0.19. The tenth is Snijders TAB (2010) in Cluster #3 with the centrality of 0.19.

Table 12.6 lists top articles sorted in terms of sigma. "Sigma indicator measures the combined strength of structural and temporal properties of a node, namely, its

Table 12.6 Top articles sorted in terms of sigma

Cited references	Sigma	Cluster #
Kaplan AM, 2010, Bus Horizons, V53, P59	5819401151475.16	2
Newman MEJ, 2003, Siam Rev, V45, P167	27277977080.97	1
Albert R, 2002, Rev Mod Phys, V74, P47	79866981.70	1
Centola D, 2010, Science, V329, P1194	66555034.25	0
Bond RM, 2012, Nature, V489, P295	8847176.59	0
Onnela JP, 2007, P Natl Acad Sci USA, V104, P7332	40726.42	0
Newman MEJ, 2010, Networks Intro, V, P	25355.81	0
Kossinets G, 2006, Science, V311, P88	24045.82	0
McPherson M, 2001, Annu Rev Sociol, V27, P415	19341.76	5
Amaral LAN, 2000, P Natl Acad Sci USA, V97, P11149	1555.13	1

betweenness centrality and citation burst" [15]. The top-ranked item by sigma is Kaplan AM (2010) in Cluster #2 with a sigma of 5819401151475.16. The second one is Newman MEJ (2003) in Cluster #1 with a sigma of 27277977080.97. The third is Albert R (2002) in Cluster #1 with a sigma of 79866981.70. The fourth is Centola D (2010) in Cluster #0 with a sigma of 66555034.25. The fifth is Bond RM (2012) in Cluster #0 with a sigma of 8847176.59. The sixth is Onnela JP (2007) in Cluster #0 with a sigma of 40726.42. The seventh is Newman MEJ (2010) in Cluster #0 with a sigma of 25355.81. The eighth is Kossinets G (2006) in Cluster #0 with a sigma of 24045.82. The ninth is McPherson M (2001) in Cluster #5 with a sigma of 19341.76. The tenth is Amaral LAN (2000) in Cluster #1 with a sigma of 1555.13.

After a detailed visualization of cited references co-citation network, our next target is to identify largest connected cluster in the network of cited references.

12.4.1.1 Identification of largest cluster in cited references

Our next analysis is the identification of the giant component in the document co-citation network. Figure 12.5 shows the clustered view of the document co-citation network. The network of cited references is decomposed into 22 co-citation clusters. These clusters are labeled by index terms from their own citers. The largest connected component has 6 clusters comprising 122 documents which are 85% of entire articles in the network. The largest connected cluster is summarized in Table 12.7.

There are several turning points and pivot points in the giant component playing the brokerage role between different components. "Turning points refer to the revolutionary articles identified by domain experts, whereas pivotal points refer to articles that share similar topological properties in the network generated by CiteSpace" [8]. The articles Kaplan AM (2010) and Bond RM (2012) are the key turning points joining Cluster #0 and Cluster #2. Similarly, McPherson M (2001) and Kossinets (2005) are the turning points connecting Cluster #2 and Cluster #5, whereas Amaral Lan (2000) and Watts D (1999) are the pivot points joining Cluster #1 and Cluster #4. Similarly, Newman MEJ (2002) is the pivot point joining Cluster #1 and Cluster #5.

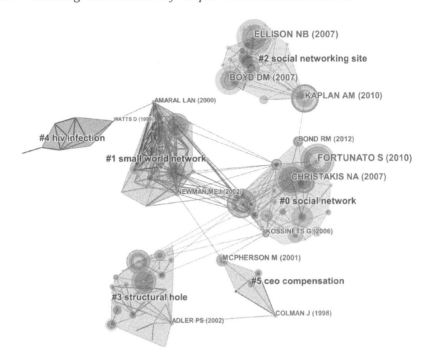

Figure 12.5   Clusters of cited references co-citation network

Table 12.7   Summary of the largest connected cluster of the cited references of co-citation network

Cluster ID	Size	Silhouette	Average (Citee year)	Label		
				TFIDF[a]	LLR[a]	MI[a]
0	30	0.825	2008	Application | complex network	Social network	Statistical analysis
1	28	0.933	2001	Role | infection	Small world network	Assortative interaction
2	26	0.998	2009	Feelings | educational method	Social networking site	Job search
3	25	0.846	2004	Coauthorship network | physical activity	Structural hole	Acute myocardial infarction
4	8	0.969	1996	Practice | perspective	HIV infection	Social network
5	5	0.945	1999	Dyadic stability	CEO compensation	Social network

[a]Labelling algorithms: Term Frequency Inverse Document Frequency (TFIDF), Log-Likelihood Ratio (LLR), and Mutual Information (MI).

The largest cluster (#0) has 30 members and a silhouette value of 0.825. The silhouette score suggests that the homogeneity of the underlying cluster is relatively high. It is labeled as "social network" by LLR, "application | complex network" by TFIDF, and "statistical analysis" by MI. The most active citer to the cluster is 0.17 Pei, S (2013) [16].

The second largest cluster (#1) has 28 members and a silhouette value of 0.933 indicating high homogeneity score. The cluster is labeled as "small world network" by LLR, "role | infection" by TFIDF, and "assortative interaction" by MI. The most active citer to the cluster is 0.29 Newman, MEJ (2001) [17].

The third largest cluster (#2) has 26 members and a silhouette value of 0.998, which indicates higher homogeneity score. The cluster is labeled as "social networking site" by LLR, "feelings | educational method" by TFIDF, and "job search" by MI. The most active citer to the cluster is 0.23 Krasnova, H (2015) [18].

The fourth largest cluster (#3) has 25 members and a silhouette score of 0.846. It is labeled as "structural hole" by LLR, "coauthorship network | physical activity" by TFIDF, and "acute myocardial infarction" by MI. The most active citer to the cluster is 0.24 Hoppe, B (2010) [19].

The fifth largest cluster (#4) has eight members and a silhouette value of 0.969. It is labeled as "HIV infection" by LLR, "practice | perspective" by TFIDF, and "social network" by MI. The most active citer to the cluster is 0.62 Riolo, CS (2001) [20].

The sixth largest cluster (#5) has five members and a silhouette value of 0.945. It is labeled as "CEO compensation" by LLR, "dyadic stability" by TFIDF, and "social network" by MI. The most active citer to the cluster is 0.4 Denner, J (2001) [21].

After identifying the largest component in the cited reference co-citation network, our focus is the analysis of author collaboration network.

12.4.2 Authors collaboration network analysis

Here we present visual analysis of coauthors network to identify core authors of the "social networks" domain. As shown in Figure 12.6, the merged network of coauthors comprises 1,103 authors and 1,231 coauthorship links. The selection criterion for this network is the top 50 authors per 1-year time slice in the timespan of 2001–18. The largest connected component comprises 274 authors who are 24% of the entire network.

The thickness of the links in the network represents the strength of the coauthorship. The link color represents the time slice when the first article was coauthored. The size of a specific node is proportional to the overall publication count of that author. Red color indicates the publication burst of authors. The concentric circles around author nodes represent the separation of publication years. The blue color represents the older publications, whereas the red color represents recent publications.

It is pertinent to note here that bibliometric data of authors suffer from initial-based disambiguation of author names. "The name disambiguation, i.e. identifying whether a set of name strings refers to one or more real-world persons" [22].

The broader picture of coauthors network is given in tabular form below.

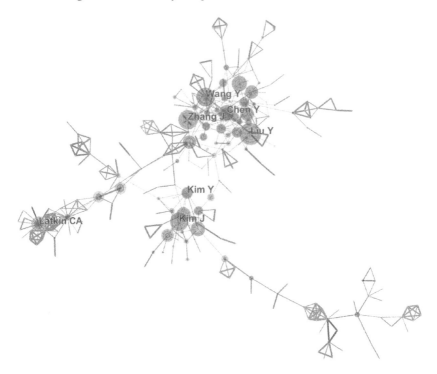

Figure 12.6   The network of collaborative authors

Table 12.8 demonstrates top collaborative authors in terms of publication counts. The top-ranked author by publication counts is "Liu Y" (2012) with publication frequency of 73. The second one is "Zhang Y" (2009) with publication frequency of 64. The third is "Zhang J" (2008) with publication frequency of 64. The fourth is "Kim J" (2011) with publication frequency of 61. The fifth is "Wang Y" (2013) with publication frequency of 61. The sixth is "Latkin CA" (2001) with publication frequency of 57. The seventh is "Chen Y" (2009) with publication frequency of 56. The eighth is "Lee S" (2010) with publication frequency of 52. The ninth is "Lee J" (2010) with publication frequency of 49. The tenth is "Chen L" (2013) with publication frequency of 48.

Table 12.9 presents top authors listed in terms of burstness of publications. The top-ranked author in terms of bursts is "Wang Y" (2013) with bursts of 17.79. The second one is "Zhang Y" (2009) with bursts of 17.69. The third is "Li Y" (2015) with bursts of 17.12. The fourth is "Liu Y" (2012) with bursts of 15.51. The fifth is "Zhang J" (2008) with bursts of 14.17. The sixth is "Chen L" (2013) with bursts of 13.64. The seventh is "Wu J" (2013) with bursts of 12.78. The eighth is "Christakis NA" (2008) with bursts of 10.54. The ninth is "Kim J" (2011) with bursts of 10.34. The tenth is "Chen X" (2014) with bursts of 10.23.

Table 12.8   Top articles sorted in terms of publication frequency

Publication counts	Author	Year
73	Liu Y	2012
64	Zhang Y	2009
64	Zhang J	2008
61	Kim J	2011
61	Wang Y	2013
57	Latkin CA	2001
56	Chen Y	2009
52	Lee S	2010
49	Lee J	2010
48	Chen L	2013

Table 12.9   Top articles sorted in terms of publication burst

Publication bursts	Author	Year
17.79	Wang Y	2013
17.69	Zhang Y	2009
17.12	Li Y	2015
15.51	Liu Y	2012
14.17	Zhang J	2008
13.64	Chen L	2013
12.78	Wu J	2013
10.54	Christakis NA	2008
10.34	Kim J	2011
10.23	Chen X	2014

Table 12.10 presents top authors sorted in terms of betweenness centrality. The top-ranked author by betweenness centrality is "Zhang Y" (2009) with centrality score of 0.04. The second one is "Kim Y" (2011) with centrality score of 0.04. The third is "Newman MEJ" (2001) with centrality score of 0.03. The fourth is "Lee J" (2010) with centrality score of 0.03. The fifth is "Wang L" (2007) with centrality score of 0.03. The sixth is "Holme P" (2003) with centrality score of 0.03. The seventh is "Fowler JH" (2007) with centrality score of 0.03. The eighth is "Park J" (2004) with centrality score of 0.03. The ninth is "Liljeros F" (2004) with centrality score of 0.03. The tenth is "Christakis NA" (2008) with centrality score of 0.02.

Table 12.11 presents top authors sorted in terms of sigma score. The top-ranked author by sigma is "Zhang Y" (2009) with a sigma of 1.98. The second one is "Newman MEJ" (2001) with a sigma of 1.33. The third is "Kim Y" (2011) with a sigma of 1.31. The fourth is "Lee J" (2010) with a sigma of 1.29. The fifth is "Wang Y" (2013) with a sigma of 1.23. The sixth is "Wang L" (2007) with a sigma of 1.21. The seventh

Table 12.10   Top authors sorted in terms of betweenness centrality

Betweenness centrality	Author	Year
0.04	Zhang Y	2009
0.04	Kim Y	2011
0.03	Newman MEJ	2001
0.03	Lee J	2010
0.03	Wang L	2007
0.03	Holme P	2003
0.03	Fowler JH	2007
0.03	Park J	2004
0.03	Liljeros F	2004
0.02	Christakis NA	2008

Table 12.11   Top authors sorted in terms of sigma score

Sigma	Author	Year
1.98	Zhang Y	2009
1.33	Newman MEJ	2001
1.31	Kim Y	2011
1.29	Lee J	2010
1.23	Wang Y	2013
1.21	Wang L	2007
1.21	Zhang J	2008
1.18	Holme P	2003
1.17	Christakis NA	2008
1.16	Park J	2004

is "Zhang J" (2008) with a sigma of 1.21. The eighth is "Holme P" (2003) with a sigma of 1.18. The ninth is "Christakis NA" (2008) with a sigma of 1.17. The tenth is "Park J" (2004) with a sigma of 1.16.

After analyzing coauthors collaboration network, we move toward visual analysis of the institution's network.

12.4.3   Institution collaboration network analysis

This section demonstrates the visualization of the institution–institution network to identify most productive institutions in the domain of "social networks" from the period of 2001 to 2018. Top 50 institutions per time slice of 1 year are selected to develop the network shown in Figure 12.7. The merged network consists of 256 institutions and 1,280 collaboration links. The largest connected component comprises 274 nodes. Five percent of the institutions are labeled in the network.

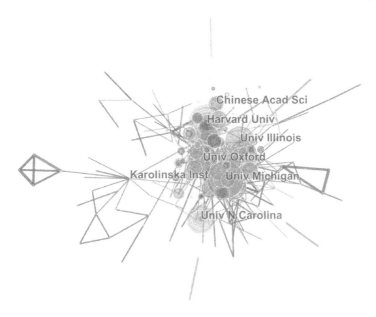

Figure 12.7   The collaborative network of institutions

Table 12.12   Top institutions sorted in terms of publication frequency

Publications count	Institution	Country	Year	World ranking 2018
452	Harvard University	USA	2001	6
427	University of Michigan	USA	2001	21
388	University of Illinois	USA	2001	37
374	University of Oxford	UK	2002	1
368	University of North Carolina	USA	2001	56
351	Pennsylvania State University	USA	2001	77
346	University of California, Los Angeles	USA	2001	15
335	Arizona State University	USA	2001	126
328	Columbia University	USA	2001	14
293	University of Toronto	Canada	2001	22

The thickness of purple trim around Harvard University indicates that it is the most central institution, and the largest radius around Harvard indicate that it is also the landmark node. The red highlights on Chinese Academy of Science indicate that it is the most active institution with the highest burst.

The detailed analysis is presented in the tabular form below.

Table 12.12 demonstrates the publication count of institutions. The top-ranked institution by publication frequency is the "Harvard University, USA" (2001) with

Table 12.13   Top institutions sorted in terms of betweenness centrality score

Betweenness centrality	Institution	Country	Year	World ranking 2018
0.18	Harvard University	USA	2001	6
0.14	Karolinska Institute	Sweden	2001	38
0.09	University of Michigan	USA	2001	21
0.08	Stockholm University	Sweden	2001	134
0.07	University of Oxford	UK	2002	1
0.07	Pennsylvania State University	USA	2001	77
0.07	University of Toronto	USA	2001	22
0.07	University of Melbourne	Australia	2001	32
0.07	University of California, Berkeley	USA	2001	18
0.06	University of Maryland	USA	2001	69

publication frequency of 452. It is ranked sixth in World University Rankings. The second one is the "University of Michigan, USA" (2001) with publication frequency of 427. It is ranked 21st in World University Rankings. The third is the "University of Illinois" (2001) with publication frequency of 388. It is ranked 37th in World University Rankings. The fourth is "University of Oxford, UK" (2002) with publication frequency of 374. It is ranked first in World University Rankings. The fifth is "University of North Carolina" (2001) with publication frequency of 368. It is ranked 56th in World University Rankings. The sixth is the "Pennsylvania State University, USA" (2001) with publication frequency of 351. It is ranked 77th in World University Rankings. The seventh is the "University of California, Los Angeles, USA" (2001) with publication frequency of 346. It is ranked 15th in World University Rankings. The eighth is the "Arizona State University, USA" (2001) with publication frequency of 335. It is ranked 126th in World University Rankings. The ninth is the "Columbia University, USA" (2001) with publication frequency of 328. It is ranked 14th in World University Rankings. The tenth is the "University of Toronto, Canada" (2001) with publication frequency of 293. It is ranked 22nd in World University Rankings.

Table 12.13 lists top institutions sorted in terms of betweenness centrality. The top-ranked institution in terms of betweenness centrality is "Harvard University, USA" (2001) with the centrality of 0.18. It is ranked sixth in World University Rankings. The second one is "Karolinska Institute, Sweden" (2001) with the centrality of 0.14. It is ranked 38th in World University Rankings. The third is "University of Michigan, USA" (2001) with the centrality of 0.09. It is ranked 21st in World University Rankings. The fourth is "Stockholm University, Sweden" (2001) with the centrality of 0.08. It is ranked 134th in World University Rankings. The fifth is the "University of Oxford, UK" (2002) with the centrality of 0.07. It is ranked first in World University Rankings. The sixth is "Pennsylvania State University, USA" (2001) with the centrality of 0.07. It is ranked 77th in World University Rankings. The seventh is "University of Toronto, Canada" (2001) with the centrality of 0.07. It is ranked 22nd in World University Rankings. The eighth is "University of Melbourne, Australia"

Table 12.14   Top Institutions sorted in terms of sigma score

Sigma	Institution	Country	Year	World ranking 2018
1.00	Harvard University	USA	2001	6
1.00	Karolinska Institute	Sweden	2001	38
1.00	University of Michigan	USA	2001	21
1.00	Stockholm University	Sweden	2001	134
1.00	University of Oxford	UK	2002	1
1.00	Pennsylvania State University	USA	2001	77
1.00	University of Toronto	Canada	2001	22
1.00	University of Melbourne	Australia	2001	32
1.00	University of California, Berkeley	USA	2001	18
1.00	University of Maryland	USA	2001	69

(2001) with the centrality of 0.07. It is ranked 32nd in World University Rankings. The ninth is "University of California, Berkeley, USA" (2001) with the centrality of 0.07. It is ranked 18th in World University Rankings. The tenth is "University of Maryland, USA" (2001) with the centrality of 0.06. It is ranked 69th in World University Rankings.

Table 12.14 presents top institutions sorted in terms of sigma score. The top-graded institution by sigma is the "Harvard University, USA" (2001) with a sigma of 1.00. It is ranked sixth in World University Rankings. The second one is the "Karolinska Institute, Sweden" (2001) with a sigma of 1.00. It is ranked 38th in World University Rankings. The third is the "University of Michigan, USA" (2001) with a sigma of 1.00. It is ranked 21st in World University Rankings. The fourth is the "Stockholm University, Sweden" (2001) with a sigma of 1.00. It is ranked 134th in World University Rankings. The fifth is the "University of Oxford, UK" (2002) with a sigma of 1.00. It is ranked first in World University Rankings. The sixth is "Pennsylvania State University, USA" (2001) with a sigma of 1.00. It is ranked 77th in World University Rankings. The seventh is the "University of Toronto, Canada" (2001) with a sigma of 1.00. It is ranked 22nd in World University Rankings. The eighth is the "University of Melbourne, Australia" (2001) with a sigma of 1.00. It is ranked 32nd in World University Rankings. The ninth is the "University of California, Berkeley, USA" (2001) with a sigma of 1.00. It is ranked 18th in World University Rankings. The tenth is the "University of Maryland, USA" (2001) with a sigma of 1.00. It is ranked 69th in World University Rankings.

Table 12.15 presents top institutions listed in terms of burst of publications. The top-ranked institution in terms of bursts is "University of Chinese Academy of Sciences, China" (2014) with bursts of 40.52. It is ranked 189th in World University Rankings. The second one is "University of Southern California, USA" (2016) with bursts of 36.5. It is ranked 66th in World University Rankings. The third is "Tsinghua University, China" (2013) with bursts of 29.49. It is ranked 30th in World University Rankings. The fourth is "Nanyang Technological University, Singapore" (2005) with

Table 12.15   The top institute sorted in terms of burstness

Institution	Country	Year	Burst	World ranking 2018
University of Chinese Academy of Sciences	China	2014	40.52	189
University of Southern California	United States	2016	36.5	66
Tsinghua University	China	2013	29.49	30
Nanyang Technological University	Singapore	2005	27.79	52
Nanyang Technological University	Singapore	2016	26.76	52

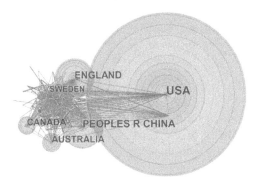

Figure 12.8   The network of collaborative countries

bursts of 27.79. It is ranked 52nd in World University Rankings. The fifth is the "Nanyang Technological University, Singapore" (2016) with bursts of 26.76. It is ranked 52nd in World University Rankings.

After giving an overview of the collaborative institution network analysis, we present an overview of the analysis of the country–country network.

12.4.4   Country collaboration network analysis

Here we demonstrate a visualization of the country collaboration network to visualize the spread of research in the "social networks" research community. As shown in Figure 12.8, the merged network consists of 84 countries and 677 collaboration links. It is a dense network. We have selected top 50 countries per 1-year slice length in the timespan of 2001–18. The largest connected component of the network contains 274 countries out of which 5.0% nodes are labeled.

The largest diameter of the United States indicates that it is the landmark node which indicates that key publications originate from the United States. The purple trims around England indicate that it is the most central country of the domain. The red marks exhibiting the United States indicates that the articles originating from the United States have garnered the interest of its research community.

Table 12.16 presents the top-ranked countries by publication frequency. The United States (2001) is on top with publication frequency of 14,490. The second

Table 12.16   Top countries sorted in terms of publication frequency

Country	Publication counts	Year
USA	14,490	2001
People's Republic of China	4,045	2001
England	3,637	2001
Australia	2,071	2001
Canada	1,955	2001
Spain	1,592	2001
Germany	1,522	2001
Netherlands	1,480	2001
Italy	1,216	2001
South Korea	1,156	2001

Table 12.17   Top countries sorted in terms of burstness

Country	Bursts	Year
USA	52.20	2001
Sweden	51.65	2001
India	38.31	2002
Iran	26.02	2010
Saudi Arabia	23.92	2013
Pakistan	21.89	2011
Egypt	13.04	2002
Croatia	10.57	2009
Israel	8.67	2001
Estonia	7.04	2008

one is People's Republic of China (2001) with publication frequency of 4,045. The third is England (2001) with publication frequency of 3,637. The fourth is Australia (2001) with publication frequency of 2,071. The fifth is Canada (2001) with publication frequency of 1,955. The sixth is Spain (2001) with publication frequency of 1,592. The seventh is Germany (2001) with publication frequency of 1,522. The eighth is Netherlands (2001) with publication frequency of 1,480. The ninth is Italy (2001) with publication frequency of 1,216. The tenth is South Korea (2001) with publication frequency of 1,156.

Table 12.17 lists the top countries ranked in terms of burstness. The top country ranked by bursts is the United States (2001) with bursts of 52.20. The second one is Sweden (2001) with bursts of 51.65. The third is India (2002) with bursts of 38.31. The fourth is Iran (2010) with bursts of 26.02. The fifth is Saudi Arabia (2013) with bursts of 23.92. The sixth is Pakistan (2011) with bursts of 21.89. The seventh is Egypt (2002) with bursts of 13.04. The eighth is Croatia (2009) with bursts of 10.57. The ninth is Israel (2001) with bursts of 8.67. The tenth is Estonia (2008) with bursts of 7.04.

Table 12.18   Top countries sorted in terms of betweenness centrality

Betweenness centrality	Country	Year
0.25	England	2001
0.12	USA	2001
0.11	Canada	2001
0.11	Spain	2001
0.10	France	2001
0.08	Denmark	2001
0.08	Switzerland	2001
0.08	Chile	2008
0.07	Finland	2001
0.07	Germany	2001

Table 12.19   Top countries sorted in terms of sigma

Sigma	Country	Year
426.27	USA	2001
6.14	Sweden	2001
2.09	Saudi Arabia	2013
1.52	India	2002
1.42	Norway	2001
1.37	Denmark	2001
1.35	Finland	2001
1.27	Hungary	2002
1.11	Israel	2001
1.10	Iran	2010

Table 12.18 presents the central regions in terms of betweenness centrality. The top-ranked country in terms of centrality is England (2001) with the centrality of 0.25. The second one is the United States (2001) with the centrality of 0.12. The third is Canada (2001) with the centrality of 0.11. The fourth is Spain (2001) with the centrality of 0.11. The fifth is France (2001) with the centrality of 0.10. The sixth is Denmark (2001) with the centrality of 0.08. The seventh is Switzerland (2001) with the centrality of 0.08. The eighth is Chile (2008) with the centrality of 0.08. The ninth is Finland (2001) with the centrality of 0.07. The tenth is Germany (2001) with the centrality of 0.07.

Table 12.19 contains top countries sorted in terms of sigma score. The top-graded country by sigma is the United States (2001) with a sigma of 426.27. The second one is Sweden (2001) with a sigma of 6.14. The third is Saudi Arabia (2013) with a sigma of 2.09. The fourth is India (2002) with a sigma of 1.52. The fifth is Norway (2001) with a sigma of 1.42. The sixth is Denmark (2001) with a sigma of 1.37. The seventh is Finland (2001) with a sigma of 1.35. The eighth is Hungary (2002) with a sigma of 1.27. The ninth is Israel (2001) with a sigma of 1.11. The tenth is Iran (2010) with a sigma of 1.10.

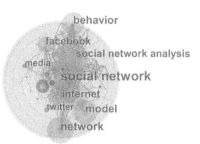

Figure 12.9   The keyword co-occurrence network

Table 12.20   Top keywords in terms of co-occurrence frequency

Co-occurrence counts	Keywords	Year
11,151	Social network	2001
2,983	Network	2001
2,679	Model	2001
2,452	Behavior	2001
1,801	Social network analysis	2005
1,740	Internet	2002
1,689	Health	2001
1,639	Facebook	2010
1,616	Community	2001
1,557	Performance	2001

After analyzing the network of collaborative countries, we focus on the analysis of keywords co-occurrence network.

12.4.5   Keywords co-occurrence network analysis

Here we present a visualization of the keywords co-occurrence network to identify popular keywords associated with the "social networks" domain. The merged network comprises 135 keywords and 887 co-occurrence links. It is a dense network with 5.0% nodes labeled. For this analysis, we have selected top 50 nodes per time slice of the year in the timespan of 2001–18.

As shown in Figure 12.9, the diameter of a node is proportional to the frequency of occurrence. The keyword "social network" with the largest diameter is the landmark node. The thickness of a node corresponds to the high betweenness centrality score. The keyword "social network" with purple trims is also the most central node. The detailed analysis is given underneath in the tables.

Table 12.20 demonstrates top keywords sorted in terms of frequency of co-occurrence. The top-ranked keyword based on a co-occurrence frequency is "social

Table 12.21   Top keywords sorted in terms of
co-occurrence burst

Co-occurrence burst	Keywords	Year
135.87	Media	2015
101.12	Children	2001
95.19	Stress	2001
89.61	Social media	2012
80.67	Twitter	2014
77.78	Mortality	2001
70.61	Weak ty	2001
70.20	Quality of life	2001
64.17	Life	2001
63.19	Care	2001

network" (2001) with a co-occurrence frequency of 11,151. The second one is "network" (2001) with a co-occurrence frequency of 2,983. The third is "model" (2001) with a co-occurrence frequency of 2,679. The fourth is "behavior" (2001) with a co-occurrence frequency of 2,452. The fifth is "social network analysis" (2005) with a co-occurrence frequency of 1,801. The sixth is "Internet" (2002) with co-occurrence frequency of 1,740. The seventh is "health" (2001) with a co-occurrence frequency of 1,689. The eighth is "Facebook" (2010) with a co-occurrence frequency of 1,639. The ninth is "community" (2001) with a co-occurrence frequency of 1,616. The tenth is "performance" (2001) with co-occurrence frequency of 1,557.

Table 12.21 presents top keywords based on co-occurrence burst. The top-graded keyword by bursts is "media" (2015) with bursts of 135.87. The second one is "children" (2001) with bursts of 101.12. The third is "stress" (2001) with bursts of 95.19. The fourth is "social media" (2012) with bursts of 89.61. The fifth is "Twitter" (2014) with bursts of 80.67. The sixth is "mortality" (2001) with bursts of 77.78. The seventh is "weak ty" (2008) with bursts of 70.61. The eighth is "quality of life" (2001) with bursts of 70.20. The ninth is "life" (2001) with bursts of 64.17. The tenth is "care" (2001) with bursts of 63.19.

Table 12.22 illustrates top keywords listed in terms of betweenness centrality. The top-ranked keyword by betweenness centrality is "social network" (2001) with the centrality of 0.50. The second one is "behavior" (2001) with the centrality of 0.13. The third is "support" (2001) with the centrality of 0.09. The fourth is "Facebook" (2010) with the centrality of 0.09. The fifth is "mortality" (2001) with the centrality of 0.08. The sixth is "health" (2001) with the centrality of 0.08. The seventh is "community" (2001) with the centrality of 0.08. The eighth is "predictor" (2001) with the centrality of 0.07. The ninth is "children" (2001) with the centrality of 0.06. The tenth is "population" (2001) with the centrality of 0.06

Table 12.23 lists top keywords sorted in terms of sigma. The top-ranked keyword by sigma is "mortality" (2001) with a sigma of 389.96. The second one is "children" (2001) with a sigma of 234.65. The third is "support" (2001) with a sigma of 13.08. The fourth is "predictor" (2001) with a sigma of 7.48. The fifth is "social support"

Table 12.22   Top keywords sorted in terms of betweenness centrality

Betweenness centrality	Keywords	Year
0.50	Social network	2001
0.13	Behavior	2001
0.09	Support	2001
0.09	Facebook	2010
0.08	Mortality	2001
0.08	Health	2001
0.08	Community	2001
0.07	Predictor	2001
0.06	Children	2001
0.06	Population	2001

Table 12.23   Top keywords sorted in terms of sigma

Sigma	Keyword	Year
389.96	Mortality	2001
234.65	Children	2001
13.08	Support	2001
7.48	Predictor	2001
6.44	Social support	2001
5.33	Women	2001
5.11	Privacy	2014
3.65	Stress	2001
3.45	HIV	2001
3.09	Population	2001

(2001) with a sigma of 6.44. The sixth is "women" (2001) with a sigma of 5.33. The seventh is "privacy" (2014) with a sigma of 5.11. The eighth is "stress" (2001) with a sigma of 3.65. The ninth is "HIV" (2001) with a sigma of 3.45. The tenth is "population" (2001) with a sigma of 3.09.

After giving an overview of the popular keywords of the domain, we present an overview of the key subject categories of the domain.

12.4.6   Category co-occurrence network analysis

This section presents a visual analysis of the category co-occurrence network to iden-tify articles associated with key subject categories of the "social network" domain. The merged network contains 87 categories and 432 co-occurrence links. We have selected top 50 articles as per slice length of 1 year for the period of 2001–18. The largest connected component is composed of 274 nodes.

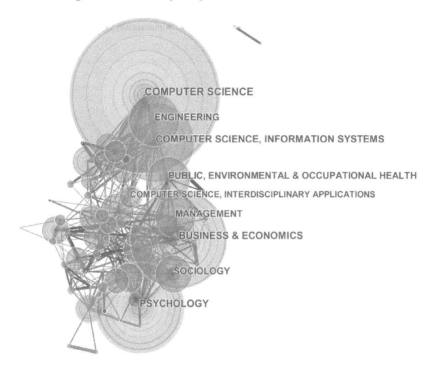

Figure 12.10   The category co-occurrence network

Table 12.24   The top categories sorted in terms of co-occurrence frequency

Co-occurrence counts	Categories	Year
6,939	Computer Science	2001
4,510	Business & Economics	2001
4,097	Psychology	2001
3,695	Computer Science, Information Systems	2002
2,423	Engineering	2003
2,296	Public, Environmental & Occupational Health	2001
2,141	Management	2001
2,011	Sociology	2001
1,834	Information Science & Library Science	2001
1,667	Business	2001

In Figure 12.10, it can be noted that "computer science" is the most occurred category, and it is the most influential category of the "social network" domain. A detailed analysis is listed below in the tabular form.

Table 12.24 presents top categories sorted in terms of co-occurrence frequency. The top-ranked subject category by co-occurrence frequency is "Computer Science"

Table 12.25 *The top categories sorted in terms of co-occurrence bursts*

Co-occurrence bursts	Categories	Year
77.40	Sociology	2001
63.49	Gerontology	2001
60.05	Nursing	2001
56.93	Geriatrics & Gerontology	2001
48.15	Hospitality, Leisure, Sport & Tourism	2016
44.72	Psychiatry	2001
44.70	Psychology, Developmental	2001
44.45	Demography	2001
42.99	Physics, Mathematical	2001
40.51	Psychology, Applied	2001

(2001) with a co-occurrence frequency of 6,939. The second one is "Business & Economics" (2001) with a co-occurrence frequency of 4,510. The third is "Psychology" (2001) with a co-occurrence frequency of 4,097. The fourth is "Computer Science, Information Systems" (2002) with a co-occurrence frequency of 3,695. The fifth is "Engineering" (2003) with a co-occurrence frequency of 2,423. The sixth is "Public, Environmental & Occupational Health" (2001) with a co-occurrence frequency of 2,296. The seventh is "Management" (2001) with a co-occurrence frequency of 2,141. The eighth is "Sociology" (2001) with a co-occurrence frequency of 2,011. The ninth is "Information Science & Library Science" (2001) with a co-occurrence frequency of 1,834. The tenth is "Business" (2001) with a co-occurrence frequency of 1,667.

Table 12.25 presents top categories sorted in terms of co-occurrence bursts. The top-ranked item by bursts is "Sociology" (2001) with bursts of 77.40. The second one is "Gerontology" (2001) with bursts of 63.49. The third is "Nursing" (2001) with bursts of 60.05. The fourth is "Geriatrics & Gerontology" (2001) with bursts of 56.93. The fifth is "Hospitality, Leisure, Sport & Tourism" (2016) with bursts of 48.15. The sixth is "Psychiatry" (2001) with bursts of 44.72. The seventh is "Psychology, Developmental" (2001) with bursts of 44.70. The eighth is "Demography" (2001) with bursts of 44.45. The ninth is "Physics, Mathematical" (2001) with bursts of 42.99. The tenth is "Psychology, Applied" (2001) with bursts of 40.51.

Table 12.26 presents top categories sorted in terms of betweenness centrality. The top-ranked item by centrality is "Psychology" (2001) with the centrality of 0.21. The second one is "Computer Science, Interdisciplinary Applications" (2002) with the centrality of 0.20. The third is "Health Care Sciences & Services" (2001) with the centrality of 0.20. The fourth is "Mathematics" (2001) with the centrality of 0.18. The fifth is "Social Sciences—Other Topics" (2001) with the centrality of 0.16. The sixth is "Public, Environmental & Occupational Health" (2001) with the centrality of 0.15. The seventh is "Environmental Sciences & Ecology" (2001) with the centrality of 0.13. The eighth is "Engineering" (2003) with the centrality of 0.11.

Table 12.26   The top categories sorted in terms of betweenness centrality

Betweenness centrality	Subject categories	Year
0.21	Psychology	2001
0.20	Computer Science, Interdisciplinary Applications	2002
0.20	Health Care Sciences & Services	2001
0.18	Mathematics	2001
0.16	Social Sciences—Other Topics	2001
0.15	Public, Environmental & Occupational Health	2001
0.13	Environmental Sciences & Ecology	2001
0.11	Engineering	2003
0.11	Social Sciences, Interdisciplinary	2001
0.10	Psychology, Multidisciplinary	2001

Table 12.27   Top categories based on sigma

Sigma	Categories	Year
350.69	Sociology	2001
14.41	Psychology	2001
9.19	Psychiatry	2001
8.71	Physics, Mathematical	2001
6.38	Social Sciences, Biomedical	2001
6.38	Biomedical Social Sciences	2001
5.45	Rehabilitation	2001
5.35	Public, Environmental & Occupational Health	2001
3.22	Physics	2001
2.44	Mathematics, Interdisciplinary Applications	2001

The ninth is "Social Sciences, Interdisciplinary" (2001) with the centrality of 0.11. The tenth is "Psychology, Multidisciplinary" (2001) with the centrality of 0.10.

Table 12.27 presents top categories sorted in terms of sigma. The top-ranked item by sigma is "Sociology" (2001) with a sigma of 350.69. The second one is "Psychology" (2001) with a sigma of 14.41. The third is "Psychiatry" (2001) with a sigma of 9.19. The fourth is "Physics, Mathematical" (2001) with a sigma of 8.71. The fifth is "Social Sciences, Biomedical" (2001) with a sigma of 6.38. The sixth is "Biomedical Social Sciences" (2001) with a sigma of 6.38. The seventh is "Rehabilitation" (2001) with a sigma of 5.45. The eighth is "Public, Environmental & Occupational Health" (2001) with a sigma of 5.35. The ninth is "Physics" (2001) with a sigma of 3.22. The tenth is "Mathematics, Interdisciplinary Applications" (2001) with a sigma of 2.44.

After analyzing the key subject categories, we perform our final analysis of journal co-citation network.

Table 12.28   Top journals sorted in terms of frequency of publications

Source title	Record count	% of 36,616 of total	Impact factor
Computers in Human Behavior	781	2.133	3.536
PLoS One	636	1.737	2.806
Physica A Statistical Mechanics and Its Applications	442	1.207	2.132
Social Networks	408	1.114	2.530
Social Science Medicine	279	0.76	3.007
Physical Review E	252	0.688	2.366
Cyberpsychology Behavior and Social Networking	193	0.527	2.689
Scientometrics	192	0.524	2.173
Scientific Reports	186	0.508	4.609
Expert Systems with Applications	171	0.467	3.711

12.4.7   Journal co-citation network analysis

In this section, we are visualizing journal co-citation network to identify core journals in the "social networks" domain. The detailed analysis of journal co-citation is given below in the tabular form.

Table 12.28 demonstrates top-core journals sorted in terms of frequency of publications. The top-ranked journal is "Computers in Human Behavior" with the publication frequency of 781. It has an impact factor of 3.536. The second one is "PLoS One" with the publication frequency of 636. It has an impact factor of 2.806. The third is "Physica A Statistical Mechanics and Its Applications" with the publication frequency of 442. It has an impact factor of 2.132. The fourth is "Social Networks" with the publication frequency of 408. It has an impact factor of 2.530. The fifth is "Social Science Medicine" with the publication frequency of 279. It has an impact factor of 3.007. The sixth is "Physical Review E" with the publication frequency of 252. It has an impact factor of 2.366. The seventh is the "Cyberpsychology Behavior and Social Networking" with the publication frequency of 193. It has an impact factor of 2.689. The eighth is "Scientometrics" with the publication frequency of 192. It has an impact factor of 2.173. The ninth is "Scientific Reports" with the publication frequency of 186. It has an impact factor of 4.609. The tenth is "Expert Systems with Applications" with the publication frequency of 171. It has an impact factor of 3.71.

12.5   Summary of results

In this manuscript, CiteSpace—a key visual analytic tool is used for the visualization of scientometric data retrieved from WoS, devoted to the "social networks" over last 18 years in the timespan of 2001–18. The retrieved dataset contains 36,469 unique records. Here we provide highlights of the key findings of this scientometric study.

First, we observed the popularity of the domain over past 10 years and revealed that starting from a few citations in 2007, the "social network" has risen to 18,000 citations only in the year 2017. Similarly, the number of publications in the domain has risen from 314 publications in the year 2001 to 5,710 publications in the year 2017.

In the successive analysis of cited reference co-citation network, we identified that the article "Ellison NB (2007)" is the landmark node with a citation frequency of 558, the article "Newman MEJ (2003)" has strongest citation burst of strength 140.95, the top-ranked article in terms of betweenness centrality is "Centola D (2010)," and the top-ranked article in terms of sigma is "Kaplan AM (2010)." We have also identified key turning points in the largest connected component of cited reference co-citation network which are "Kaplan AM (2010)," "Bond RM (2012)," "McPherson M (2001)," "Watts D (1999)," and "Newman MEJ (2002)" and the key pivot points which are "Amaral Lan (2000)," Watts D (1999)," and "Newman MEJ (2000)."

In the analysis of collaborative author network, we observed that the author "Liu Y (2012)" is the landmark node with the highest number of publications. We also observed that "Wang Y (2013)" has the strongest publication burst. We also observed that "Zhang Y (2009)" is the most central author in the domain and also has the highest burst.

In the institution–institution network analysis, it is revealed that Harvard is the most productive and the most centrally organization in the domain. It also has highest sigma score. We also observed that the "University of Chinese Academy of Sciences" has the strongest burst.

In the visualization of keywords co-occurrence network, we observed that "social network" is the most popular and most central keyword in the domain. We also observed that the keyword "media" has the highest co-occurrence burst and the keyword "mortality" has a highest sigma score.

In the category co-occurrence network, we observed that "computer science" is the most frequently occurred subject category of the domain. We also observed that "sociology" has highest co-occurrence burst and highest sigma score. We also found that "psychology" is the most central category.

Subsequently, in the analysis of the journal co-citation network, we also noted that "computers in human behavior" journal has highest publication frequency and its impact factor is 3.536.

12.6   Conclusions and future work

In this study, CiteSpace—a key visual analytic tool is used to trace the emerging trends, temporal patterns, and development in the domain of "social networks." To this end, we use a methodology involving the use of complex networks to visually analyze recent data from WoS in the social network domain. We took bibliographic records from the period of 2001 to 2018 and applied analysis on the data. We have visualized co-citation networks of cited references and journals, collaboration networks of authors, institutions, and countries and co-occurrence networks of keywords and subject categories.

We have identified key turning points, pivot nodes, landmark publications, central articles, active articles, key journals, most productive and influential organizations, active organizations, core countries, central countries, most productive countries, popular keywords, central keywords, active keywords, key subject categories, central subject categories, and bursty subject categories. Additionally, major clusters in the cited reference co-citation network are also revealed.

A future research may conduct a visual analysis of hybrid networks, such as author–country network, institution–country network, author–institution network, author–document network, and document–category network.

References

[1]   M. Newman, *Networks: An introduction*: Oxford University Press, New York, USA, 2010.

[2]   M. Newman, *Networks: An introduction*: New York, NY: Oxford University Press Inc., 2010, pp. 1–2.

[3]   D. J. D. S. Price, "Networks of scientific papers," *Science,* USA, Vol. 149, No. 3683, pp. 510–515, 1965.

[4]   J. R. Clough and T. S. Evans, "What is the dimension of citation space?," *Physica A: Statistical Mechanics and its Applications,* Netherlands, vol. 448, pp. 235–247, 2016.

[5]   C. Chen, *CiteSpace: A practical guide for mapping scientific literature*: Nova Science Publishers, Incorporated, New York, USA, 2016.

[6]   M. E. Newman, "Coauthorship networks and patterns of scientific collaboration," *Proceedings of the National Academy of Sciences,* vol. 101, pp. 5200–5205, 2004.

[7]   P. Pradhan, *Science mapping and visualization tools used in bibliometric & scientometric studies: An overview,* INFLIBNET Centre, Gandhinagar, Series 23, Report no.4, ISSN 0971-9849 2017.

[8]   C. Chen, "CiteSpace II: Detecting and visualizing emerging trends and transient patterns in scientific literature," *Journal of the Association for Information Science and Technology,* vol. 57, pp. 359–377, 2006.

[9]   O. Persson, R. Danell, and J. W. Schneider, "How to use BibExcel for various types of bibliometric analysis," In: Åström, F., Danell, R., Larsen, B., Wiborg-Schneider, J. (Eds.), *Celebrating scholarly communication studies: A Festschrift for Olle Persson at his 60th Birthday,* vol. 5, pp. 9–24, 2009.

[10] A. Thor, W. Marx, L. Leydesdorff, and L. Bornmann, "Introducing CitedReferencesExplorer (CRExplorer): A program for reference publication year spectroscopy with cited references standardization," *Journal of Informetrics,* vol. 10, pp. 503–515, 2016.

[11] N. J. Van Eck and L. Waltman, "CitNetExplorer: A new software tool for analyzing and visualizing citation networks," *Journal of Informetrics,* vol. 8, pp. 802–823, 2014.

[12]   S. Grauwin and P. Jensen, "Mapping scientific institutions," *Scientometrics,* vol. 89, pp. 943–954, 2011.

[13]   J. Kleinberg, "Bursty and hierarchical structure in streams," *Data Mining and Knowledge Discovery,* vol. 7, pp. 373–397, 2003.

[14]   U. Brandes, "A faster algorithm for betweenness centrality," *Journal of Mathematical Sociology,* vol. 25, pp. 163–177, 2001.

[15]   C. Chen, Y. Chen, M. Horowitz, H. Hou, Z. Liu, and D. Pellegrino, "Towards an explanatory and computational theory of scientific discovery," *Journal of Informetrics,* vol. 3, pp. 191–209, 2009.

[16]   S. Pei and H. A. Makse, "Spreading dynamics in complex networks," *Journal of Statistical Mechanics: Theory and Experiment,* vol. 2013, p. P12002, 2013.

[17]   M. E. Newman, S. H. Strogatz, and D. J. Watts, "Random graphs with arbitrary degree distributions and their applications," *Physical Review E,* vol. 64, p. 026118, 2001.

[18]   H. Krasnova, T. Widjaja, P. Buxmann, H. Wenninger, and I. Benbasat, "Research note—why following friends can hurt you: An exploratory investigation of the effects of envy on social networking sites among college-age users," *Information Systems Research,* vol. 26, pp. 585–605, 2015.

[19]   B. Hoppe and C. Reinelt, "Social network analysis and the evaluation of leadership networks," *The Leadership Quarterly,* vol. 21, pp. 600–619, 2010.

[20]   C. S. Riolo, J. S. Koopman, and S. E. Chick, "Methods and measures for the description of epidemiologic contact networks," *Journal of Urban Health,* vol. 78, pp. 446–457, 2001.

[21]   J. Denner, D. Kirby, K. Coyle, and C. Brindis, "The protective role of social capital and cultural norms in Latino communities: A study of adolescent births," *Hispanic Journal of Behavioral Sciences,* vol. 23, pp. 3–21, 2001.

[22]   J. Kim, H. Kim, and J. Diesner, "The impact of name ambiguity on properties of coauthorship networks," *Journal of Information Science Theory and Practice,* vol. 2, pp. 6–15, 2014.

Index